Introduction to Seismology

Second Edition

This book provides an approachable and concise introduction to seismic theory, designed as a first course for graduate students or advanced undergraduate students. It clearly explains the fundamental concepts, emphasizing intuitive understanding over lengthy derivations.

Incorporating over 30% new material, this second edition includes all the topics needed for a one-semester course in seismology. Additional material has been added throughout, including numerical methods, 3-D ray tracing, earthquake location, attenuation, normal modes, and receiver functions. The chapter on earthquakes and source theory has been extensively revised and enlarged, and now includes details on non-double-couple sources, earthquake scaling, radiated energy, and finite slip inversions.

Each chapter includes worked problems and detailed exercises that give students the opportunity to apply the techniques they have learned to compute results of interest and to illustrate the Earth's seismic properties. Computer subroutines and data sets for use in the exercises are available on the book's website.

Peter M. Shearer is a Professor of Geophysics at the Scripps Institution of Oceanography, University of California, San Diego. He has written over 100 scientific papers on various aspects of seismology and is currently the President-Elect of the seismology section of the American Geophysical Union. He has taught the introductory seismology class at Scripps for over 15 years; this book is based on material and problems sets that were developed for this class.

INTRODUCTION TO
Seismology
SECOND EDITION

Peter M. Shearer

Scripps Institution of Oceanography
University of California, San Diego

CAMBRIDGE
UNIVERSITY PRESS

CAMBRIDGE UNIVERSITY PRESS

Cambridge, New York, Melbourne, Madrid, Cape Town, Singapore, São Paulo, Delhi

Cambridge University Press
The Edinburgh Building, Cambridge CB2 8RU, UK

Published in the United States of America by Cambridge University Press, New York

www.cambridge.org
Information on this title: www.cambridge.org/Shearer

First published 2009

Printed in the United Kingdom at the University Press, Cambridge

A catalog record for this publication is available from the British Library

ISBN 978-0-521-88210-1 hardback
ISBN 978-0-521-70842-5 paperback

Cover illustration: the seismic wavefield resulting from an isotropic source
located 600 km beneath the North Pole, as computed using a finite-difference
method with a grid spacing of 2 to 4 kilometers. This snapshot shows the radial
component of motion 871 seconds after the source, which has a dominant
period of about 6 seconds. Image provided courtesy of Gunnar Jahnke.

CONTENTS

Preface to the first edition   *page* xi
Preface to the second edition   xiii
Acknowledgment   xiv

1 Introduction   **1**
  1.1 A brief history of seismology   2
  1.2 Exercises   15

2 Stress and strain   **17**
  2.1 The stress tensor   17
     2.1.1 Example: Computing the traction vector   19
     2.1.2 Principal axes of stress   20
     2.1.3 Example: Computing the principal axes   22
     2.1.4 Deviatoric stress   23
     2.1.5 Values for stress   24
  2.2 The strain tensor   25
     2.2.1 Values for strain   29
     2.2.2 Example: Computing strain for a seismic wave   29
  2.3 The linear stress–strain relationship   30
     2.3.1 Units for elastic moduli   32
  2.4 Exercises   33

3 The seismic wave equation   **39**
  3.1 Introduction: The wave equation   39
  3.2 The momentum equation   40
  3.3 The seismic wave equation   42
     3.3.1 Potentials   46
  3.4 Plane waves   46
     3.4.1 Example: Harmonic plane wave equation   48
  3.5 Polarizations of P and S waves   48

3.6   Spherical waves   50
3.7   Methods for computing synthetic seismograms[†]   51
3.8   The future of seismology?[†]   53
3.9   Equations for 2-D isotropic finite differences[†]   56
3.10   Exercises   61

4   Ray theory: Travel times   **65**
4.1   Snell's law   65
4.2   Ray paths for laterally homogeneous models   67
       4.2.1 Example: Computing $X(p)$ and $T(p)$ 70
       4.2.2 Ray tracing through velocity gradients   71
4.3   Travel time curves and delay times   72
       4.3.1 Reduced velocity   73
       4.3.2 The $\tau(p)$ function   73
4.4   Low-velocity zones   76
4.5   Summary of 1-D ray tracing equations   77
4.6   Spherical-Earth ray tracing   80
4.7   The Earth-flattening transformation   82
4.8   Three-dimensional ray tracing[†]   83
4.9   Ray nomenclature   86
       4.9.1 Crustal phases   86
       4.9.2 Whole Earth phases   87
       4.9.3 *PKJKP*: The Holy Grail of body wave seismology   88
4.10   Global body-wave observations   89
4.11   Exercises   98

5   Inversion of travel time data   **103**
5.1   One-dimensional velocity inversion   103
5.2   Straight-line fitting   106
       5.2.1 Example: Solving for a layer-cake model   108
       5.2.2 Other ways to fit the $T(X)$ curve   109
5.3 $\tau(p)$ Inversion   110
       5.3.1 Example: The layer-cake model revisited   111
       5.3.2 Obtaining $\tau(p)$ constraints   112
5.4   Linear programming and regularization methods   115
5.5   Summary: One-dimensional velocity inversion   117
5.6   Three-dimensional velocity inversion   117
       5.6.1 Setting up the tomography problem   118
       5.6.2 Solving the tomography problem   122
       5.6.3 Tomography complications   124
       5.6.4 Finite frequency tomography   125
5.7   Earthquake location   127
       5.7.1 Iterative location methods   133

	5.7.2	Relative event location methods	134
	5.8	Exercises	135
6	**Ray theory: Amplitude and phase**		**139**
	6.1	Energy in seismic waves	139
	6.2	Geometrical spreading in 1-D velocity models	142
	6.3	Reflection and transmission coefficients	144
	6.3.1	SH-wave reflection and transmission coefficients	145
	6.3.2	Example: Computing SH coefficients	149
	6.3.3	Vertical incidence coefficients	149
	6.3.4	Energy-normalized coefficients	151
	6.3.5	Dependence on ray angle	152
	6.4	Turning points and Hilbert transforms	156
	6.5	Matrix methods for modeling plane waves†	159
	6.6	Attenuation	163
	6.6.1	Example: Computing intrinsic attenuation	164
	6.6.2	t^* and velocity dispersion	165
	6.6.3	The absorption band model†	168
	6.6.4	The standard linear solid†	171
	6.6.5	Earth's attenuation	173
	6.6.6	Observing Q	175
	6.6.7	Non-linear attenuation	176
	6.6.8	Seismic attenuation and global politics	177
	6.7	Exercises	177
7	**Reflection seismology**		**181**
	7.1	Zero-offset sections	182
	7.2	Common midpoint stacking	184
	7.3	Sources and deconvolution	188
	7.4	Migration	191
	7.4.1	Huygens' principle	192
	7.4.2	Diffraction hyperbolas	193
	7.4.3	Migration methods	195
	7.5	Velocity analysis	197
	7.5.1	Statics corrections	198
	7.6	Receiver functions	199
	7.7	Kirchhoff theory†	202
	7.7.1	Kirchhoff applications	208
	7.7.2	How to write a Kirchhoff program	210
	7.7.3	Kirchhoff migration	210
	7.8	Exercises	211

8   Surface waves and normal modes   **215**

8.1   Love waves   215
       8.1.1   Solution for a single layer   218
8.2   Rayleigh waves   219
8.3   Dispersion   224
8.4   Global surface waves   226
8.5   Observing surface waves   228
8.6   Normal modes   231
8.7   Exercises   238

9   Earthquakes and source theory   **241**

9.1   Green's functions and the moment tensor   241
9.2   Earthquake faults   245
       9.2.1   Non-double-couple sources   248
9.3   Radiation patterns and beach balls   251
       9.3.1   Example: Plotting a focal mechanism   259
9.4   Far-field pulse shapes   260
       9.4.1   Directivity   262
       9.4.2   Source spectra   265
       9.4.3   Empirical Green's functions   267
9.5   Stress drop   268
       9.5.1   Self-similar earthquake scaling   271
9.6   Radiated seismic energy   273
       9.6.1   Earthquake energy partitioning   277
9.7   Earthquake magnitude   280
       9.7.1   The b value   288
       9.7.2   The intensity scale   290
9.8   Finite slip modeling   291
9.9   The heat flow paradox   293
9.10   Exercises   297

10   Earthquake prediction   **301**

10.1   The earthquake cycle   301
10.2   Earthquake triggering   309
10.3   Searching for precursors   314
10.4   Are earthquakes unpredictable?   316
10.5   Exercises   318

11   Instruments, noise, and anisotropy   **321**

11.1   Instruments   321
       11.1.1 Modern seismographs   327
11.2   Earth noise   330
11.3   Anisotropy[†]   332

11.3.1 Snell's law at an interface   337
11.3.2 Weak anisotropy   337
11.3.3 Shear-wave splitting   339
11.3.4 Hexagonal anisotropy   341
11.3.5 Mechanisms for anisotropy   343
11.3.6 Earth's anisotropy   344
11.4 Exercises   346

Appendix A   The PREM model   **349**

Appendix B   Math review   **353**
B.1 Vector calculus   353
B.2 Complex numbers   358

Appendix C   The eikonal equation   **361**

Appendix D   Fortran subroutines   **367**

Appendix E   Time series and Fourier transforms   **371**
E.1 Convolution   371
E.2 Fourier transform   373
E.3 Hilbert transform   373

Bibliography   377
Index   391

PREFACE TO THE FIRST EDITION

Why another book on seismology? Several excellent texts already exist that cover most parts of the field. None, however, is ideal for the purposes of an introductory class. Most simply present far more material than can be adequately covered in a single quarter or semester. My goal for this book is to produce a readable, concise introduction to the quantitative aspects of seismology that is designed specifically for classroom instruction. The result is not as rigorous or comprehensive as Aki and Richards (1980) or Lay and Wallace (1995), but I hope that it is more suited for teaching an overview of seismology within a limited time period.

To quicken the pace, many results are described without detailed proofs or derivations of equations. In these cases, the reader is usually referred to Aki and Richards or other sources for more complete explanations. Generally I have attempted to provide practical descriptions of the main concepts and how they are used to study Earth structure. Some knowledge of physics and vector calculus is assumed, but in an effort to make the book self-contained most of the key concepts are reviewed in the Appendices.

Any book to some extent reflects the prejudices of its author. In this regard, I have perhaps included more material on ray theory and body wave travel times, and less on surface waves and normal modes, than a truly balanced book would require. In my defense, it can be argued that a large fraction of current seismological research continues to rely on travel times, and that ray theory provides a good starting point for students as it is intuitively easier to understand than more advanced theories. Although some current research results are presented, I have concentrated more on fundamental principles and key data sets in an effort to avoid rapid obsolescence after this book goes to press.

The emphasis in the student exercises is not on deriving equations (which few seismologists spend much time doing anyway), but on using techniques explained in the text to compute results of interest and to illustrate some of Earth's seismic properties. Since computer programming skills are often a necessity for performing seismology research, I have included a number of computer-based assignments. These are designed to give a taste of real research problems, while requiring only a moderate level of programming ability. Subroutines to assist in the exercises are listed in Appendix D.

PREFACE TO THE SECOND EDITION

During the last ten years, I have continued teaching the beginning seismology class at University of California, San Diego, and have received feedback from my students, as well as other instructors who have been using the book. The second edition is my attempt to expand on some subjects, clarify parts of the book that have proven confusing, and update the discussion of current research results. The biggest changes are to the Source Theory chapter, which now provides a more complete discussion of non-double-couple sources, stress drop, earthquake scaling, radiated energy, energy partitioning, and magnitude saturation. However, I have also tried to remain concise enough that the book can still be used for a one-quarter or one-semester class, although depending upon the pace of the class it may be necessary to skip some of the material. Sections flagged with a † are suggestions for possible areas to skip without much compromise in understanding of the remaining subjects.

The computer subroutines and data for some of the exercises can now be obtained from www.cambridge.org/shearer, which also contains links to any errors found in the text and other supplemental information that I plan to add in the future.

ACKNOWLEDGMENT

This book began as a series of lecture notes that I developed while teaching the beginning seismology class to first year graduate students in geophysics at University of California, San Diego. Some of the material in Chapters 4–5 and the section on the eikonal equation is derived from notes that John Orcutt wrote for a similar class. The stacked images in Chapter 4 were produced in collaboration with Luciana Astiz and Paul Earle. I am grateful to Steve Day, Dick Hilt, Youshun Sun, and Ruedi Widmer-Schnidrig for alerting me to some mistakes in the first edition; to Heidi Houston, Cliff Thurber, Bob Nowack, and Arthur Snoke for their suggestions for the second edition; and to Emily Brodsky, Heidi Houston, Heiner Igel, and John Vidale for their comments on drafts of the second edition.

1

Introduction

Every day there are about fifty earthquakes worldwide that are strong enough to be felt locally, and every few days an earthquake occurs that is capable of damaging structures. Each event radiates seismic waves that travel throughout Earth, and several earthquakes per day produce distant ground motions that, although too weak to be felt, are readily detected with modern instruments anywhere on the globe. Seismology is the science that studies these waves and what they tell us about the structure of Earth and the physics of earthquakes. It is the primary means by which scientists learn about Earth's deep interior, where direct observations are impossible, and has provided many of the most important discoveries regarding the nature of our planet. It is also directly concerned with understanding the physical processes that cause earthquakes and seeking ways to reduce their destructive impacts on humanity.

Seismology occupies an interesting position within the more general fields of geophysics and Earth sciences. It presents fascinating theoretical problems involving analysis of elastic wave propagation in complex media, but it can also be applied simply as a tool to examine different areas of interest. Applications range from studies of Earth's core, thousands of kilometers below the surface, to detailed mapping of shallow crustal structure to help locate petroleum deposits. Much of the underlying physics is no more advanced than Newton's second law ($F = ma$), but the complications introduced by realistic sources and structures have motivated sophisticated mathematical treatments and extensive use of powerful computers. Seismology is driven by observations, and improvements in instrumentation and data availability have often led to breakthroughs both in seismology theory and in our understanding of Earth structure.

The information that seismology provides has widely varying degrees of uncertainty. Some parameters, such as the average compressional wave travel time through the mantle, are known to a fraction of a percent, while others, such as the degree of damping of seismic energy within the inner core, are known only very approximately. The average radial seismic velocity structure of Earth has been known fairly well for over fifty years, and the locations and seismic radiation patterns of earthquakes are now routinely mapped, but many important aspects of the physics of earthquakes themselves remain a mystery.

1.1 A brief history of seismology

Seismology is a comparatively young science that has only been studied quantitatively for about 100 years. Reviews of the history of seismology include Dewey and Byerly (1969) and Agnew (2002). Early thinking about earthquakes was, as one might expect, superstitious and not very scientific. It was noted that earthquakes and volcanoes tended to go together, and explanations for earthquakes involving underground explosions were common. In the early 1800s the theory of elastic wave propagation began to be developed by Cauchy, Poisson, Stokes, Rayleigh, and others who described the main wave types to be expected in solid materials. These include compressional and shear waves, termed *body waves* since they travel through solid volumes, and *surface waves*, which travel along free surfaces. Since compressional waves travel faster than shear waves and arrive first, they are often called primary or P waves, whereas the later arriving shear waves are called secondary or S waves. At this time theory was ahead of seismic observations, since these waves were not identified in Earth until much later.

In 1857 a large earthquake struck near Naples. Robert Mallet, an Irish engineer interested in earthquakes, traveled to Italy to study the destruction caused by the event. His work represented the first significant attempt at observational seismology and described the idea that earthquakes radiate seismic waves away from a focus point (now called the *hypocenter*) and that they can be located by projecting these waves backward to the source. Mallet's analysis was flawed since he assumed that earthquakes are explosive in origin and only generate compressional waves. Nevertheless, his general concept was sound, as were his suggestions that observatories be established to monitor earthquakes and his experiments on measuring seismic velocities using artificial sources.

Early seismic instrumentation was based on undamped pendulums, which did not continuously record time, although sometimes an onset time was measured. The first time-recording seismograph was built in Italy by Filippo Cecchi in 1875. Soon after this, higher-quality instruments were developed by the British in Japan, beginning with a horizontal pendulum design by James Ewing that recorded on a rotating disk of smoked glass. The first observation of a distant earthquake, or *teleseism*, was made in Potsdam in 1889 for a Japanese event. In 1897 the first North American seismograph was installed at Lick Observatory near San Jose in California; this device was later to record the 1906 San Francisco earthquake. These early instruments were undamped, and they could provide accurate estimates of ground motion only for a short time at the beginning of shaking. In 1898 E. Wiechert introduced the first seismometer with viscous damping, capable of producing useful records for the entire duration of an earthquake. The first electromagnetic seismographs, in which a moving pendulum is used to generate an electric current in a coil, were developed in the early 1900s, by B. B. Galitzen, who established a chain of stations across Russia. All modern seismographs are electromagnetic, since these instruments have numerous advantages over the purely mechanical designs of the earliest instruments.

The availability of seismograms recorded at a variety of ranges from earthquakes led to rapid progress in determining Earth's seismic velocity structure. By 1900 Richard Oldham reported the identification of P, S, and surface waves on seismograms, and later (1906) he detected the presence of Earth's core from the absence of direct P and S arrivals at source–receiver distances beyond about 100°. In 1909 Andrija Mohorovičić reported observations showing the existence of a velocity discontinuity separating the crust and mantle (this interface is now generally referred to, somewhat irreverently, as the "Moho"). Tabulations of arrival times led to the construction of travel time tables (arrival time as a function of distance from the earthquake); the first widely used tables were produced by Zöppritz in 1907. Beno Gutenberg published tables in 1914 with core phases (waves that penetrate or reflect off the core) and reported the first accurate estimate for the depth of Earth's fluid core (2900 km, very close to the modern value of 2889 km). In 1936, Inge Lehmann discovered the solid inner core, and in 1940 Harold Jeffreys and K. E. Bullen published the final version of their travel time tables for a large number of seismic phases. The JB tables are still in use today and contain times that differ by only a few seconds from current models.

The travel times of seismic arrivals can be used to determine Earth's average velocity versus depth structure, and this was largely accomplished over fifty years ago. The crust varies from about 6 km in thickness under the oceans to 30–50 km beneath continents. The deep interior is divided into three main layers: the mantle, the outer core, and the inner core (Fig. 1.1). The mantle is the solid rocky outer shell that makes up 84% of our planet's volume and 68% of the mass. It is characterized by a fairly rapid velocity increase in the upper mantle between about 300 and 700 km depth, a region termed the *transition zone*, where several mineralogical phase changes are believed to occur (including those at the 410 and 660 km seismic discontinuities, shown as the dashed arcs in Fig. 1.1). Between about 700 km to near the core–mantle boundary (CMB), velocities increase fairly gradually with depth; this increase is in general agreement with that expected from the changes in pressure and temperature on rocks of uniform composition and crystal structure.

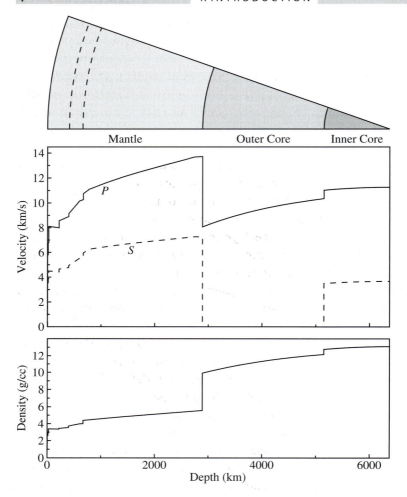

Figure 1.1 Earth's P velocity, S velocity, and density as a function of depth. Values are plotted from the Preliminary Reference Earth Model (PREM) of Dziewonski and Anderson (1981); except for some differences in the upper mantle, all modern Earth models are close to these values. PREM is listed as a table in Appendix A.

At the CMB, the P velocity drops dramatically from almost 14 km/s to about 8 km/s and the S velocity goes from about 7 km/s to zero. This change (larger than the velocity contrast at Earth's surface!) occurs at a sharp interface that separates the solid mantle from the fluid outer core. Within the outer core, the P velocity again increases gradually, at a rate consistent with that expected for a well-mixed fluid. However, at a radius of about 1221 km the core becomes solid, the P velocities increase slightly, and non-zero shear velocities are present. Earth's core is believed to be composed mainly of iron, and the inner-core boundary (ICB) is thought to represent a phase change in iron to a different crystal structure.

Earth's internal density distribution is much more difficult to determine than the velocity structure, since P and S travel times provide no direct constraints on density. However, by using probable velocity versus density scaling relationships and Earth's known mass and moment of inertia, K. E. Bullen showed that it is possible to infer a density profile similar to that shown in Figure 1.1. Modern results from normal mode seismology, which provides more direct constraints on density (although with limited vertical resolution), have generally proven consistent with the older density profiles.

Seismic surveying using explosions and other artificial sources was developed during the 1920s and 1930s for prospecting purposes in the oil-producing regions of Mexico and the United States. Early work involved measuring the travel time versus distance of P waves to determine seismic velocity at depth. Later studies focused on reflections from subsurface layering (reflection seismology), which can achieve high resolution when instruments are closely spaced. The common-midpoint (CMP) stacking method for reflection seismic data was patented in 1956, leading to reduced noise levels and higher-quality profiles. The Vibroseis method, also developed in the 1950s, applies signal-processing techniques to data recorded using a long-duration, vibrating source.

The increasing number of seismic stations established in the early 1900s enabled large earthquakes to be routinely located, leading to the discovery that earthquakes are not randomly distributed but tend to occur along well-defined belts (Fig. 1.2). However, the significance of these belts was not fully appreciated until the 1960s, as part of the plate tectonics revolution in the Earth sciences. At that time, it was recognized that Earth's surface features are largely determined by the motions of a small number of relatively rigid plates that drift slowly over geological time (Fig. 1.3). The relative motions between adjacent plates give rise to earthquakes along the plate boundaries. The plates are spreading apart along the mid-oceanic ridges, where new oceanic lithosphere is being formed. This has caused the splitting apart and separation of Europe and Africa from the Americas (the "continental drift" hypothesized by Alfred Wegener in 1915). The plates are recycled back into the mantle in the trenches and subduction zones around the Pacific margin. Large shear faults, such as the San Andreas Fault in California, are a result of transverse motion between plates. Plate boundaries across continents are often more diffuse and marked by distributed seismicity, such as occurs in the Himalayan region between the northward moving Indian Plate and the Eurasian Plate.

In the 1960s, seismologists were able to show that the *focal mechanisms* (the type of faulting as inferred from the radiated seismic energy) of most global earthquakes are consistent with that expected from plate tectonic theory, thus helping to validate the still emerging paradigm. However, considering the striking similarity between Figures 1.2 and 1.3, why didn't seismologists begin to develop the theory of plate

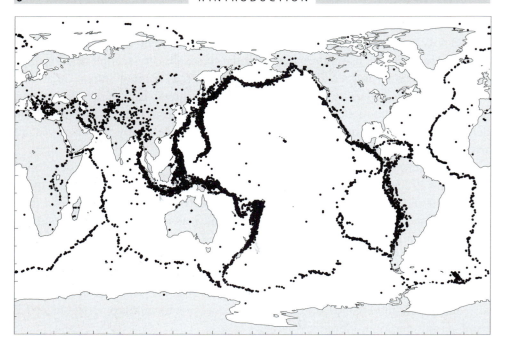

Figure 1.2 Selected global earthquake locations from 1977 to 1994 (taken from the PDE and ISC catalogs). Earthquakes occur along well-defined belts of seismicity; these are particularly prominent around the Pacific rim and along mid-oceanic ridges. We now know that these belts define the edges of the tectonic plates within Earth's rigid outermost layer (see Fig. 1.3).

tectonics much earlier? In part, this can be attributed to the lower resolution of the older earthquake locations compared to more modern results. However, a more important reason was that seismologists, like most geophysicists at the time, did not feel that ideas of continental drift had a sound physical basis. Thus they were unable to fully appreciate the significance and implications of the earthquake locations, and tended to interpret their results in terms of local and regional tectonics, rather than a unifying global theory.

In 1923, H. Nakano introduced the theory for the seismic radiation from a double-couple source (two pairs of opposing point forces). For about the next forty years, a controversy would rage over the question of whether a single- or double-couple source is the most appropriate for earthquakes, despite the fact that theory shows that single-couple sources are physically impossible. In 1928, Kiyoo Wadati reported the first convincing evidence for deep focus earthquakes (below 100 km depth). A few years earlier, H. H. Turner had located some earthquakes at significant depth, but his analyses were not generally accepted (particularly since he also located some events in the air above the surface!). Deep focus events are typically observed along dipping planes of seismicity (often termed Wadati–Benioff zones) that can extend

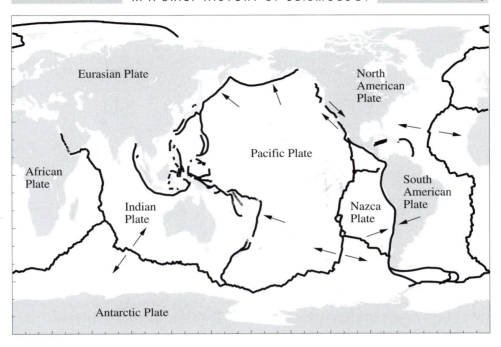

Figure 1.3 Earth's major tectonic plates. The arrows indicate relative plate motions at some of the plate boundaries. The plates are pulling apart along spreading centers, such as the Mid-Atlantic Ridge, where new crust is being formed. Along the subduction zones in the western Pacific, the Pacific Plate is sliding back down into the mantle. The San Andreas Fault in California is a result of shear between the Pacific and North American Plates.

to almost 700 km depth; these mark the locations of subducting slabs of oceanic lithosphere that are found surrounding much of the Pacific Ocean. Figure 1.4 shows a cross-section of the earthquake locations in the Tonga subduction zone in the southwest Pacific, the world's most active area of deep seismicity. The existence of deep events was a surprising discovery because the high pressures and temperatures that exist at these depths should make most materials deform ductilely, without the sudden brittle failure that causes shallow earthquakes in the crust. Even today the physical mechanism for deep events is not well understood and is a continuing source of controversy.

In 1946, an underwater nuclear explosion near Bikini Atoll led to the first detailed seismic recordings of a nuclear bomb. Perhaps a more significant development, at least for western government funding for seismology, was the 1949 testing of a Soviet nuclear bomb. This led to an intense interest by the US military in the ability of seismology to detect nuclear explosions, estimate yields, and discriminate between explosions and earthquakes. A surge in funding for seismology resulted, helping to improve seismic instrumentation and expand government and university

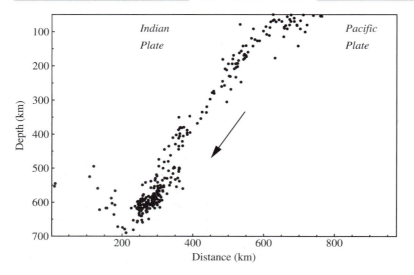

Figure 1.4 A vertical west–east cross-section of the deep seismicity in the Tonga subduction zone, showing selected earthquakes from the PDE and ISC catalogs between 1977 and 1994. The seismicity marks where the lithosphere of the Pacific Plate is sinking down into the mantle.

seismology programs. In 1961 the Worldwide Standardized Seismograph Network (WWSSN) was established, consisting of well-calibrated instruments with both short- and long-period seismometers. The ready availability of records from these seismographs led to rapid improvements in many areas of seismology, including the production of much more complete and accurate catalogs of earthquake locations and the long overdue recognition that earthquake radiation patterns are consistent with double-couple sources.

Records obtained from the great Chilean earthquake of 1960 were the first to provide definitive observations of Earth's *free oscillations*. Any finite solid will resonate only at certain vibration frequencies, and these *normal modes* provide an alternative to the traveling wave representation for characterizing the deformations in the solid. Earth "rings" for several days following large earthquakes, and its normal modes are seen as peaks in the power spectrum of seismograms. The 1960s and 1970s saw the development of the field of normal mode seismology, which gives some of the best constraints on the large-scale structure, particularly in density, of Earth's interior. Analyses of normal mode data also led to the development of many important ideas in geophysical inverse theory, providing techniques for evaluating the uniqueness and resolution of Earth models obtained from indirect observations.

Between 1969 and 1972, seismometers were placed on the Moon by the Apollo astronauts, and the first lunar quakes were recorded. These include surface impacts, shallow quakes within the top 100 km, and deeper quakes at roughly 800 to 1,000 km depth. Lunar seismograms appear very different from those on Earth,

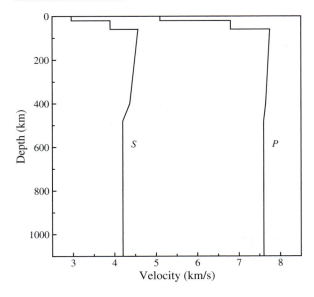

Figure 1.5 An approximate seismic velocity model derived for the Moon from observations of quakes and surface impacts (from Goins *et al.*, 1981). Velocities at greater depths (the lunar radius is 1737 km) are largely unconstrained owing to a lack of deep seismic waves in the *Apollo* data set.

with lengthy wavetrains of high-frequency scattered energy. This has complicated their interpretation, but a lunar crust and mantle have been identified, with a crustal thickness of about 60 km (see Fig. 1.5). A seismometer placed on Mars by the *Viking 2* probe in 1976 was hampered by wind noise, and only one possible Mars quake was identified.

Although it is not practical to place seismometers on the Sun, it is possible to detect oscillations of the solar surface by measuring the Doppler shift of spectral lines. Such oscillations were first observed in 1960 by Robert Leighton, who discovered that the Sun's surface vibrates continually at a period of about five minutes and is incoherent over small spatial wavelengths. These oscillations were initially interpreted as resulting from localized gas movements near the solar surface, but in the late 1960s several researchers proposed that the oscillations resulted from acoustic waves trapped within the Sun. This idea was confirmed in 1975 when it was shown that the pattern of observed vibrations is consistent with that predicted for the free oscillations of the Sun, and the field of *helioseismology* was born. Analysis is complicated by the fact that, unlike Earth, impulsive sources analogous to earthquakes are rarely observed; the excitation of acoustic energy is a continuous process. However, many of the analysis techniques developed for normal mode seismology can be applied, and the radial velocity structure of the Sun is now well constrained (Fig. 1.6). Continuing improvements in instrumentation and dedicated experiments promise further breakthroughs, including resolution of spatial and temporal variations in solar velocity structure. In only a few decades, helioseismology has become one of the most important tools for examining the structure of the Sun.

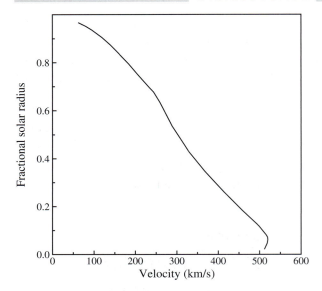

Figure 1.6 The velocity of sound within the Sun (adapted from Harvey, 1995).

The advent of computers in the 1960s changed the nature of terrestrial seismology, by enabling analyses of large data sets and more complicated problems, and led to the routine calculation of earthquake locations. The first complete theoretical seismograms for complicated velocity structures began to be computed at this time. The computer era also has seen the rapid expansion of seismic imaging techniques using artificial sources that have been applied extensively by the oil industry to map shallow crustal structure. Beginning in 1976, data started to become available from global seismographs in digital form, greatly facilitating quantitative waveform comparisons. In recent years, many of the global seismic stations have been upgraded to broadband, high dynamic range seismometers, and new instruments have been deployed to fill in gaps in the global coverage. Large numbers of portable instruments have also become available for specialized experiments in particular regions. Seismic records are now far easier to obtain, with centralized archives providing online data access in standard formats.

Earth's average radial velocity and density structures were well established by 1970, including the existence of minor velocity discontinuities near 410- and 660-km depth in the upper mantle. Attention then shifted to resolving lateral differences in velocity structure, first by producing different velocity versus depth profiles for different regions, and more recently by inverting seismic data directly for three-dimensional velocity structures. The latter methods have been given the name "tomography" by analogy to medical imaging techniques. During recent years, tomographic methods of increasing resolution have begun to provide spectacular images of the structure of Earth's crust and mantle at a variety of scale lengths. Local earthquake tomography at scales from tens to hundreds of kilometers has imaged details of crustal structure in many different regions, including the slow seismic velocities found in sedimentary basins and the sharp velocity changes that can occur near active fault zones.

Figure 1.7 shows seismic velocity perturbations in the mantle, as recently imaged using whole-Earth tomographic methods. Note that the velocity anomalies are strongest at the top and bottom of the mantle, with high velocities beneath the continents in the uppermost mantle and in a ring surrounding the Pacific in the lowermost mantle. Many, but not all, geophysicists ascribe these fast velocities near

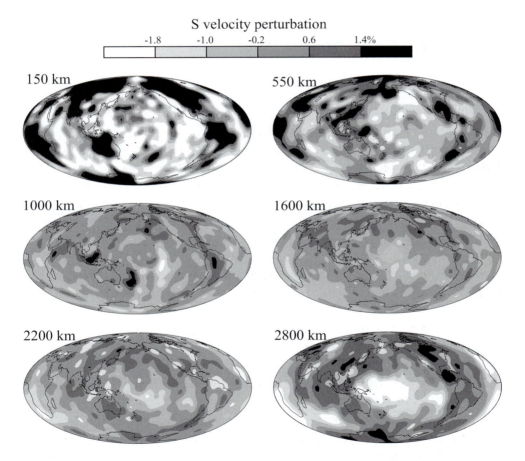

Figure 1.7 Lateral variations in *S* velocity at depths of 150, 550, 1000, 1600, 2200, and 2800 km in the mantle from Manners and Masters (2008). Velocity perturbations are contoured as shown, with black indicating regions that are more than 1.4% faster than average, and white indicating velocities over 1.8% slower than average.

the core–mantle boundary to the pooling of cold descending slabs from current and past subduction zones around the Pacific. The slow lowermost mantle S velocities seen beneath the south-central Pacific have often been interpreted as a warm region that may feed plumes and oceanic island volcanism, but differences between P- and S-wave tomography models now indicate that the anomaly is largely compositional in origin (e.g., Masters *et al.*, 2000). Other features include ponding of slabs in the transition zone between the 410- and 660-km discontinuties (see 550 km slice) as well as some evidence for slabs in the midmantle beneath Tonga and South America (see 1000 km slice).

At shallower depths, reflection seismic experiments using controlled sources have led to detailed images of crustal structure, both on land and beneath the oceans (Fig. 1.8). The ability to image three-dimensional structures has greatly expanded the power of seismology to help resolve many outstanding problems in

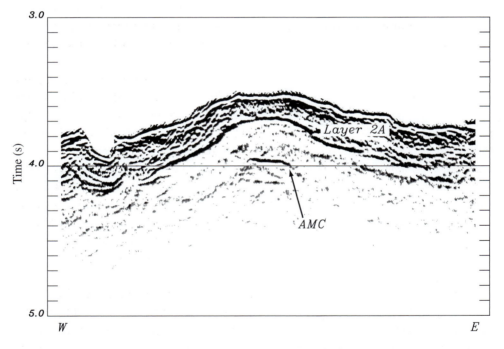

Figure 1.8 An image of the axial magma chamber (AMC) beneath the East Pacific Rise near 14°15′S obtained through migration processing of reflection seismic data (from Kent et al., 1994). The profile is about 7 km across, with the vertical axis representing the two-way travel time of compressional waves between the surface of the ocean and the reflection point. The sea floor is the reflector at about 3.5 s in the middle of the plot, while the magma chamber appears at about 4.0 s and is roughly 750 m wide. Shallow axial magma chambers are commonly seen beneath fast-spreading oceanic ridges, such as those in the eastern Pacific, but not beneath slow-spreading ridges, such as the Mid-Atlantic Ridge.

the Earth sciences. These include the structure of fault zones at depth, the deep roots of continents, the properties of mineralogical phase changes in the mantle, the fate of subducting slabs, the structure of oceanic spreading centers, the nature of convection within the mantle, the complicated details of the core–mantle boundary region, and the structure of the inner core.

Most of the preceding discussion is concerned with *structural seismology*, or using records of seismic waves to learn about Earth's internal structure. Progress has also been made in learning about the physics of earthquakes themselves. The turning point came with the investigations following the 1906 San Francisco earthquake. H. F. Reid, an American engineer, studied survey lines across the fault taken before and after the earthquake. His analysis led to the *elastic rebound* theory for the origin of earthquakes in which a slow accumulation of shear stress and strain is suddenly released by movement along a fault. Subsequent work has confirmed that this mechanism is the primary cause of tectonic earthquakes in the crust and is capable of quickly releasing vast amounts of energy. Today, observations of large-scale deformations following large earthquakes, using land- and satellite-based surveying methods, are widely used to constrain the distribution of slip on subsurface faults.

The first widely used measure of earthquake size was the magnitude scale developed for earthquakes in southern California by Charles Richter and Beno Gutenberg in 1935. Because the Richter scale is logarithmic, a small range of Richter magnitudes can describe large variations in earthquake size. The smallest earthquakes that are readily felt at the surface have magnitudes of about 3, while great earthquakes such as the 1906 San Francisco earthquake have magnitudes of 8 or greater. A number of different magnitude scales, applicable to different types of seismic observations, have now been developed that are based on Richter's idea. However, most of these scales are empirical and not directly related to properties of the source. A more physically based measure of earthquake size, the *seismic moment*, was formulated by Keiiti Aki in 1966. This led to the definition of the *moment magnitude*, which remains on scale even for the earthquakes of magnitude 8 and greater.

Because catastrophic earthquakes occur rarely in any particular region, humanity often forgets how devastating these events can be. However, history should remind us of their power to suddenly kill tens of thousands of people (see Table 1.1) and of the importance of building earthquake resistant structures. Earth's rapidly increasing population, particularly in cities in seismically active regions, means that future earthquakes may be even more deadly. The great earthquake and tsunami of December 2004 killed over 250 000 people in Sumatra and around the northeast Indian ocean. This earthquake was the first magnitude 9+ earthquake recorded by modern broadband seismographs (instruments were much more primitive for the 1960 Chile and 1964 Alaskan earthquakes). The Sumatra earthquake lasted over

Table 1.1: Earthquakes with 70 000 or more deaths.			
Year	Location	Magnitude	Deaths
856	Damghan, Iran		200 000
893	Ardabil, Iran		150 000
1138	Aleppo, Syria		230 000
1290	Chihli, China		100 000
1556	Shansi, China	~8	830 000
1667	Shemakha, Caucasia		80 000
1727	Tabriz, Iran		77 000
1755	Lisbon, Portugal	8.7	70 000
1908	Messina, Italy	7.2	~85 000
1920	Gansu, China	7.8	200 000
1923	Kanto, Japan	7.9	143 000
1927	Tsinghai, China	7.9	200 000
1932	Gansu, China	7.6	70 000
1948	Ashgabat, Turkmenistan	7.3	110 000
1976	Tangshan, China	7.5	255 000
2004	Sumatra	9.1	283 106
2005	Pakistan	7.6	86 000
2008	Eastern Sichuan, China	7.9	87 652

Source: http://earthquake.usgs.gov/regional/world/most-destructive.php

8 minutes and ruptured about 1300 km of fault (see Fig. 1.9). Seismic wave displacements caused by this event were over a centimeter when its surface waves crossed the United States, over 12 000 kilometers away. The radiated seismic energy from this earthquake has been estimated as 1.4 to 3 $\times 10^{17}$ joules (Kanamori, 2006; Choy and Boatwright, 2007). Normal modes excited by this event could be observed for several months as the Earth continued to vibrate at very long periods.

During the past few decades, large networks of seismometers have been deployed in seismically active regions to map out patterns of earthquake activity, and strong motion instruments have been used to obtain on-scale recordings near large earthquakes. It has become possible to map the time–space history of the slip distribution on faults during major earthquakes. Despite these advances, many fundamental questions regarding the nature of earthquakes remain largely unanswered, including the origin of deep events and the processes by which the rupture

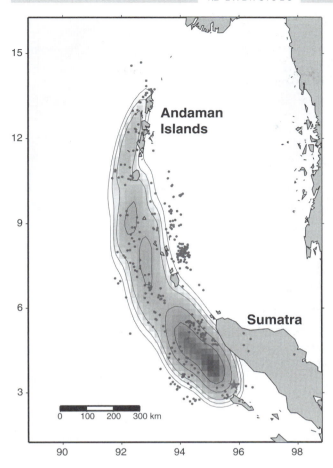

Figure 1.9 The 2004 Sumatra-Andaman earthquake as imaged by Ishii *et al.* (2005) using high-frequency data from the Japanese Hi-Net array. Note the good agreement between the 1300-km-long rupture zone and the locations of the first 35 days of aftershocks (small dots).

of a crustal fault initiates, propagates, and eventually comes to a halt. It is perhaps in these areas of earthquake physics that some of seismology's most important future discoveries remain to be made.

1.2 Exercises

1. The radii of the Earth, Moon, and Sun are 6371 km, 1738 km, and 6.951×10^5 km, respectively. From Figures 1.1, 1.5, and 1.6, make a rough estimate of how long it takes a P wave to traverse the diameter of each body.

2. Assuming that the P velocity in the ocean is 1.5 km/s, estimate the minimum and maximum water depths shown in Figure 1.8. If the crustal P velocity is 5 km/s, what is the depth to the top of the magma chamber from the sea floor?

3. Assume that the S velocity perturbations plotted at 150 km depth in Figure 1.7 extend throughout the uppermost 300 km of the mantle. Estimate how many seconds earlier a vertically upgoing S-wave will arrive at a seismic station in the middle of Canada, compared to a station in the eastern Pacific. Ignore any topographic or crustal thickness differences between the sites; consider only the integrated travel time difference through the upper mantle.

4. Earthquake moment is defined as $M_0 = \mu DA$, where μ is the shear modulus, D is the average displacement on the fault, and A is the fault area that slipped. The moment of the 2004 Sumatra-Andaman earthquake has been estimated to be about 1.0×10^{23} N m. Assuming that the fault is horizontal, crudely estimate the slip area from the image shown in Figure 1.9. Assuming that the shear modulus $\mu = 3.0 \times 10^{10}$ N/m^2, then compute the average displacement on the fault.

5. Do some research on the web to find the energy release of the following: (a) a 1 megaton nuclear explosion, (b) the yearly electricity consumption in the United States, (c) yearly dissipation of tidal energy in Earth's oceans, and (d) the daily energy release of a typical hurricane. Express all your answers in joules (J) and compare these numbers to the seismic energy release of the 2004 Sumatra earthquake (see text). Note that the total energy release (including heat generated on the fault, etc.) of the Sumatra earthquake may be significantly greater than the seismically radiated energy. This is discussed in Chapter 9.

Stress and strain

Any quantitative description of seismic wave propagation or of earthquake physics requires the ability to characterize the internal forces and deformations in solid materials. We now begin a brief review of those parts of stress and strain theory that will be needed in subsequent chapters. Although this section is intended to be self-contained, we will not derive many equations, and the reader is referred to any continuum mechanics text (Malvern, 1969, is a classic but there are many others) for further details.

Deformations in three-dimensional materials are termed strain; internal forces between different parts of the medium are called stress. Stress and strain do not exist independently in materials; they are linked through the constitutive relationships that describe the nature of elastic solids.

2.1 The stress tensor

Consider an infinitesimal plane of arbitrary orientation within a homogeneous elastic medium in static equilibrium. The orientation of the plane may be specified by its unit normal vector, $\hat{\mathbf{n}}$. The force per unit area exerted by the side in the direction of $\hat{\mathbf{n}}$ across this plane is termed the *traction* and is represented by the vector $\mathbf{t}(\hat{\mathbf{n}}) = (t_x, t_y, t_z)$. If $\mathbf{t}$ acts in the direction shown 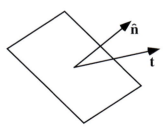 here, then the traction force is pulling the opposite side toward the interface. This definition is the usual convention in seismology and results in extensional forces being positive and compressional forces being negative. In some other fields, such as rock mechanics, the definition is reversed and compressional forces are positive. There is an equal and opposite force exerted by the side opposing $\hat{\mathbf{n}}$, such that

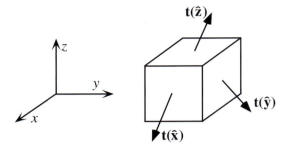

Figure 2.1 The traction vectors $\mathbf{t}(\hat{\mathbf{x}})$, $\mathbf{t}(\hat{\mathbf{y}})$, and $\mathbf{t}(\hat{\mathbf{z}})$ describe the forces on the faces of an infinitesimal cube in a Cartesian coordinate system.

$\mathbf{t}(-\hat{\mathbf{n}}) = -\mathbf{t}(\hat{\mathbf{n}})$. The part of $\mathbf{t}$ which is normal to the plane is termed the *normal stress*, that which is parallel is called the *shear stress*. In the case of a fluid, there are no shear stresses and $\mathbf{t} = -P\hat{\mathbf{n}}$, where P is the pressure.

In general, the magnitude and direction of the traction vector will vary as a function of the orientation of the infinitesimal plane. Thus, to fully describe the internal forces in the medium, we need a general method for determining $\mathbf{t}$ as a function of $\hat{\mathbf{n}}$. This is accomplished with the *stress tensor*, which provides a linear mapping between $\hat{\mathbf{n}}$ and $\mathbf{t}$. The stress tensor, $\boldsymbol{\tau}$, in a Cartesian coordinate system (Fig. 2.1) may be defined[1] by the tractions across the yz, xz, and xy planes:

$$\boldsymbol{\tau} = \begin{bmatrix} t_x(\hat{\mathbf{x}}) & t_x(\hat{\mathbf{y}}) & t_x(\hat{\mathbf{z}}) \\ t_y(\hat{\mathbf{x}}) & t_y(\hat{\mathbf{y}}) & t_y(\hat{\mathbf{z}}) \\ t_z(\hat{\mathbf{x}}) & t_z(\hat{\mathbf{y}}) & t_z(\hat{\mathbf{z}}) \end{bmatrix} = \begin{bmatrix} \tau_{xx} & \tau_{xy} & \tau_{xz} \\ \tau_{yx} & \tau_{yy} & \tau_{yz} \\ \tau_{zx} & \tau_{zy} & \tau_{zz} \end{bmatrix}. \tag{2.1}$$

Because the solid is in static equilibrium, there can be no net rotation from the shear stresses. For example, consider the shear stresses in the xz plane. To balance the torques, $\tau_{xz} = \tau_{zx}$. Similarly, $\tau_{xy} = \tau_{yx}$ and $\tau_{yz} = \tau_{zy}$, and the stress tensor $\boldsymbol{\tau}$ is symmetric, that is,

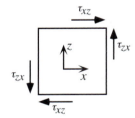

$$\boldsymbol{\tau} = \boldsymbol{\tau}^T = \begin{bmatrix} \tau_{xx} & \tau_{xy} & \tau_{xz} \\ \tau_{xy} & \tau_{yy} & \tau_{yz} \\ \tau_{xz} & \tau_{yz} & \tau_{zz} \end{bmatrix}. \tag{2.2}$$

The stress tensor $\boldsymbol{\tau}$ contains only six independent elements, and these are sufficient to completely describe the state of stress at a given point in the medium.

---

[1] Often the stress tensor is defined as the transpose of (2.1) so that the first subscript of $\boldsymbol{\tau}$ represents the surface normal direction. In practice, it makes no difference as $\boldsymbol{\tau}$ is symmetric.

The traction across any arbitrary plane of orientation defined by $\hat{\mathbf{n}}$ may be obtained by multiplying the stress tensor by $\hat{\mathbf{n}}$, that is,

$$
\mathbf{t}(\hat{\mathbf{n}}) = \boldsymbol{\tau}\hat{\mathbf{n}} = \begin{bmatrix} t_x(\hat{\mathbf{n}}) \\ t_y(\hat{\mathbf{n}}) \\ t_z(\hat{\mathbf{n}}) \end{bmatrix} = \begin{bmatrix} \tau_{xx} & \tau_{xy} & \tau_{xz} \\ \tau_{xy} & \tau_{yy} & \tau_{yz} \\ \tau_{xz} & \tau_{yz} & \tau_{zz} \end{bmatrix} \begin{bmatrix} \hat{n}_x \\ \hat{n}_y \\ \hat{n}_z \end{bmatrix}. \tag{2.3}
$$

This can be shown by summing the forces on the surfaces of a tetrahedron (the *Cauchy tetrahedron*) bounded by the plane normal to $\hat{\mathbf{n}}$ and the xy, xz, and yz planes.

The stress tensor is simply the linear operator that produces the traction vector $\mathbf{t}$ from the normal vector $\hat{\mathbf{n}}$, and, in this sense, the stress tensor exists independent of any particular coordinate system. In seismology we almost always write the stress tensor as a 3×3 matrix in a Cartesian geometry. Note that the symmetry requirement reduces the number of independent parameters in the stress tensor to six from the nine that are present in the most general form of a second-order tensor (scalars are considered zeroth-order tensors, vectors are first order, etc.).

The stress tensor will normally vary with position in a material; it is a measure of the forces acting on infinitesimal planes at each point in the solid. Stress provides a measure only of the forces exerted across these planes and has units of force per unit area. However, other forces may be present (e.g., gravity); these are termed *body forces* and have units of force per unit volume or mass.

2.1.1 Example: Computing the traction vector

Suppose we are given that the horizontal components of the stress tensor are

$$
\boldsymbol{\tau} = \begin{bmatrix} \tau_{xx} & \tau_{xy} \\ \tau_{xy} & \tau_{yy} \end{bmatrix} = \begin{bmatrix} -40 & -10 \\ -10 & -60 \end{bmatrix} \text{MPa.}
$$

Assuming this is a two-dimensional problem, let us compute the forces acting across a fault oriented at $45°$ (clockwise) from the x-axis. We typically assume that the x-axis points east and the y-axis points north, so in this case the fault is trending from the northwest to the southeast. To compute the traction vector from equation (2.3), we need the normal vector $\hat{\mathbf{n}}$. This vector is perpendicular to the fault and thus points to the northeast, or parallel to the vector $(1,1)$ in our (x, y) coordinate system. However, remember that $\hat{\mathbf{n}}$ is a *unit* vector and thus we must normalize its length to obtain

$$
\hat{\mathbf{n}} = \begin{bmatrix} 1/\sqrt{2} \\ 1/\sqrt{2} \end{bmatrix} = \begin{bmatrix} 0.7071 \\ 0.7071 \end{bmatrix}.
$$

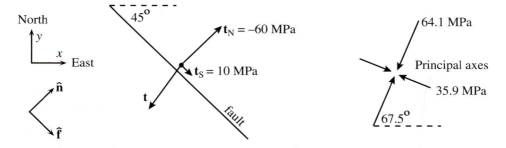

Figure 2.2 The fault tractions and principal stresses for Examples 2.1.1 and 2.1.3.

Substituting into (2.3), we have

$$\mathbf{t}(\hat{\mathbf{n}}) = \tau\hat{\mathbf{n}} = \begin{bmatrix} -40 & -10 \\ -10 & -60 \end{bmatrix} \begin{bmatrix} 1/\sqrt{2} \\ 1/\sqrt{2} \end{bmatrix} = \begin{bmatrix} -50/\sqrt{2} \\ -70/\sqrt{2} \end{bmatrix} \approx \begin{bmatrix} -35.4 \\ -49.4 \end{bmatrix} \text{ MPa.}$$

Note that the traction vector points approximately southwest (see Fig. 2.2). This is the force exerted by the northeast side of the fault (i.e., in the direction of our normal vector) on the southwest side of the fault. Thus we see that there is fault normal compression on the fault. To resolve the normal and shear stress on the fault, we compute the dot products with unit vectors perpendicular ($\hat{\mathbf{n}}$) and parallel ($\hat{\mathbf{f}}$) to the fault

$$\mathbf{t}_N = \mathbf{t} \cdot \hat{\mathbf{n}} = (-50/\sqrt{2}, -70/\sqrt{2}) \cdot (1/\sqrt{2}, 1/\sqrt{2}) = -60 \text{ MPa}$$
$$\mathbf{t}_S = \mathbf{t} \cdot \hat{\mathbf{f}} = (-50/\sqrt{2}, -70/\sqrt{2}) \cdot (1/\sqrt{2}, -1/\sqrt{2}) = 10 \text{ MPa.}$$

The fault normal compression is 60 MPa. The shear stress is 10 MPa.

2.1.2 Principal axes of stress

For any stress tensor, it is always possible to find a direction $\hat{\mathbf{n}}$ such that there are no shear stresses across the plane normal to $\hat{\mathbf{n}}$, that is, $\mathbf{t}(\hat{\mathbf{n}})$ points in the $\hat{\mathbf{n}}$ direction. In this case

$$\mathbf{t}(\hat{\mathbf{n}}) = \lambda\hat{\mathbf{n}} = \tau\hat{\mathbf{n}},$$
$$\tau\hat{\mathbf{n}} - \lambda\hat{\mathbf{n}} = 0, \tag{2.4}$$
$$(\boldsymbol{\tau} - \lambda\mathbf{I})\hat{\mathbf{n}} = 0,$$

where $\mathbf{I}$ is the identity matrix and λ is a scalar. This is an eigenvalue problem that has a non-trivial solution only when the determinant vanishes

$$\det[\boldsymbol{\tau} - \lambda\mathbf{I}] = 0. \tag{2.5}$$

This is a cubic equation with three solutions, the eigenvalues λ_1, λ_2, and λ_3 (do not confuse these with the Lamé parameter λ that we will discuss later). Since $\boldsymbol{\tau}$ is symmetric and real, the eigenvalues are real. Corresponding to the eigenvalues are the eigenvectors $\hat{\mathbf{n}}^{(1)}$, $\hat{\mathbf{n}}^{(2)}$, and $\hat{\mathbf{n}}^{(3)}$. The eigenvectors are orthogonal and define the *principal axes* of stress. The planes perpendicular to these axes are termed the *principal planes*. We can rotate $\boldsymbol{\tau}$ into the $\hat{\mathbf{n}}^{(1)}$, $\hat{\mathbf{n}}^{(2)}$, $\hat{\mathbf{n}}^{(3)}$ coordinate system by applying a similarity transformation (see Appendix B for details about coordinate rotations and transformation tensors):

$$\boldsymbol{\tau}^R = \mathbf{N}^{\mathrm{T}}\boldsymbol{\tau}\mathbf{N} = \begin{bmatrix} \tau_1 & 0 & 0 \\ 0 & \tau_2 & 0 \\ 0 & 0 & \tau_3 \end{bmatrix}, \tag{2.6}$$

where $\boldsymbol{\tau}^R$ is the rotated stress tensor and τ_1, τ_2, and τ_3 are the *principal stresses* (identical to the eigenvalues λ_1, λ_2, and λ_3). Here $\mathbf{N}$ is the matrix of eigenvectors

$$\mathbf{N} = \begin{bmatrix} n_x^{(1)} & n_x^{(2)} & n_x^{(3)} \\ n_y^{(1)} & n_y^{(2)} & n_y^{(3)} \\ n_z^{(1)} & n_z^{(2)} & n_z^{(3)} \end{bmatrix}, \tag{2.7}$$

with $\mathbf{N}^{\mathrm{T}} = \mathbf{N}^{-1}$ for orthogonal eigenvectors normalized to unit length.

By convention, the three principal stresses are sorted by size, such that $|\tau_1| > |\tau_2| > |\tau_3|$. The maximum shear stress occurs on planes at $45°$ to the maximum and minimum principle stress axes. In the principal axes coordinate system, one of these planes has normal vector $\hat{\mathbf{n}} = (1/\sqrt{2}, 0, 1/\sqrt{2})$. The traction vector for the stress across this plane is

$$\mathbf{t}(45°) = \begin{bmatrix} \tau_1 & 0 & 0 \\ 0 & \tau_2 & 0 \\ 0 & 0 & \tau_3 \end{bmatrix} \begin{bmatrix} 1/\sqrt{2} \\ 0 \\ 1/\sqrt{2} \end{bmatrix} = \begin{bmatrix} \tau_1/\sqrt{2} \\ 0 \\ \tau_3/\sqrt{2} \end{bmatrix}. \tag{2.8}$$

This can be decomposed into normal and shear stresses on the plane:

$$\mathbf{t}_N(45°) = \mathbf{t}(45°) \cdot (1/\sqrt{2},\ 0,\ 1/\sqrt{2}) = (\tau_1 + \tau_3)/2 \tag{2.9}$$

$$\mathbf{t}_S(45°) = \mathbf{t}(45°) \cdot (1/\sqrt{2},\ 0,\ -1/\sqrt{2}) = (\tau_1 - \tau_3)/2 \tag{2.10}$$

and we see that the maximum shear stress is $(\tau_1 - \tau_3)/2$.

If $\tau_1 = \tau_2 = \tau_3$, then the stress field is called *hydrostatic* and there are no planes of any orientation in which shear stress exists. In a fluid the stress tensor can be written

$$\tau = \begin{bmatrix} -P & 0 & 0 \\ 0 & -P & 0 \\ 0 & 0 & -P \end{bmatrix}, \qquad (2.11)$$

where P is the pressure.

2.1.3 Example: Computing the principal axes

Let us compute the principal axes for our previous example, for which the 2-D stress tensor is given by

$$\tau = \begin{bmatrix} -40 & -10 \\ -10 & -60 \end{bmatrix} \text{MPa}$$

From equation (2.5), we have

$$\det \begin{bmatrix} -40 - \lambda & -10 \\ -10 & -60 - \lambda \end{bmatrix} = 0$$

or

$$(-40 - \lambda)(-60 - \lambda) - (-10)^2 = 0$$
$$\lambda^2 + 100\lambda + 2300 = 0 \qquad \frac{-b \pm \sqrt{b^2 - 4ac}}{2a}$$

This quadratic equation has roots $\lambda_1 = -64.14$ and $\lambda_2 = -35.86$. Substituting into equation (2.4), we have two eigenvector equations

$$\begin{bmatrix} 24.14 & -10 \\ -10 & 4.14 \end{bmatrix} \begin{bmatrix} n_x^{(1)} \\ n_y^{(1)} \end{bmatrix} = 0 \quad \text{and} \quad \begin{bmatrix} -4.14 & -10 \\ -10 & -24.14 \end{bmatrix} \begin{bmatrix} n_x^{(2)} \\ n_y^{(2)} \end{bmatrix} = 0$$

with solutions for the two eigenvectors (normalized to unit length) of

$$\hat{\mathbf{n}}^{(1)} = \begin{bmatrix} 0.3827 \\ 0.9239 \end{bmatrix} \quad \text{and} \quad \hat{\mathbf{n}}^{(2)} = \begin{bmatrix} -0.9239 \\ 0.3827 \end{bmatrix}.$$

Note that these vectors are orthogonal ($\hat{\mathbf{n}}^{(1)} \cdot \hat{\mathbf{n}}^{(2)} = 0$) and define the principal stress axes. The maximum compressive stress is in the direction $\hat{\mathbf{n}}^{(1)}$, or at an angle to 67.5° with the x axis (see Fig. 2.2). The eigenvector matrix is

$$
\mathbf{N} = \begin{bmatrix} n_x^{(1)} & n_x^{(2)} \\ n_y^{(1)} & n_y^{(2)} \end{bmatrix} = \begin{bmatrix} 0.383 & -0.924 \\ 0.924 & 0.383 \end{bmatrix}
$$

which we can use to rotate $\boldsymbol{\tau}$ into the principal stress coordinate system:

$$
\boldsymbol{\tau}^R = \mathbf{N}^T \boldsymbol{\tau} \mathbf{N} = \begin{bmatrix} 0.383 & 0.924 \\ -0.924 & 0.383 \end{bmatrix} \begin{bmatrix} -40 & -10 \\ -10 & -60 \end{bmatrix} \begin{bmatrix} 0.383 & -0.924 \\ 0.924 & 0.383 \end{bmatrix}
$$

$$
= \begin{bmatrix} -64.14 & 0 \\ 0 & -35.86 \end{bmatrix} \text{MPa.}
$$

As expected, the principal stresses are simply the eigenvalues, λ_1 and λ_2. In practice, matrix eigenvector problems are most easily solved using software such as Matlab or Mathematica, or an appropriate computer subroutine. A Matlab script to solve this example is given in the supplemental web material.

2.1.4 Deviatoric stress

Stresses in the deep Earth are dominated by the large compressive stress from the hydrostatic pressure. Often it is convenient to consider only the much smaller *deviatoric* stresses, which are computed by subtracting the mean normal stress (given by the average of the principle stresses, that is $\tau_m = (\tau_1 + \tau_2 + \tau_3)/3$) from the diagonal components of the stress tensor, thus defining the deviatoric stress tensor

$$
\boldsymbol{\tau}_D = \begin{bmatrix} \tau_{xx} - \tau_m & \tau_{xy} & \tau_{xz} \\ \tau_{xy} & \tau_{yy} - \tau_m & \tau_{yz} \\ \tau_{xz} & \tau_{yz} & \tau_{zz} - \tau_m \end{bmatrix} \tag{2.12}
$$

It should be noted that the trace of the stress tensor is invariant with respect to rotation, so the mean stress τ_m can be computed by averaging the diagonal elements of $\boldsymbol{\tau}$ without computing the eigenvalues (i.e., $\tau_m = (\tau_{11} + \tau_{22} + \tau_{33})/3$). In addition, the deviatoric stress tensor has the same principal stress axes as the original stress tensor.

The stress tensor can then be written as the sum of two parts, the hydrostatic stress tensor $\tau_m \mathbf{I}$ and the deviatoric stress tensor $\boldsymbol{\tau}_D$

$$
\boldsymbol{\tau} = \tau_m \mathbf{I} + \boldsymbol{\tau}_D = \begin{bmatrix} -p & 0 & 0 \\ 0 & -p & 0 \\ 0 & 0 & -p \end{bmatrix} + \begin{bmatrix} \tau_{xx} + p & \tau_{xy} & \tau_{xz} \\ \tau_{xy} & \tau_{yy} + p & \tau_{yz} \\ \tau_{xz} & \tau_{yz} & \tau_{zz} + p \end{bmatrix} \tag{2.13}
$$

Table 2.1: Pressure versus depth inside Earth.		
Depth (km)	Region	Pressure (GPa)
0–24	Crust	0–0.6
24–400	Upper mantle	0.6–13.4
400–670	Transition zone	13.4–23.8
670–2891	Lower mantle	23.8–135.8
2891–5150	Outer core	135.8–328.9
5150–6371	Inner core	328.9–363.9

where $p = -\tau_m$ is the mean normal pressure. For isotropic materials (see Section 2.3), hydrostatic stress produces volume change without any change in the shape; it is the deviatoric stress that causes shape changes.

2.1.5 Values for stress

Stress has units of force per unit area. In SI units

$$1 \text{ pascal (Pa)} = 1 \text{ N m}^{-2}.$$

Recall that 1 newton (N) $= 1 \text{ kg m s}^{-2} = 10^5$ dyne. Another commonly used unit for stress is the *bar*:

$$1 \text{ bar } = 10^5 \text{ Pa},$$
$$1 \text{ kbar } = 10^8 \text{ Pa } = 100 \text{ MPa},$$
$$1 \text{ Mbar } = 10^{11} \text{ Pa} = 100 \text{ GPa}.$$

Pressure increases rapidly with depth in Earth, as shown in Table 2.1 using values taken from the reference model PREM (Dziewonski and Anderson, 1981). Pressures reach 13.4 GPa at 400 km depth, 136 GPa at the core–mantle boundary, and 329 GPa at the inner-core boundary. In contrast, the pressure at the center of the Moon is only about 4.8 GPa, a value reached in Earth at 150 km depth (Latham *et al.*, 1969). This is a result of the much smaller mass of the Moon.

These are the hydrostatic pressures inside Earth; shear stresses at depth are much smaller in magnitude and include stresses associated with mantle convection and the dynamic stresses caused by seismic wave propagation. Static shear stresses can be maintained in the upper, brittle part of the crust. Measuring shear stress in the crust is a topic of current research and the magnitude of the stress is a subject of some controversy. Crustal shear stress is probably between about 100 and 1000 bars (10 to 100 MPa), with a tendency for lower stresses to occur close to active faults (which act to relieve the stress).

2.2 The strain tensor

Now let us consider how to describe changes in the positions of points within a continuum. The location of a particular particle at time t relative to its position at a reference time t_0 can be expressed as a vector field, that is, the displacement field $\mathbf{u}$ is given by

$$\mathbf{u}(\mathbf{r}_0, t) = \mathbf{r} - \mathbf{r}_0, \tag{2.14}$$

where $\mathbf{r}$ is the position of the point at time t and $\mathbf{r}_0$ is the reference location of the point. This approach of following the displacements of particles specified by their original positions at some reference time is called the *Lagrangian description* of motion in a continuum and is almost always the most convenient formulation in seismology.[2] As we will discuss in Chapter 11, seismometers respond to the motion of the particles in the Earth connected to the instrument and thus provide a record of Lagrangian motion. The particle displacement is $\mathbf{u}(t)$, the particle velocity is $\partial\mathbf{u}/\partial t$, and the particle acceleration is $\partial^2\mathbf{u}/\partial t^2$.

The displacement field, $\mathbf{u}$, is an important concept and we will refer to it often in this book. It is an absolute measure of position changes. In contrast, *strain* is a local measure of relative changes in the displacement field, that is, the spatial gradients in the displacement field. Strain is related to the deformation, or change in shape, of a material rather than any absolute change in position. For example, *extensional strain* is defined as the change in length with respect to length. If a 100 m long string is fixed at one end and uniformly stretched to a length of 101 m, then the displacement field varies from 0 to 1 m along the string, whereas the strain field is constant at 0.01 (1%) everywhere in the string.

Consider the displacement $\mathbf{u} = (u_x, u_y, u_z)$ at position $\mathbf{x}$, a small distance away from a reference position $\mathbf{x}_0$:

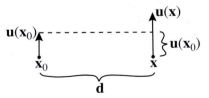

[2] The alternative approach of examining whatever particle happens to occupy a specified location is termed the *Eulerian description* and is often used in fluid mechanics.

We can expand $\mathbf{u}$ in a Taylor series to obtain

$$\mathbf{u}(\mathbf{x}) = \begin{bmatrix} u_x \\ u_y \\ u_z \end{bmatrix} = \mathbf{u}(\mathbf{x}_0) + \begin{bmatrix} \frac{\partial u_x}{\partial x} & \frac{\partial u_x}{\partial y} & \frac{\partial u_x}{\partial z} \\ \frac{\partial u_y}{\partial x} & \frac{\partial u_y}{\partial y} & \frac{\partial u_y}{\partial z} \\ \frac{\partial u_z}{\partial x} & \frac{\partial u_z}{\partial y} & \frac{\partial u_z}{\partial z} \end{bmatrix} \begin{bmatrix} d_x \\ d_y \\ d_z \end{bmatrix} = \mathbf{u}(\mathbf{x}_0) + \mathbf{Jd}, \qquad (2.15)$$

where $\mathbf{d} = \mathbf{x} - \mathbf{x}_0$. We have ignored higher-order terms in the expansion by assuming that the partials, $\partial u_x / \partial x$, $\partial u_y / \partial x$, etc., are small enough that their products can be ignored (the basis for *infinitesimal strain theory*). Seismology is fortunate that actual Earth strains are almost always small enough that this approximation is valid. We can separate out rigid rotations by dividing $\mathbf{J}$ into symmetric and antisymmetric parts:

$$\mathbf{J} = \begin{bmatrix} \frac{\partial u_x}{\partial x} & \frac{\partial u_x}{\partial y} & \frac{\partial u_x}{\partial z} \\ \frac{\partial u_y}{\partial x} & \frac{\partial u_y}{\partial y} & \frac{\partial u_y}{\partial z} \\ \frac{\partial u_z}{\partial x} & \frac{\partial u_z}{\partial y} & \frac{\partial u_z}{\partial z} \end{bmatrix} = \mathbf{e} + \mathbf{\Omega}, \qquad (2.16)$$

where the *strain tensor*, $\mathbf{e}$, is symmetric ($e_{ij} = e_{ji}$) and is given by

$$\mathbf{e} = \begin{bmatrix} \frac{\partial u_x}{\partial x} & \frac{1}{2}\left(\frac{\partial u_x}{\partial y} + \frac{\partial u_y}{\partial x}\right) & \frac{1}{2}\left(\frac{\partial u_x}{\partial z} + \frac{\partial u_z}{\partial x}\right) \\ \frac{1}{2}\left(\frac{\partial u_y}{\partial x} + \frac{\partial u_x}{\partial y}\right) & \frac{\partial u_y}{\partial y} & \frac{1}{2}\left(\frac{\partial u_y}{\partial z} + \frac{\partial u_z}{\partial y}\right) \\ \frac{1}{2}\left(\frac{\partial u_z}{\partial x} + \frac{\partial u_x}{\partial z}\right) & \frac{1}{2}\left(\frac{\partial u_z}{\partial y} + \frac{\partial u_y}{\partial z}\right) & \frac{\partial u_z}{\partial z} \end{bmatrix}, \qquad (2.17)$$

and the *rotation tensor*, $\mathbf{\Omega}$, is antisymmetric ($\Omega_{ij} = -\Omega_{ji}$) and is given by

$$\mathbf{\Omega} = \begin{bmatrix} 0 & \frac{1}{2}\left(\frac{\partial u_x}{\partial y} - \frac{\partial u_y}{\partial x}\right) & \frac{1}{2}\left(\frac{\partial u_x}{\partial z} - \frac{\partial u_z}{\partial x}\right) \\ -\frac{1}{2}\left(\frac{\partial u_x}{\partial y} - \frac{\partial u_y}{\partial x}\right) & 0 & \frac{1}{2}\left(\frac{\partial u_y}{\partial z} - \frac{\partial u_z}{\partial y}\right) \\ -\frac{1}{2}\left(\frac{\partial u_x}{\partial z} - \frac{\partial u_z}{\partial x}\right) & -\frac{1}{2}\left(\frac{\partial u_y}{\partial z} - \frac{\partial u_z}{\partial y}\right) & 0 \end{bmatrix}. \qquad (2.18)$$

The reader should verify that $\mathbf{e} + \mathbf{\Omega} = \mathbf{J}$.

The effect of $\mathbf{e}$ and $\mathbf{\Omega}$ may be illustrated by considering what happens to an infinitesimal cube (Fig. 2.3). The off-diagonal elements of $\mathbf{e}$ cause shear strain; for example, in two dimensions, if $\mathbf{\Omega} = \mathbf{0}$ and we assume $\partial u_x / \partial x = \partial u_z / \partial z = 0$, then

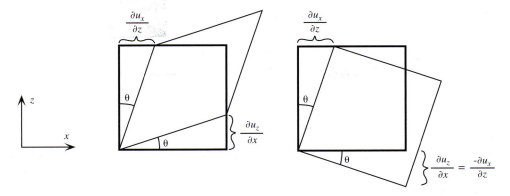

Figure 2.3 The different effects of the strain tensor **e** and the rotation tensor Ω are illustrated by the deformation of a square in the $x - z$ plane. The off-diagonal components of **e** cause shear deformation (left square), whereas Ω causes rigid rotation (right square). The deformations shown here are highly exaggerated compared to those for which infinitesimal strain theory is valid.

$\partial u_x/\partial z = \partial u_z/\partial x$, and

$$\mathbf{J} = \mathbf{e} = \begin{bmatrix} 0 & \theta \\ \theta & 0 \end{bmatrix} = \begin{bmatrix} 0 & \frac{\partial u_x}{\partial z} \\ \frac{\partial u_z}{\partial x} & 0 \end{bmatrix}, \qquad (2.19)$$

where θ is the angle (in radians, not degrees!) through which each side rotates. Note that the total change in angle between the sides is 2θ.

In contrast, the Ω matrix causes rigid rotation, for example, if $\mathbf{e} = \mathbf{0}$, then $\partial u_x/\partial z = -\partial u_z/\partial x$ and

$$\mathbf{J} = \Omega = \begin{bmatrix} 0 & \theta \\ -\theta & 0 \end{bmatrix} = \begin{bmatrix} 0 & \frac{\partial u_x}{\partial z} \\ \frac{\partial u_z}{\partial x} & 0 \end{bmatrix}. \qquad (2.20)$$

In both of these cases there is no volume change in the material. The relative volume increase, or *dilatation*, $\Delta = (V - V_0)/V_0$, is given by the sum of the extensions in the x, y, and z directions:

$$\Delta = \frac{\partial u_x}{\partial x} + \frac{\partial u_y}{\partial y} + \frac{\partial u_z}{\partial z} = \text{tr}[\mathbf{e}] = \nabla \cdot \mathbf{u}, \qquad (2.21)$$

where $\text{tr}[\mathbf{e}] = e_{11} + e_{22} + e_{33}$, the *trace* of **e**. Note that the dilatation is given by the divergence of the displacement field.

What about the curl of the displacement field? Recall the definition of the curl of a vector field:

$$\nabla \times \mathbf{u} = \left(\frac{\partial u_z}{\partial y} - \frac{\partial u_y}{\partial z} \right) \hat{\mathbf{x}} + \left(\frac{\partial u_x}{\partial z} - \frac{\partial u_z}{\partial x} \right) \hat{\mathbf{y}} + \left(\frac{\partial u_y}{\partial x} - \frac{\partial u_x}{\partial y} \right) \hat{\mathbf{z}}. \qquad (2.22)$$

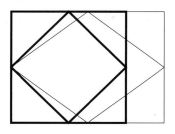

Figure 2.4 Simple extensional strain in the x direction results in shear strain; internal angles are not preserved.

A comparison of this equation with (2.18) shows that $\nabla \times \mathbf{u}$ is nonzero only if $\boldsymbol{\Omega}$ is non-zero and the displacement field contains some rigid rotation.

The strain tensor, like the stress tensor, is symmetric and contains six independent parameters. The *principal axes* of strain may be found by computing the directions $\hat{\mathbf{n}}$ for which the displacements are in the same direction, that is,

$$\mathbf{u} = \lambda\hat{\mathbf{n}} = e\hat{\mathbf{n}}. \tag{2.23}$$

This is analogous to the case of the stress tensor discussed in the previous section. The three eigenvalues are the *principal strains*, e_1, e_2, and e_3, while the eigenvectors define the principal axes. Note that, except in the case $e_1 = e_2 = e_3$ (*hydrostatic strain*), there is always some shear strain present.

For example, consider a two-dimensional square with extension only in the x direction (Fig. 2.4), so that $\mathbf{e}$ is given by

$$\mathbf{e} = \begin{bmatrix} e_1 & 0 \\ 0 & 0 \end{bmatrix} = \begin{bmatrix} \frac{\partial u_x}{\partial x} & 0 \\ 0 & 0 \end{bmatrix}. \tag{2.24}$$

Angles between lines parallel to the coordinate axes do not change, but lines at intermediate angles are seen to rotate. The angle changes associated with shearing become obvious if we consider the diagonal lines at 45° with respect to the square. If we rotate the coordinate system (see Appendix B) by 45° as defined by the unit vectors $(1/\sqrt{2}, 1/\sqrt{2})$ and $(-1/\sqrt{2}, 1/\sqrt{2})$ we obtain

$$\mathbf{e}' = \begin{bmatrix} 1/\sqrt{2} & 1/\sqrt{2} \\ -1/\sqrt{2} & 1/\sqrt{2} \end{bmatrix} \begin{bmatrix} e_1 & 0 \\ 0 & 0 \end{bmatrix} \begin{bmatrix} 1/\sqrt{2} & -1/\sqrt{2} \\ 1/\sqrt{2} & 1/\sqrt{2} \end{bmatrix} = \begin{bmatrix} e_1/2 & -e_1/2 \\ -e_1/2 & e_1/2 \end{bmatrix}$$

$$\tag{2.25}$$

and we see that the strain tensor now has off-diagonal terms. As we shall see in the next chapter, the type of deformation shown in Figure 2.4 would be produced by a seismic *P* wave traveling in the *x* direction; our discussion here shows how *P* waves involve both compression and shearing.

In subsequent sections, we shall find it helpful to express the strain tensor using index notation. Equation (2.17) can be rewritten as

$$e_{ij} = \tfrac{1}{2}(\partial_i u_j + \partial_j u_i), \tag{2.26}$$

where i and j are assumed to range from 1 to 3 (for the x, y, and z directions) and we are using the notation $\partial_x u_y = \partial u_y / \partial x$.

2.2.1 Values for strain

Strain is dimensionless since it represents a change in length divided by length. Dynamic strains associated with the passage of seismic waves in the far field are typically less than 10^{-6}.

2.2.2 Example: Computing strain for a seismic wave

A seismic plane shear wave is traveling through a solid with displacement that can locally be approximated as

$$u_z = A \sin\left[2\pi f(t - x/c)\right]$$

where A is the amplitude, f is the frequency, and c is the velocity of the wave. What is the maximum strain for this wave?

The non-zero partial derivative from equation (2.17) is

$$\frac{\partial u_z}{\partial x} = \frac{-2\pi f A}{c} \cos\left[2\pi f(t - x/c)\right].$$

The maximum occurs when $\cos = -1$ and is thus

$$\left(\frac{\partial u_z}{\partial x}\right)_{\text{max}} = \frac{2\pi f A}{c}.$$

For example, if $f = 2\,\text{Hz}$, $c = 3.14\,\text{km/s}$ $(3140\,\text{m/s})$ and $A = 1\,\text{mm}$ $(10^{-3}\,\text{m})$, then $(\partial u_z/\partial x)_{\text{max}} = 4 \times 10^{-6}$ and the strain tensor is given by

$$\mathbf{e}_{\text{max}} = \begin{bmatrix} 0 & 0 & 2 \times 10^{-6} \\ 0 & 0 & 0 \\ 2 \times 10^{-6} & 0 & 0 \end{bmatrix}.$$

2.3 The linear stress–strain relationship

Stress and strain are linked in elastic media by a stress–strain or *constitutive* relationship. The most general linear relationship between the stress and strain tensors can be written

$$\tau_{ij} = c_{ijkl}e_{kl} \equiv \sum_{k=1}^{3}\sum_{l=1}^{3}c_{ijkl}e_{kl}, \qquad (2.27)$$

where c_{ijkl} is termed the *elastic tensor*. Here we begin using the *summation convention* in our index notation. Any repeated index in a product indicates that the sum is to be taken as the index varies from 1 to 3. Equation (2.27) is sometimes called the *generalized Hooke's law* and assumes perfect elasticity; there is no energy loss or attenuation as the material deforms in response to the applied stress (sometimes these effects are modeled by permitting c_{ijkl} to be complex). A solid obeying (2.27) is called *linearly elastic*. Non-linear behavior is sometimes observed in seismology (examples include the response of some soils to strong ground motions and the fracturing of rock near earthquakes and explosions) but the non-linearity greatly complicates the mathematics. In this chapter we only consider linearly elastic solids, deferring a discussion of anelastic behavior and attenuation until Chapter 6. Note that stress is not sensitive to the rotation tensor Ω; stress changes are caused by changes in the volume or shape of solids, as defined by the strain tensor, rather than by rigid rotations.

Equation (2.27) should not be applied to compute the strain for the large values of hydrostatic stress that are present within Earth's interior (see Table 2.1). These strains, representing the compression of rocks under high pressure, are too large for linear stress–strain theory to be valid. Instead, this equation applies to perturbations in stress, termed incremental stresses, with respect to an initial state of stress at which the strain is assumed to be zero. This is standard practice in seismology and we will assume throughout this section that stress is actually defined in terms of incremental stress.

The elastic tensor, c_{ijkl}, is a fourth-order tensor with 81 (3^4) components. However, because of the symmetry of the stress and strain tensors and thermodynamic considerations, only 21 of these components are independent. These 21 components are necessary to specify the stress–strain relationship for the most general form of elastic solid. The properties of such a solid may vary with direction; if they do, the material is termed *anisotropic*. In contrast, the properties of an *isotropic* solid are the same in all directions. Isotropy has proven to be a reasonable first-order approximation for much of the Earth's interior, but in some regions anisotropy has been observed and this is an important area of current research (see Section 11.3 for more about anisotropy).

If we assume isotropy (c_{ijkl} is invariant with respect to rotation), it can be shown that the number of independent parameters is reduced to two:

$$c_{ijkl} = \lambda \delta_{ij}\delta_{kl} + \mu(\delta_{il}\delta_{jk} + \delta_{ik}\delta_{jl}), \qquad (2.28)$$

where λ and μ are called the *Lamé parameters* of the material and δ_{ij} is the *Kronecker delta* ($\delta_{ij} = 1$ for $i = j$, $\delta_{ij} = 0$ for $i \neq j$). Thus, for example, $C_{1111} = \lambda + 2\mu$, $C_{1112} = 0$, $C_{1122} = \lambda$, $C_{1212} = \mu$, etc. As we shall see, the Lamé parameters, together with the density, will eventually determine the seismic velocities of the material. The stress–strain equation (2.27) for an isotropic solid is

$$
\begin{aligned}
\tau_{ij} &= [\lambda \delta_{ij}\delta_{kl} + \mu(\delta_{il}\delta_{jk} + \delta_{ik}\delta_{jl})]e_{kl} \\
&= \lambda \delta_{ij}e_{kk} + 2\mu e_{ij},
\end{aligned}
\qquad (2.29)
$$

where we have used $e_{ij} = e_{ji}$ to combine the μ terms. Note that $e_{kk} = \mathrm{tr}[\mathbf{e}]$, the sum of the diagonal elements of $\mathbf{e}$. Using this equation, we can directly write the components of the stress tensor in terms of the strains:

$$
\tau = \begin{bmatrix}
\lambda\,\mathrm{tr}[\mathbf{e}] + 2\mu e_{11} & 2\mu e_{12} & 2\mu e_{13} \\
2\mu e_{21} & \lambda\,\mathrm{tr}[\mathbf{e}] + 2\mu e_{22} & 2\mu e_{23} \\
2\mu e_{31} & 2\mu e_{32} & \lambda\,\mathrm{tr}[\mathbf{e}] + 2\mu e_{33}
\end{bmatrix}. \qquad (2.30)
$$

The two Lamé parameters completely describe the linear stress–strain relation within an isotropic solid. μ is termed the *shear modulus* and is a measure of the resistance of the material to shearing. Its value is given by half of the ratio between the applied shear stress and the resulting shear strain, that is, $\mu = \tau_{xy}/2e_{xy}$. The other Lamé parameter, λ, does not have a simple physical explanation. Other commonly used elastic constants for isotropic solids include:

Young's modulus E: The ratio of extensional stress to the resulting extensional strain for a cylinder being pulled on both ends. It can be shown that

$$E = \frac{(3\lambda + 2\mu)\mu}{\lambda + \mu}. \qquad (2.31)$$

Bulk modulus κ: The ratio of hydrostatic pressure to the resulting volume change, a measure of the incompressibility of the material. It can be expressed as

$$\kappa = \lambda + \tfrac{2}{3}\mu. \qquad (2.32)$$

Poisson's ratio σ: The ratio of the lateral contraction of a cylinder (being pulled on its ends) to its longitudinal extension. It can be expressed as

$$\sigma = \frac{\lambda}{2(\lambda + \mu)}. \tag{2.33}$$

In seismology, we are mostly concerned with the compressional (P) and shear (S) velocities. As we will show later, these can be computed from the elastic constants and the density, ρ:

P velocity, α, can be expressed as

$$\alpha = \sqrt{\frac{\lambda + 2\mu}{\rho}}. \tag{2.34}$$

S velocity, β, can be expressed as

$$\beta = \sqrt{\frac{\mu}{\rho}}. \tag{2.35}$$

Poisson's ratio σ is often used as a measure of the relative size of the P and S velocities; it can be shown that

$$\sigma = \frac{\alpha^2 - 2\beta^2}{2(\alpha^2 - \beta^2)} = \frac{(\alpha/\beta)^2 - 2}{2(\alpha/\beta)^2 - 2}. \tag{2.36}$$

Note that σ is dimensionless and varies between 0 and 0.5 with the upper limit representing a fluid ($\mu = 0$). For a *Poisson solid*, $\lambda = \mu$, $\sigma = 0.25$, and $\alpha/\beta = \sqrt{3}$ and this is a common approximation in seismology for estimating the S velocity from the P velocity and vice versa. Note that the minimum possible P-to-S velocity ratio for an isotropic solid is $\sqrt{2}$, which occurs when $\lambda = \sigma = 0$. Most crustal rocks have Poisson's ratios between 0.25 and 0.30.

Although many different elastic parameters have been defined, it should be noted that two parameters and density are sufficient to give a complete description of isotropic elastic properties. In seismology, these parameters are often simply the P and S velocities. Other elastic parameters can be computed from the velocities, assuming the density is also known.

2.3.1 Units for elastic moduli

The Lamé parameters, Young's modulus, and the bulk modulus all have the same units as stress (i.e., pascals). Recall that

$$1\,\text{Pa} = 1\,\text{N}\,\text{m}^{-2} = 1\,\text{kg}\,\text{m}^{-1}\,\text{s}^{-2}.$$

Note that when this is divided by density $(\mathrm{kg\,m^{-3}})$ the result is units of velocity squared (appropriate for equations (2.34) and (2.35)).

2.4 Exercises

1. Assume that the horizontal components of the 2-D stress tensor are

$$\boldsymbol{\tau} = \begin{bmatrix} \tau_{xx} & \tau_{xy} \\ \tau_{yx} & \tau_{yy} \end{bmatrix} = \begin{bmatrix} -30 & -20 \\ -20 & -40 \end{bmatrix} \mathrm{MPa}$$

   (a) Compute the normal and shear stresses on a fault that strikes $10°$ east of north.

   (b) Compute the principal stresses, and give the azimuths (in degrees east of north) of the maximum and minimum compressional stress axes.

2. The principal stress axes for a 2-D geometry are oriented at N45° E and N135° E, corresponding to principal stresses of -15 and -10 MPa. What are the four components of the 2-D stress tensor in a (x = east, y = north) coordinate system?

3. Figure 2.5 shows a vertical-component seismogram of the 1989 Loma Prieta earthquake recorded in Finland. Make an estimate of the *maximum* strain recorded at this site. Hints: 1 micron $= 10^{-6}$ m, note that the time axis is in 100s of seconds, assume the Rayleigh surface wave phase velocity at the dominant period is 3.9 km/s, remember that strain is proportional to $\partial u_z / \partial x$, Table 3.1 may be helpful.

4. Using equations (2.4), (2.23), and (2.30), show that the principal stress axes always coincide with the principal strain axes for isotropic media. In other words, show that if $\mathbf{x}$ is an eigenvector of $\mathbf{e}$, then it is also an eigenvector of $\boldsymbol{\tau}$.

5. From equations (2.34) and (2.35) derive expressions for the Lamé parameters in terms of the seismic velocities and density.

6. Seismic observations of S velocity can be directly related to the shear modulus μ. However, P velocity is a function of both the shear and bulk moduli. For this reason, sometimes seismologists will compute the *bulk sound speed*, defined as:

$$V_c = \sqrt{\frac{\kappa}{\rho}} \qquad (2.37)$$

   which isolates the sensitivity to the bulk modulus κ. Derive an equation for V_c in terms of the P and S velocities.

7. What is the P/S velocity ratio for a rock with a Poisson's ratio of 0.30?

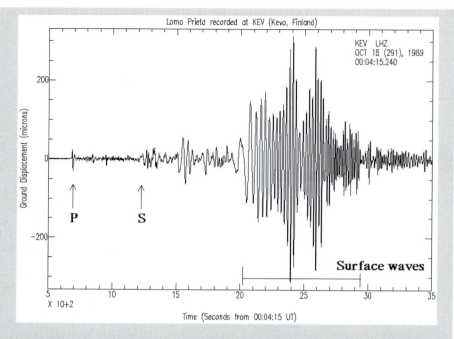

Figure 2.5 A vertical component seismogram of the 1989 Loma Prieta earthquake in California, recorded in Finland. This plot was taken from the Princeton Earth Physics Project, PEPP, website at www.gns.cri.nz/outreach/qt/quaketrackers/curr/seismic_waves.htm.

8. A sample of granite in the laboratory is observed to have a P velocity of 5.5 km/s and a density of 2.6 Mg/m^3. Assuming it is a Poisson solid, obtain values for the Lamé parameters, Young's modulus, and the bulk modulus. Express your answers in pascals.

9. Using values from the PREM model (Appendix A), compute values for the bulk modulus on both sides of (a) the core–mantle boundary (CMB) and (b) the inner-core boundary (ICB). Express your answers in pascals.

10. Figure 2.6 shows surface displacement rates as a function of distance from the San Andreas Fault in California.

    (a) Consider this as a 2-D problem with the x-axis perpendicular to the fault and the y-axis parallel to the fault. From these data, estimate the yearly strain (**e**) and rotation (**Ω**) tensors for a point on the fault. Express your answers as 2×2 matrices.

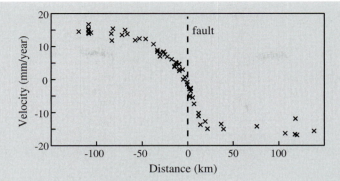

Figure 2.6 Geodetically determined displacement rates near the San Andreas Fault in central California. Velocities are in mm per year for motion parallel to the fault; distances are measured perpendicular to the fault. Velocities are normalized to make the velocity zero at the fault. Data points courtesy of Duncan Agnew.

(b) Assuming the crustal shear modulus is 27 GPa, compute the yearly change in the stress tensor. Express your answer as a 2×2 matrix with appropriate units.

(c) If the crustal shear modulus is 27 GPa, what is the shear stress across the fault after 200 years, assuming zero initial shear stress?

(d) If large earthquakes occur every 200 years and release all of the accumulated strain by movement along the fault, what, if anything, can be inferred about the *absolute* level of shear stress?

(e) What, if anything, can be learned about the fault from the observation that most of the deformation occurs within a zone less than 50 km wide?

(f) Note: The asymmetry in the deformation pattern is a long-standing puzzle. To learn more, see Schmalzle *et al.* (2006).

11. Do some research on the observed density of the Sun. Are the high sound velocities in the Sun (see Fig. 1.6) compared to Earth's P velocities caused primarily by low solar densities compared to the Earth, a higher bulk modulus or some combination of these factors?

12. The University of California, San Diego, operates the Piñon Flat Observatory (PFO) in the mountains northeast of San Diego (near Anza). Instruments include high-quality strain meters for measuring crustal deformation.

(a) Assume, at 5 km depth beneath PFO, the seismic velocities are $\alpha = 6$ km/s and $\beta = 3.5$ km/s and the density is $\rho = 2.7$ Mg/m^3. Compute values for the Lamé parameters, λ and μ, from these numbers. Express your answer in units of pascals.

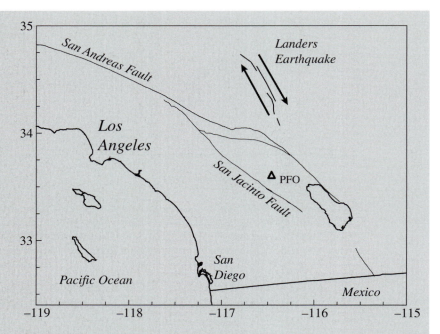

Figure 2.7 The 1992 Landers earthquake ($M_S = 7.3$) in southern California produced measurable strain changes at PFO observatory, located about 80 km south of the event.

(b) Following the 1992 Landers earthquake ($M_S = 7.3$), located in southern California 80 km north of PFO (Fig. 2.7), the PFO strain meters measured a large static change in strain compared to values before the event. Horizontal components of the strain tensor changed by the following amounts: $e_{11} = -0.26 \times 10^{-6}$, $e_{22} = 0.92 \times 10^{-6}$, $e_{12} = -0.69 \times 10^{-6}$. In this notation 1 is east, 2 is north, and extension is positive. You may assume that this strain change occurred instantaneously at the time of the event. Assuming these strain values are also accurate at depth, use the result you obtained in part (a) to determine the change in stress due to the Landers earthquake at 5 km, that is, compute the change in τ_{11}, τ_{22}, and τ_{12}. Treat this as a two-dimensional problem by assuming there is no strain in the vertical direction and no depth dependence of the strain.

(c) Compute the orientations of the principal strain axes (horizontal) for the response at PFO to the Landers event. Express your answers as azimuths (degrees east of north).

(d) A steady long-term change in strain at PFO has been observed to occur in which the changes in one year are: $e_{11} = 0.101 \times 10^{-6}$, $e_{22} = -0.02 \times 10^{-6}$, $e_{12} = 0.005 \times 10^{-6}$. Note that the long-term strain change is close to simple E–W extension. Assuming that this strain rate has occurred steadily for the last 1000 years, from an initial state of zero stress, compute the components of the stress tensor at 5 km depth. (Note: This is probably not a very realistic assumption!) Don't include the large hydrostatic component of stress at 5 km depth.

(e) Farmer Bob owns a $1 \, \text{km}^2$ plot of land near PFO that he has fenced and surveyed with great precision. How much land does Farmer Bob gain or lose each year? How much did he gain or lose as a result of the Landers earthquake? Express your answers in m^2.

(f) (COMPUTER) Write a computer program that computes the stress across vertical faults at azimuths between 0 and 170 degrees (east from north, at 10 degree increments). For the stress tensors that you calculated in (b) and (d), make a table that lists the fault azimuth and the corresponding shear stress and normal stress across the fault (for Landers these are the stress changes, not absolute stresses). At what azimuths are the maximum shear stresses for each case?

(g) (COMPUTER) Several studies (e.g., Stein *et al.*, 1992, 1994; Harris and Simpson, 1992; Harris *et al.*, 1995; Stein, 1999; Harris, 2002) have modeled the spatial distribution of events following large earthquakes by assuming that the likelihood of earthquake rupture along a fault is related to the *Coulomb failure function* (CFF). Ignoring the effect of pore fluid pressure, the change in CFF may be expressed as:

$$\Delta\text{CFF} = \Delta|\tau_s| + \mu_s \Delta\tau_n,$$

where τ_s is the shear stress (traction), τ_n is the normal stress, and μ_s is the coefficient of static friction (don't confuse this with the shear modulus!). Note that CFF increases as the shear stress increases, and as the compressional stress on the fault is reduced (recall in our sign convention that extensional stresses are positive and compressional stresses are negative). Assume that $\mu_s = 0.2$ and modify your computer program to compute ΔCFF for each fault orientation. Make a table of the yearly change in ΔCFF due to the long-term strain change at each fault azimuth.

(h) (COMPUTER) Now assume that the faults will fail when their long-term CFF reaches some critical threshold value. The change in time to the next earthquake may be expressed as

$$\Delta t = \frac{\text{CFF}_{1000+L} - \text{CFF}_{1000}}{\text{CFF}_a},$$

where CFF_a is the annual change in CFF, CFF_{1000} is the thousand year change in CFF, and CFF_{1000+L} is the thousand year + Landers change in CFF (note that $\text{CFF}_{1000+L} \neq \text{CFF}_{1000} + \text{CFF}_L$). Compute the effect of the Landers earthquake in terms of advancing or retarding the time until the next earthquake for each fault orientation. Express your answer in years, using the sign convention of positive time for advancement of the next earthquake and negative time for retardation. (Warning: This is tricky.) Check your answer against the values of shear stress on the fault. Generally (but not always) the earthquake time should advance when the long-term and Landers shear changes agree in sign (either both positive or both negative), and the time should be delayed when the shear stress changes disagree in sign.

(i) No increase in seismicity (small earthquake activity) has been observed near PFO following the Landers event. Does this say anything about the validity of the threshold CFF model?

Hint: Getting the signs correct in parts (f)–(h) can be complicated, particularly for part (h). Stresses can be either positive or negative. To help get it right, define two unit vectors for each fault azimuth, one parallel to the fault ($\hat{\mathbf{f}}$) and one perpendicular to the fault ($\hat{\mathbf{p}}$). Compute the traction vector by multiplying the stress tensor by $\hat{\mathbf{p}}$. Then resolve the traction vector into shear stress and normal stress by computing the dot product with $\hat{\mathbf{f}}$ and $\hat{\mathbf{p}}$, respectively. Naturally, $\hat{\mathbf{f}}$ and $\hat{\mathbf{p}}$ must be of unit length for this to work.

3

The seismic wave equation

Using the stress and strain theory developed in the previous chapter, we now construct and solve the seismic wave equation for elastic wave propagation in a uniform whole space. We will show that two types of solutions are possible, corresponding to compressional (P) and shear (S) waves, and we will derive the equations for their velocities that we presented in the last chapter. This will involve vector calculus and complex numbers; some of the mathematics is reviewed in Appendix B. For simplicity, in this chapter we assume perfect elasticity with no energy loss in the seismic waves from any intrinsic attenuation.

3.1 Introduction: The wave equation

To motivate our discussion, consider the one-dimensional wave equation

$$\frac{\partial^2 u}{\partial t^2} = c^2 \frac{\partial^2 u}{\partial x^2} \tag{3.1}$$

and its general solution

$$u(x, t) = f(x \pm ct), \tag{3.2}$$

which represents waves of arbitrary shape propagating at velocity c in the positive and negative x directions. This is a very common equation in physics and can be used to describe, for example, the vibrations of a string or acoustic waves in a pipe. The velocity of the wave is determined by the physical properties of the material through which it propagates. In the case of a vibrating string, $c^2 = F/\rho$ where F is the string tension force and ρ is the density.

The wave equation is classified as a *hyperbolic equation* in the theory of linear partial differential equations. Hyperbolic equations are among the most challenging to solve because sharp features in their solutions will persist and can reflect off boundaries. Unlike, for example, the diffusion equation, solutions will be smooth only if the initial conditions are smooth. This complicates both analytical and numerical solution methods.

As we shall see, the seismic wave equation is more complicated than equation (3.1) because it is three dimensional and the link between force and displacement involves the full stress–strain relationship for an elastic solid. However, the *P* and *S* seismic wave solutions share many characteristics with the solutions to the 1-D wave equation. They involve pulses of arbitrary shape that travel at speeds determined by the elastic properties and density of the medium, and these pulses are often decomposed into harmonic wave solutions involving sine and cosine functions. Stein and Wysession (2003, Section 2.2) provide a useful review of the 1-D wave equation as applied to a vibrating string, with analogies to seismic wave propagation in the Earth.

3.2 The momentum equation

In the previous chapter, the stress, strain, and displacement fields were considered in static equilibrium and unchanging with time. However, because seismic waves are time-dependent phenomena that involve velocities and accelerations, we need to account for the effect of momentum. We do this by applying Newton's law ($F = ma$ from your freshman physics class) to a continuous medium.

Consider the forces on an infinitesimal cube in an (x_1, x_2, x_3) coordinate system (Fig. 3.1). The forces on each surface of the cube are given by the product of the traction vector and the surface area. For example, the force on the plane normal to x_1 is given by

$$
\begin{aligned}
\mathbf{F}(\hat{\mathbf{x}}_1) &= \mathbf{t}(\hat{\mathbf{x}}_1)\, dx_2\, dx_3 \\[2mm]
&= \boldsymbol{\tau}\hat{\mathbf{x}}_1\, dx_2\, dx_3 \\[2mm]
&= \begin{bmatrix} \tau_{11} \\ \tau_{21} \\ \tau_{31} \end{bmatrix} dx_2\, dx_3,
\end{aligned}
\tag{3.3}
$$

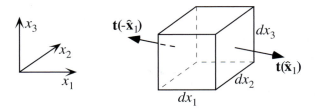

Figure 3.1 The force on the (x_2, x_3) face of an infinitesimal cube is given by $\mathbf{t}(\hat{\mathbf{x}}_1) \, dx_2 \, dx_3$, the product of the traction vector and the surface area.

where $\mathbf{F}$ is the force vector, $\mathbf{t}$ is the traction vector, and $\boldsymbol{\tau}$ is the stress tensor. In the case of a homogeneous stress field, there is no net force on the cube since the forces on opposing sides will cancel out, that is, $\mathbf{F}(-\hat{\mathbf{x}}_1) = -\mathbf{F}(\hat{\mathbf{x}}_1)$. Net force will only be exerted on the cube if spatial gradients are present in the stress field. In this case, the net force from the planes normal to x_1 is

$$\mathbf{F}(\hat{\mathbf{x}}_1) = \frac{\partial}{\partial x_1} \begin{bmatrix} \tau_{11} \\ \tau_{21} \\ \tau_{31} \end{bmatrix} dx_1 \, dx_2 \, dx_3, \tag{3.4}$$

and we can use index notation and the summation convention to express the total force from the stress field on all the faces of the cube as

$$F_i = \sum_{j=1}^{3} \frac{\partial \tau_{ij}}{\partial x_j} dx_1 \, dx_2 \, dx_3$$

$$= \partial_j \tau_{ij} \, dx_1 \, dx_2 \, dx_3. \tag{3.5}$$

The $d_j \tau_{ij}$ term is the divergence of the stress tensor (recall that the summation convention means that this term is summed over $j = 1, 2, 3$). There may also exist a body force on the cube that acts in proportion to the volume of material, that is,

$$F_i^{\text{body}} = f_i \, dx_1 \, dx_2 \, dx_3. \tag{3.6}$$

The mass of our infinitesimal cube is given by

$$m = \rho \, dx_1 \, dx_2 \, dx_3, \tag{3.7}$$

where ρ is the density. The acceleration of the cube is given by the second time derivative of the displacement $\mathbf{u}$. Substituting (3.5)–(3.7) into $F = ma$ and canceling the common factor of $dx_1 \, dx_2 \, dx_3$, we obtain[1]

$$\rho \frac{\partial^2 u_i}{\partial t^2} = \partial_j \tau_{ij} + f_i. \tag{3.8}$$

This is the fundamental equation that underlies much of seismology. It is called the *momentum equation* or the *equation of motion* for a continuum. Each of the terms u_i, τ_{ij}, and f_i is a function of position $\mathbf{x}$ and time. The body force term $\mathbf{f}$ generally consists of a gravity term $\mathbf{f}_g$ and a source term $\mathbf{f}_s$. Gravity is an important factor at very low frequencies in normal mode seismology, but it can generally be neglected for body- and surface-wave calculations at typically observed wavelengths. We will consider the effects of the source term $\mathbf{f}_s$ later in this book (Chapter 9). In the absence of body forces, we have the *homogeneous equation of motion*

$$\rho \frac{\partial^2 u_i}{\partial t^2} = \partial_j \tau_{ij}, \tag{3.9}$$

which governs seismic wave propagation outside of seismic source regions. Generating solutions to (3.8) or (3.9) for realistic Earth models is an important part of seismology; such solutions provide the predicted ground motion at specific locations at some distance from the source and are commonly termed *synthetic seismograms*.

   If, on the other hand, we assume that the acceleration term in (3.8) is zero, the result is the *static equilibrium equation*

$$\partial_j \tau_{ij} = -f_i. \tag{3.10}$$

in which the body forces are balanced by the divergence of the stress tensor. This equation is applicable to static deformation problems in geodesy, engineering, and many other fields.

3.3 The seismic wave equation

In order to solve (3.9) we require a relationship between stress and strain so that we can express $\boldsymbol{\tau}$ in terms of the displacement $\mathbf{u}$. Recall the linear, isotropic stress–strain

---

[1] In expressing the acceleration term, we approximate the *total* derivatives of $\mathbf{u}$ with respect to time with the *partial* derivatives of $\mathbf{u}$ with respect to time. That is, we make the small-deformation approximation such that the terms in the total derivative containing the spatial derivatives of $\mathbf{u}$ can be ignored. This is generally assumed valid in seismology, but the spatial derivatives (advection terms) are very important in fluid mechanics.

relationship,

$$\tau_{ij} = \lambda \delta_{ij} e_{kk} + 2\mu e_{ij}, \tag{3.11}$$

where λ and μ are the Lamé parameters and the strain tensor is defined as

$$e_{ij} = \tfrac{1}{2}(\partial_i u_j + \partial_j u_i). \tag{3.12}$$

Substituting for e_{ij} in (3.11), we obtain

$$\tau_{ij} = \lambda \delta_{ij} \partial_k u_k + \mu(\partial_i u_j + \partial_j u_i). \tag{3.13}$$

Equations (3.9) and (3.13) provide a coupled set of equations for the displacement and stress. These equations are sometimes used directly at this point to model wave propagation in computer calculations by applying finite-difference techniques. In these methods, the stresses and displacements are computed at a series of grid points in the model, and the spatial and temporal derivatives are approximated through numerical differencing. The great advantage of finite-difference schemes is their relative simplicity and ability to handle Earth models of arbitrary complexity. However, they are extremely computationally intensive and do not necessarily provide physical insight regarding the behavior of the different wave types.

In the equations that follow, we will switch back and forth between vector notation and index notation. A brief review of vector calculus is given in Appendix B. If we substitute (3.13) into (3.9), we obtain

$$
\begin{aligned}
\rho \frac{\partial^2 u_i}{\partial t^2} &= \partial_j [\lambda \delta_{ij} \partial_k u_k + \mu(\partial_i u_j + \partial_j u_i)] \\
&= \partial_i \lambda \partial_k u_k + \lambda \partial_i \partial_k u_k + \partial_j \mu(\partial_i u_j + \partial_j u_i) + \mu \partial_j \partial_i u_j + \mu \partial_j \partial_j u_i \\
&= \partial_i \lambda \partial_k u_k + \partial_j \mu(\partial_i u_j + \partial_j u_i) + \lambda \partial_i \partial_k u_k + \mu \partial_i \partial_j u_j + \mu \partial_j \partial_j u_i.
\end{aligned} \tag{3.14}
$$

Defining $\ddot{\mathbf{u}} = \partial^2 \mathbf{u}/\partial t^2$, we can write this in vector notation as

$$\rho \ddot{\mathbf{u}} = \nabla \lambda (\nabla \cdot \mathbf{u}) + \nabla \mu \cdot [\nabla \mathbf{u} + (\nabla \mathbf{u})^T] + (\lambda + \mu)\nabla\nabla \cdot \mathbf{u} + \mu \nabla^2 \mathbf{u}. \tag{3.15}$$

We now use the vector identity

$$\nabla \times \nabla \times \mathbf{u} = \nabla\nabla \cdot \mathbf{u} - \nabla^2 \mathbf{u} \tag{3.16}$$

to change this to a more convenient form. We have

$$\nabla^2 \mathbf{u} = \nabla\nabla \cdot \mathbf{u} - \nabla \times \nabla \times \mathbf{u}. \tag{3.17}$$

Substituting this into (3.15), we obtain

$$\rho\ddot{\mathbf{u}} = \nabla\lambda(\nabla\cdot\mathbf{u}) + \nabla\mu\cdot[\nabla\mathbf{u} + (\nabla\mathbf{u})^T] + (\lambda + 2\mu)\nabla\nabla\cdot\mathbf{u} - \mu\nabla\times\nabla\times\mathbf{u}. \qquad (3.18)$$

This is one form of the *seismic wave equation*. The first two terms on the right-hand side (r.h.s.) involve gradients in the Lamé parameters themselves and are non-zero whenever the material is inhomogeneous (i.e., contains velocity gradients). Most non-trivial Earth models for which we might wish to compute synthetic seismograms contain such gradients. However, including these factors makes the equations very complicated and difficult to solve efficiently. Thus, most practical synthetic seismogram methods ignore these terms, using one of two different approaches.

First, if velocity is only a function of depth, then the material can be modeled as a series of homogeneous layers. Within each layer, there are no gradients in the Lamé parameters and so these terms go to zero. The different solutions within each layer are linked by calculating the reflection and transmission coefficients for waves at both sides of the interface separating the layers. The effects of a continuous velocity gradient can be simulated by considering a "staircase" model with many thin layers. As the number of layers increases, these results can be shown to converge to the continuous gradient case (more layers are needed at higher frequencies). This approach forms the basis for many techniques for computing predicted seismic motions from one-dimensional Earth models; we will term these *homogeneous-layer methods*. They are particularly useful for studying surface waves and low- to medium-frequency body waves. However, at high frequencies they become relatively inefficient because large numbers of layers are necessary for accurate modeling.

Second, it can be shown that the strength of these gradient terms varies as $1/\omega$, where ω is frequency, and thus at high frequencies these terms will tend to zero. This approximation is made in most *ray-theoretical methods*, in which it is assumed that the frequencies are sufficiently high that the $1/\omega$ terms are unimportant. However, note that at any given frequency this approximation will break down if the velocity gradients in the material become steep enough. At velocity discontinuities between regions of shallow gradients, the approximation cannot be used directly, but the solutions above and below the discontinuities can be patched together through the use of reflection and transmission coefficients. The distinction between the homogeneous-layer and ray-theoretical approaches is often important and will be emphasized later in this book.

If we ignore the gradient terms, the momentum equation for homogeneous media becomes

$$\rho\ddot{\mathbf{u}} = (\lambda + 2\mu)\nabla\nabla\cdot\mathbf{u} - \mu\nabla\times\nabla\times\mathbf{u}. \qquad (3.19)$$

This is a standard form for the seismic wave equation in homogeneous media and forms the basis for most body-wave synthetic seismogram methods. However, it is important to remember that it is an approximate expression, which has neglected the gravity and velocity gradient terms and has assumed a linear, isotropic Earth model.

We can separate this equation into solutions for P waves and S waves by taking the divergence and curl, respectively. Taking the divergence of (3.19) and using the vector identity $\nabla \cdot (\nabla \times \mathbf{\Psi}) = 0$, we obtain:

$$\frac{\partial^2 (\nabla \cdot \mathbf{u})}{\partial t^2} = \frac{\lambda + 2\mu}{\rho} \nabla^2 (\nabla \cdot \mathbf{u}) \tag{3.20}$$

or

$$\nabla^2 (\nabla \cdot \mathbf{u}) - \frac{1}{\alpha^2} \frac{\partial^2 (\nabla \cdot \mathbf{u})}{\partial t^2} = 0, \tag{3.21}$$

where the P-wave velocity, α, is given by

$$\alpha^2 = \frac{\lambda + 2\mu}{\rho}. \tag{3.22}$$

Taking the curl of (3.19) and using the vector identity $\nabla \times (\nabla \phi) = 0$, we obtain:

$$\frac{\partial^2 (\nabla \times \mathbf{u})}{\partial t^2} = -\frac{\mu}{\rho} \nabla \times \nabla \times (\nabla \times \mathbf{u}). \tag{3.23}$$

Using the vector identity (3.16) and $\nabla \cdot (\nabla \times \mathbf{u}) = 0$, this becomes

$$\frac{\partial^2 (\nabla \times \mathbf{u})}{\partial t^2} = \frac{\mu}{\rho} \nabla^2 (\nabla \times \mathbf{u}) \tag{3.24}$$

or

$$\nabla^2 (\nabla \times \mathbf{u}) - \frac{1}{\beta^2} \frac{\partial^2 (\nabla \times \mathbf{u})}{\partial t^2} = 0, \tag{3.25}$$

where the S-wave velocity, β, is given by

$$\beta^2 = \frac{\mu}{\rho}. \tag{3.26}$$

We can use (3.22) and (3.26) to rewrite the elastic wave equation (3.18) directly in terms of the P and S velocities:

$$\ddot{\mathbf{u}} = \alpha^2 \nabla \nabla \cdot \mathbf{u} - \beta^2 \nabla \times \nabla \times \mathbf{u}. \tag{3.27}$$

3.3.1 Potentials

The displacement $\mathbf{u}$ is often expressed in terms of the P-wave scalar potential ϕ and S-wave vector potential $\boldsymbol{\Psi}$, using the Helmholtz decomposition theorem (e.g., Aki and Richards, 2002, pp. 67–9), i.e.,

$$\mathbf{u} = \nabla\phi + \nabla \times \boldsymbol{\Psi}, \quad \nabla \cdot \boldsymbol{\Psi} = 0. \tag{3.28}$$

We then have

$$\nabla \cdot \mathbf{u} = \nabla^2 \phi \tag{3.29}$$

and

$$
\begin{aligned}
\nabla \times \mathbf{u} &= \nabla \times \nabla \times \boldsymbol{\Psi} \\
&= \nabla\nabla \cdot \boldsymbol{\Psi} - \nabla^2 \boldsymbol{\Psi} \quad \text{(from 3.16)} \\
&= -\nabla^2 \boldsymbol{\Psi} \quad \text{(since } \nabla \cdot \boldsymbol{\Psi} = 0 \text{).}
\end{aligned}
\tag{3.30}
$$

Motivated by (3.21) and (3.25), we require that these potentials also satisfy

$$\nabla^2 \phi - \frac{1}{\alpha^2}\frac{\partial^2 \phi}{\partial t^2} = 0, \tag{3.31}$$

$$\nabla^2 \boldsymbol{\Psi} - \frac{1}{\beta^2}\frac{\partial^2 \boldsymbol{\Psi}}{\partial t^2} = 0. \tag{3.32}$$

After solving these equations for ϕ and $\boldsymbol{\Psi}$, the P-wave displacement is given by the gradient of ϕ and the S-wave displacement is given by the curl of $\boldsymbol{\Psi}$, following (3.28).

3.4 Plane waves

At this point it is helpful to introduce the concept of a *plane wave*. This is a solution to the wave equation in which the displacement varies only in the direction of wave propagation and is constant in the directions orthogonal to the propagation direction. For example, for a plane wave traveling along the x axis, the displacement may be expressed as

$$\mathbf{u}(x, t) = \mathbf{f}(t \pm x/c), \tag{3.33}$$

where c is the velocity of the wave, $\mathbf{f}$ is any arbitrary function (a vector function is required to express the polarization of the wave), and the waves are propagating in

Table 3.1: Harmonic wave parameters.			
Angular frequency	ω	time^{-1}	$\omega = 2\pi f = \frac{2\pi}{T} = ck$
Frequency	f	time^{-1}	$f = \frac{\omega}{2\pi} = \frac{1}{T} = \frac{c}{\Lambda}$
Period	T	time	$T = \frac{1}{f} = \frac{2\pi}{\omega} = \frac{\Lambda}{c}$
Velocity	c	distance time^{-1}	$c = \frac{\Lambda}{T} = f\Lambda = \frac{\omega}{k}$
Wavelength	Λ	distance	$\Lambda = \frac{c}{f} = cT = \frac{2\pi}{k}$
Wavenumber	k	distance^{-1}	$k = \frac{\omega}{c} = \frac{2\pi}{\Lambda} = \frac{2\pi f}{c} = \frac{2\pi}{cT}$

either the $+x$ or $-x$ direction. The displacement does not vary with y or z; the wave extends to infinity in these directions. If $\mathbf{f}(t)$ is a discrete pulse, then u assumes the form of a displacement pulse traveling as a planar wavefront. More generally, displacement at position vector $\mathbf{x}$ for a plane wave propagating in the unit direction $\hat{\mathbf{s}}$ may be expressed as

$$\mathbf{u}(\mathbf{x}, t) = \mathbf{f}(t - \hat{\mathbf{s}} \cdot \mathbf{x}/c) \tag{3.34}$$

$$= \mathbf{f}(t - \mathbf{s} \cdot \mathbf{x}), \tag{3.35}$$

where $\mathbf{s} = \hat{\mathbf{s}}/c$ is the *slowness vector*, whose magnitude is the reciprocal of the velocity.

Since seismic energy is usually radiated from localized sources, seismic wavefronts are always curved to some extent; however, at sufficiently large distances from the source the wavefront becomes flat enough that a plane wave approximation becomes locally valid. Furthermore, many techniques for solving the seismic wave equation involve expressing the complete solution as a sum of plane waves of differing propagation angles. Often the time dependence is also removed from the equations by transforming into the frequency domain. In this case the displacement for a particular angular frequency ω may be expressed as

$$\mathbf{u}(\mathbf{x}, t) = \mathbf{A}(\omega)e^{-i\omega(t - \mathbf{s} \cdot \mathbf{x})} \tag{3.36}$$

$$= \mathbf{A}(\omega)e^{-i(\omega t - \mathbf{k} \cdot \mathbf{x})}, \tag{3.37}$$

where $\mathbf{k} = \omega\mathbf{s} = (\omega/c)\hat{\mathbf{s}}$ is termed the *wavenumber vector*. We will use complex numbers to represent harmonic waves throughout this book; details of how this works are reviewed in Appendix B. This may be termed a *monochromatic* plane wave; it is also sometimes called the *harmonic* or *steady-state* plane wave solution. Other parameters used to describe such a wave are the wavenumber $k = |\mathbf{k}| = \omega/c$, the frequency $f = \omega/(2\pi)$, the period $T = 1/f$, and the wavelength $\Lambda = cT$. Equations relating the various harmonic wave parameters are summarized in Table 3.1.

3.4.1 Example: Harmonic plane wave equation

What is the equation for the displacement of a 1 Hz P-wave propagating in the $+x$ direction at 6 km/s? In this case $\omega = 2\pi f$, where $f = 1$ Hz, and thus $\omega = 2\pi$. The slowness vector is in the direction of the x axis and thus $\hat{\mathbf{s}} = \hat{\mathbf{x}} = (1, 0, 0)$ and $\mathbf{s} = (1/c, 0, 0) = (1/6, 0, 0)$ s/km. We can thus express (3.36) as

$$\mathbf{u}(\mathbf{x}, t) = \mathbf{u}(x, t) = \mathbf{A}e^{-2i\pi(t-x/6)}$$

where t is in s and x is in km. As we shall see in the next section, P waves are polarized in the direction of wave propagation, so $\mathbf{u} = (u_x, 0, 0)$ and we can express this more simply as

$$u_x(x, t) = Ae^{-2i\pi(t-x/6)}.$$

In general, the coefficient A is complex to permit any desired phase at $x = 0$. As described in Appendix B, the real part must be taken for this equation to have a physical meaning. An alternative form is

$$u_x(x, t) = a \cos\left[2\pi(t - x/6) - \phi\right]$$

where a is the amplitude and ϕ is the phase at $x = 0$ (see Figure B.3).

3.5 Polarizations of P and S waves

Consider plane P waves propagating in the x direction. From (3.31) we have

$$\alpha^2 \partial_{xx}\phi = \partial_{tt}\phi. \tag{3.38}$$

A general solution to (3.38) can be written as

$$\phi = \phi_0(t \pm x/\alpha), \tag{3.39}$$

where a minus sign corresponds to propagation in the $+x$ direction and a plus sign denotes propagation in the $-x$ direction. Because $\mathbf{u} = \nabla\phi$, we have

$$\begin{aligned} u_x &= \partial_x\phi, \\ u_y &= 0, \\ u_z &= 0. \end{aligned} \tag{3.40}$$

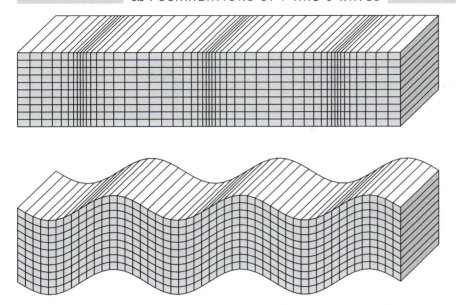

Figure 3.2 Displacements occurring from a harmonic plane P wave (top) and S wave (bottom) traveling horizontally across the page. S-wave propagation is pure shear with no volume change, whereas P waves involve both a volume change and shearing (change in shape) in the material. Strains are highly exaggerated compared to actual seismic strains in the Earth.

Note that for a plane wave propagating in the x direction there is no change in the y and z directions, and so the spatial derivatives ∂_y and ∂_z are zero. For P waves, the only displacement occurs in the direction of propagation along the x axis. Such wave motion is termed "longitudinal." Also, because $\nabla \times \nabla \phi = 0$, the motion is curl-free or "irrotational." Since P waves introduce volume changes in the material ($\nabla \cdot \mathbf{u} \neq 0$), they can also be termed "compressional" or "dilatational." However, note that P waves involve shearing as well as compression; this is why the P velocity is sensitive to both the bulk and shear moduli. Particle motion for a harmonic P wave is shown in Figure 3.2.

Now consider a plane S wave propagating in the positive x direction. The vector potential becomes

$$\mathbf{\Psi} = \Psi_x(t - x/\beta)\hat{\mathbf{x}} + \Psi_y(t - x/\beta)\hat{\mathbf{y}} + \Psi_z(t - x/\beta)\hat{\mathbf{z}}. \tag{3.41}$$

The displacement is

$$
\begin{aligned}
u_x &= (\nabla \times \mathbf{\Psi})_x = \partial_y \Psi_z - \partial_z \Psi_y = 0, \\
u_y &= (\nabla \times \mathbf{\Psi})_y = \partial_z \Psi_x - \partial_x \Psi_z = -\partial_x \Psi_z, \\
u_z &= (\nabla \times \mathbf{\Psi})_z = \partial_x \Psi_y - \partial_y \Psi_x = \partial_x \Psi_y,
\end{aligned}
\tag{3.42}
$$

where again we have used $\partial_y = \partial_z = 0$, thus giving

$$\mathbf{u} = -\partial_x \Psi_z \hat{\mathbf{y}} + \partial_x \Psi_y \hat{\mathbf{z}}. \tag{3.43}$$

The motion is in the y and z directions, perpendicular to the propagation direction. S-wave particle motion is often divided into two components: the motion within a vertical plane through the propagation vector (SV waves) and the horizontal motion in the direction perpendicular to this plane (SH waves). Because $\nabla \cdot \mathbf{u} = \nabla \cdot (\nabla \times \mathbf{\Psi}) = 0$, the motion is pure shear without any volume change (hence the name shear waves). Particle motion for a harmonic shear wave polarized in the vertical direction (SV wave) is illustrated in Figure 3.2.

3.6 Spherical waves

Another solution to the scalar wave equation (3.31) for the P-wave potential ϕ is possible if we assume spherical symmetry. In spherical coordinates, the Laplacian operator is

$$\nabla^2 \phi(r) = \frac{1}{r^2} \frac{\partial}{\partial r} \left[r^2 \frac{\partial \phi}{\partial r} \right], \tag{3.44}$$

where we have dropped the angular derivatives because of the spherical symmetry. Using this expression in (3.31), we have

$$\frac{1}{r^2} \frac{\partial}{\partial r} \left[r^2 \frac{\partial \phi}{\partial r} \right] - \frac{1}{\alpha^2} \frac{\partial^2 \phi}{\partial t^2} = 0. \tag{3.45}$$

Solutions to this equation outside the point $r = 0$ may be expressed as

$$\phi(r, t) = \frac{f(t \pm r/\alpha)}{r}. \tag{3.46}$$

Note that this is identical to the plane wave equation (3.33), except for the factor of $1/r$. Inward and outward propagating waves are specified by the $+$ and $-$ signs respectively. Since this expression is usually used to model waves radiating away from a point source, the inward propagating solution is normally ignored. In this case the $1/r$ term represents a decay in the wave amplitude with range, a geometrical spreading factor that we will explore further in Chapter 6.

Equation (3.46) is not a valid solution to (3.45) at $r = 0$. However, it can be shown (e.g., Aki and Richards, 2002, Section 4.1) that (3.46) is the solution to the inhomogeneous wave equation

$$\nabla^2\phi(r) - \frac{1}{\alpha^2}\frac{\partial^2\phi}{\partial t^2} = -4\pi\delta(r)f(t), \tag{3.47}$$

where the delta function $\delta(r)$ is zero everywhere except $r = 0$ and has a volume integral of 1. The factor $4\pi\delta(r)f(t)$ represents the source-time function at the origin. We will return to this equation when we discuss seismic source theory in Chapter 9.

3.7 Methods for computing synthetic seismograms[†]

A large part of seismology involves devising and implementing techniques for computing synthetic seismograms for realistic Earth models. In general, our goal is to calculate what would be recorded by a seismograph at a specified receiver location, given an exact specification of the seismic source and the Earth model through which the seismic waves propagate. This is a well-defined forward problem that, in principle, can be solved exactly. However, errors in the synthetic seismograms often occur in practical applications. These inaccuracies can be be separated into two parts:

1. Inaccuracies arising from approximations in the theory used to compute the synthetic seismograms. Examples of this would include many applications of ray theory which do not properly account for head waves, diffracted waves, or the coupling between different wave types at long periods. Another computational error is the grid dispersion that occurs in most finite difference schemes.
2. Errors caused by using a simplified Earth or source model. In this case the synthetic seismogram may be exact for the simplified model, but the model is an inadequate representation of the real problem. These simplifications might be necessary in order to apply a particular numerical technique, or might result from ignorance of many of the details of the model. Examples would include the use of 1-D models that do not fully account for 3-D structure, the assumption of a point source rather than a finite rupture, and neglecting the effects of attenuation or anisotropy in the calculations.

The first category of errors may be addressed by applying a more exact algorithm, although in practice limits on computing resources may prevent achieving the desired accuracy in the case of complicated models. The second category is more serious because often one simply does not know the properties of the Earth well enough to be able to model every wiggle in the observed seismograms. This is particularly true at high frequencies (0.5 Hz and above). For teleseismic arrivals, long-period body waves (15–50 s period) and surface waves (40–300 s period) can usually be fit well with current Earth models, whereas the coda of high-frequency body-wave arrivals can only be modeled statistically (fitting the envelope function but not the individual wiggles).

Because of the linearity of the problem and the superposition principle (in which distributed sources can be described as the sum of multiple point sources), there is no great difficulty in modeling even very complicated sources (inverting for these sources, is, of course, far more difficult, but here we are only concerned with the forward problem). If the source can be exactly specified, then computing synthetics for a distributed source is only slightly more complicated than for a simple point source. By far the most difficult part in computing synthetic seismograms is solving for the propogation effects through realistic velocity structures. Only for a few grossly simplified models (e.g., whole space or half-spaces) are analytical solutions possible.

The part of the solution that connects the force distribution at the source with the displacements at the receiver is termed the elastodynamic Green's function, and will be discussed in greater detail in Chapter 9. Computation of the Green's function is the key part of the synthetic seismogram calculation because this function must take into account all of the elastic properties of the material and the appropriate boundary conditions.

There are a large number of different methods for computing synthetic seismograms. Most of these fall into the following categories:

1. Finite-difference and finite-element methods that use computer power to solve the wave equation over a discrete set of grid points or model elements. These have the great advantage of being able to handle models of arbitrary complexity. Their computational cost grows with the number of required grid points; more points are required for 3-D models (vs. 2-D) and for higher frequencies. These methods are discussed in more detail in the next section.
2. Ray-theoretical methods in which ray geometries are explicitly specified and ray paths are computed. These methods include simple (or geometrical) ray theory, WKBJ, and so-called "generalized" ray theory. They are most useful at high frequencies for which the ray-theoretical approximation is most accurate. They are most simply applied to 1-D Earth models but can be generalized to 3-D models.
3. Homogeneous layer methods in which the model consists of a series of horizontal layers with constant properties within each layer. Matrix methods are then used to connect the solutions between layers. Examples of this approach include "reflectivity" and "wavenumber integration." These methods yield an exact solution but can become moderately computationally intensive at high frequencies because a large number of layers are required to accurately simulate continuous velocity gradients. Unlike finite-difference and ray-theoretical methods, homogeneous-layer techniques are restricted to 1-D Earth models. However, spherically symmetric models can be computed using the flat Earth transformation.

4. Normal mode summation methods in which the standing waves (eigenvectors) of the spherical Earth are computed and then summed to generate synthetic seismograms. This is the most natural and complete way to compute synthetic seismograms for the spherical Earth, but is computationally intensive at high frequencies. Generalization to 3-D Earth models requires including coupling between modes; this is generally done using asymptotic approximations and greatly increases the complexity of the algorithm.

There is no single "best" way to compute synthetic seismograms as each method has its own advantages and disadvantages. The method of choice will depend upon the particular problem to be addressed and the available computer power; thus it is useful to be aware of the full repertoire of techniques. This book will cover only how relatively simple ray-theoretical synthetic seismograms can be computed for 1-D Earth models. For details regarding ray-theoretical and homogeneous-layer methods, see Kennett (2001) and Chapman (2004). For normal mode methods, see Dahlen and Tromp (1998).

3.8 The future of seismology?[†]

Increasing computer capabilities now make possible ambitious numerical simulations of seismic wave propagation that were impractical only a few years ago and this trend is likely to continue for many decades. These calculations involve finite-difference or finite-element methods that approximate the continuum of elastic properties with a large number of discrete values or model elements and solve the wave equation numerically over a series of discrete time steps. They provide a complete image of the wavefield at each point in the model for every time step, as illustrated in Figure 3.3, which shows a snapshot at 10 minutes of the SH wavefield in the mantle for a source at 500 km (Thorne et al., 2007). Finite difference methods specify the model at a series of grid points and approximate the spatial and temporal derivatives by using the model values at nearby grid points. Finite-element methods divide the model into a series of volume elements with specified properties and match the appropriate boundary conditions among adjacent elements. Historically, because of their simplicity, finite-difference methods have been used in seismology more often than finite elements. However, finite-difference algorithms can have difficulty correctly handling boundary conditions at sharp interfaces, including the irregular topography at the Earth's surface, for which finite-element schemes are more naturally suited.

Discrete modeling approaches can accurately compute seismograms for complicated 3-D models of Earth structure, provided the gridding or meshing scheme has sufficient resolution. Complicated analytical techniques are not required, although

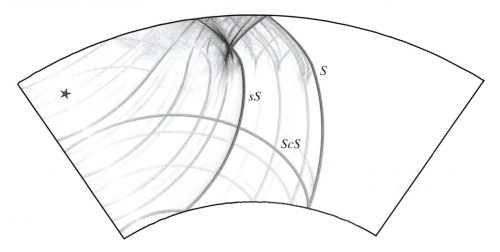

Figure 3.3 The *SH*-velocity wavefield in the mantle after 10 minutes for a source at 500 km depth (star), adapted from a figure in Thorne *et al.* (2007). This axi-symmetric 2-D finite-difference calculation used the PREM velocity model. The major seismic phases are labeled (see Chapter 4); the lower amplitude phases are mainly reflections off upper-mantle discontinuities and an assumed discontinuity 264 km above the core--mantle boundary.

the speed of the algorithm depends upon the skill of the computer programmer in developing efficient code. Typically, a certain number of grid points or model elements are required per seismic wavelength for accurate results, with the exact number depending upon the specific algorithm. In three dimensions, the number of grid points grows inversely as the cube of the grid spacing and number of required time steps normally also increases. Because of this, computational requirements increase rapidly with decreasing seismic wavelength, with the most challenging calculations being at high frequencies and the greatest required grid densities occurring in the slowest parts of the model.

Finite-difference methods vary depending upon how the temporal and spatial derivatives in these equations are calculated. Simple first-order differencing schemes for the spatial derivatives are fast and easy to program, but require more grid points per wavelength to achieve accuracy comparable to higher-order differencing schemes. Many finite-difference programs use the staggered grid approach in which the velocities and stresses are computed at different grid points.

A few general points to keep in mind:

1. Finite-difference programs run most efficiently if their arrays fit into memory and thus machines with large memories are desirable. Higher-order finite-difference schemes generally have an advantage because fewer grid points per wavelength are required for accurate results, thus reducing the size of the arrays.

2. Simple first-order differencing schemes require more grid points per wavelength than higher-order schemes. A commonly used "rule of thumb" is that first-order differencing algorithms require about 20 grid points per wavelength, but even this is not sufficient if the calculation is performed for a large model that spans many wavelengths. So called pseudo-spectral methods are equivalent to very high-order differencing methods and in principle require the smallest number of grid points per wavelength (approaching 2 in certain idealized situations). However, models with sharp velocity discontinuities often require more grid points, so much of the advantage of the spectral methods is lost in this case.

3. An important aspect of finite-difference and finite-element methods is devising absorbing boundary conditions to prevent annoying reflections from the edges of the model. This is a non-trivial problem, and many papers have been written discussing various techniques. Many of these methods work adequately for waves hitting the boundaries at near normal incidence, but have problems for grazing incidence angles.

Finite-element programs often have advantages over finite differences in applying boundary conditions. Currently the most developed finite-element program in global seismology is the implementation of the spectral-element method by Komatitsch, Tromp and coworkers (Komatitsch *et al.*, 2002, 2005), which explicitly includes the free-surface and fluid/solid boundary conditions at the core–mantle and inner-core boundaries. This program is designed to run in parallel on large high-performance computing clusters. It uses a variable size meshing scheme for the entire Earth that maintains a relatively constant number of grid points per wavelength (see Fig. 3.4). The method includes the effects of general anisotropy, anelasticity (attenuation), surface topography, ellipticity, rotation, and self-gravitation. As implemented, the method requires an average of five grid points per wavelength for many applications. The algorithm has been validated through comparisons with synthetics computed using normal mode summation.

Numbers cited by Komatitsch *et al.* (2005) for calculations on the Earth Simulator at the Japan Agency for Marine Earth Science and Technology (JAMSTEC) provide some perspective on the computational requirements. Using 48 nodes (with 64 gigaflops and 16 gigabytes of memory per node), a global simulation can model waveperiods down to 9 s in about 10 hours of computer time. Shorter periods can be reached in the same time if more nodes are used. This calculation provides synthetic seismograms from a single earthquake to any number of desired receivers. The Earth model itself can be arbitrarily complicated with calculations for general 3-D velocity and density variations taking no longer than those for 1-D reference models.

Large-scale numerical simulations are also important for modeling strong ground motions in and around sedimentary basins from large earthquakes and a number of groups are now performing these calculations (e.g., Akcelik *et al.*, 2003; Olsen *et al.*,

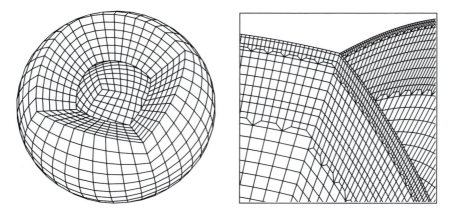

Figure 3.4 The meshing scheme used by the Komatitsch *et al.* (2002, 2005) implementation of the spectral element method. The spherical Earth is decomposed into six "cubical" blocks. The right plot shows how the blocks join and how the number of elements increases near the surface.

2006). A challenging aspect of these problems is the very slow shear velocities observed in shallow sedimentary layers. For example, in the Los Angeles basin the average shear velocity approaches 200 m/s at the surface (e.g., Magistrale *et al.*, 2000). These calculations are valuable because they show how focusing effects from rupture directivity and basin geometry can lead to large variations in expected wave amplitudes.

As computer speed and memory size increase, these numerical methods will become practical even on desktop machines and eventually it will be possible to routinely compute broadband synthetics for general 3-D Earth models. In time, these computer-intensive algorithms will probably replace many of the alternative synthetic seismogram methods. However, there will still be a need for the techniques of classical seismology, such as ray theory and surface wave dispersion analysis, in order to understand and interpret the results that the computers provide. Ultimately the challenge will be to devise new methods of data analysis and inversion that will fully exploit the computational capabilities that are rapidly coming to the field.

3.9 Equations for 2-D isotropic finite differences[†]

As an example of a discrete modeling method, this section presents equations for simple isotropic 1-D and 2-D finite differences. Much of this material is adapted from Section 13.6 of the 2nd volume of the first edition of Aki and Richards (1980).

We begin with the momentum equation:

$$\rho \frac{\partial^2 \mathbf{u}}{\partial t^2} = \nabla \cdot \boldsymbol{\tau}. \tag{3.48}$$

Now let $\mathbf{u} = (u_x, u_y, u_z) = (u, v, w)$ and recall that $(\nabla \cdot \boldsymbol{\tau})_i = \partial_j \tau_{ij}$. For the two-dimensional case of *SH* waves propagating in the xz plane, displacement only occurs in the y direction (i.e., $\mathbf{u} = (0, v, 0)$) and we can write:

$$\rho \frac{\partial^2 v}{\partial t^2} = \partial_j \tau_{yj} = \frac{\partial \tau_{yx}}{\partial x} + \frac{\partial \tau_{yz}}{\partial z}. \tag{3.49}$$

Note that $\frac{\partial}{\partial y} = 0$ for the two-dimensional problem. Now recall (3.13) which relates stress to displacement for isotropic media:

$$\tau_{ij} = \lambda \delta_{ij} \partial_k u_k + \mu (\partial_i u_j + \partial_j u_i). \tag{3.50}$$

Using this equation we can obtain expressions for τ_{yx} and τ_{yz}:

$$\tau_{yx} = \mu \frac{\partial v}{\partial x} \tag{3.51}$$

$$\tau_{yz} = \mu \frac{\partial v}{\partial z}.$$

Substituting into (3.49), we obtain:

$$\rho \frac{\partial^2 v}{\partial t^2} = \frac{\partial}{\partial x} \left[\mu \frac{\partial v}{\partial x} \right] + \frac{\partial}{\partial z} \left[\mu \frac{\partial v}{\partial z} \right]. \tag{3.52}$$

Note that for one-dimensional wave propagation in the x direction $\frac{\partial}{\partial z} = 0$ and the *SH* equation reduces to:

$$\rho(x) \frac{\partial^2 v}{\partial t^2} = \frac{\partial}{\partial x} \left[\mu(x) \frac{\partial v}{\partial x} \right]. \tag{3.53}$$

This is equivalent to equation (13.129) in Aki and Richards (1980). A similar equation exists for one-dimensional *P*-wave propagation if the $\mu(x)$ is replaced with $\lambda(x) + 2\mu(x)$ and the displacements in the y direction (v) are replaced with displacements in the x direction (u).

We can avoid the double time derivative and the space derivatives of μ if we use the particle velocity $\dot{v}$ and stress $\tau = \mu \partial v / \partial x$ as variables. We then have the simultaneous equations:

$$\frac{\partial \dot{v}}{\partial t} = \frac{1}{\rho(x)} \frac{\partial \tau}{\partial x} \tag{3.54}$$

$$\frac{\partial \tau}{\partial t} = \mu(x)\frac{\partial \dot{v}}{\partial x}.$$

A solution to these equations can be obtained directly using finite-difference approximations for the derivatives. In order to design a stable finite-difference algorithm for the wave equation, it is important to use *centered* finite-difference operators. To see this, consider the Taylor series expansion of a function $\phi(x)$

$$\phi(x + \Delta x) = \phi(x) + \frac{\partial \phi}{\partial x}\Delta x + \frac{1}{2}\frac{\partial^2 \phi}{\partial x^2}(\Delta x)^2 + \frac{1}{6}\frac{\partial^3 \phi}{\partial x^3}(\Delta x)^3 + \text{higher-order terms.}$$
$$(3.55)$$

If we solve this equation for $\partial \phi/\partial x$, we obtain

$$\frac{\partial \phi}{\partial x} = \frac{1}{\Delta x}\left[\phi(x + \Delta x) - \phi(x)\right] - \frac{1}{2}\frac{\partial^2 \phi}{\partial x^2}\Delta x - \frac{1}{6}\frac{\partial^3 \phi}{\partial x^3}(\Delta x)^2 - \cdots \qquad (3.56)$$

and we see that the simple approximation

$$\frac{\partial \phi}{\partial x} \approx \frac{1}{\Delta x}\left[\phi(x + \Delta x) - \phi(x)\right] \qquad (3.57)$$

will have a leading truncation error proportional to Δx. To obtain a better approximation, consider the expansion for $\phi(x - \Delta x)$

$$\phi(x - \Delta x) = \phi(x) - \frac{\partial \phi}{\partial x}\Delta x + \frac{1}{2}\frac{\partial^2 \phi}{\partial x^2}(\Delta x)^2 - \frac{1}{6}\frac{\partial^3 \phi}{\partial x^3}(\Delta x)^3 + \text{higher-order terms}$$
$$(3.58)$$

Solving for $\partial \phi/\partial x$, we obtain

$$\frac{\partial \phi}{\partial x} = \frac{1}{\Delta x}\left[\phi(x) - \phi(x - \Delta x)\right] + \frac{1}{2}\frac{\partial^2 \phi}{\partial x^2}\Delta x - \frac{1}{6}\frac{\partial^3 \phi}{\partial x^3}(\Delta x)^2 - \cdots \qquad (3.59)$$

Averaging (3.56) and (3.59), we obtain

$$\frac{\partial \phi}{\partial x} = \frac{1}{2\Delta x}\left[\phi(x + \Delta x) - \phi(x - \Delta x)\right] - \frac{1}{3}\frac{\partial^3 \phi}{\partial x^3}(\Delta x)^2 - \cdots \qquad (3.60)$$

and we see that the central difference formula

$$\frac{\partial \phi}{\partial x} = \frac{1}{2\Delta x}\left[\phi(x + \Delta x) - \phi(x - \Delta x)\right] \qquad (3.61)$$

has an error of order $(\Delta x)^2$. For small values of Δx, these errors will be much smaller than those obtained using (3.57). Similarly, the second derivative of ϕ can be computed by summing (3.55) and (3.58) to obtain

$$\frac{\partial^2 \phi}{\partial x^2} = \frac{1}{(\Delta x)^2}\left[\phi(x + \Delta x) - 2\phi(x) + \phi(x - \Delta x)\right], \qquad (3.62)$$

which also has error of order $(\Delta x)^2$.

To show how a centered finite-difference approach can be used to solve (3.54), consider Figure 3.5a, which shows the xt plane sampled at points $(i\Delta t,\ j\Delta x)$, where i and j are integers. We can then write

$$\frac{\dot{v}_j^{i+1} - \dot{v}_j^{i-1}}{2\Delta t} = \frac{1}{\rho_j}\frac{\tau_{j+1}^i - \tau_{j-1}^i}{2\Delta x} \qquad (3.63)$$

$$\frac{\tau_j^{i+1} - \tau_j^{i-1}}{2\Delta t} = \mu_j \frac{\dot{v}_{j+1}^i - \dot{v}_{j-1}^i}{2\Delta x}.$$

This approach will be stable provided the time-mesh interval Δt is smaller than or equal to $\Delta x/c_j$, where $c_j = \sqrt{\mu_j/\rho_j}$ is the local wave velocity.

An even better algorithm uses a *staggered-grid* approach (e.g., Virieux, 1986) in which the velocities and stresses are computed at different grid points, offset by

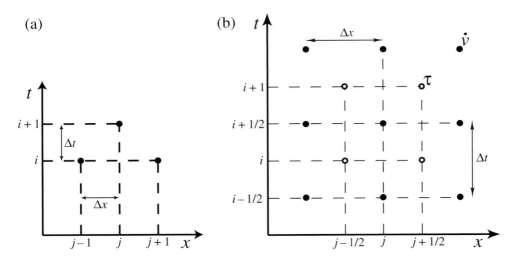

Figure 3.5 (a) A simple 1-D finite-difference gridding scheme. (b) A staggered grid in which the velocities and stresses are stored at different points.

half a grid length in both x and t (see Figure 3.5b). In this case, we have

$$\frac{\dot{v}_j^{i+\frac{1}{2}} - \dot{v}_j^{i-\frac{1}{2}}}{\Delta t} = \frac{1}{\rho_j} \frac{\tau_{j+1/2}^i - \tau_{j-1/2}^i}{\Delta x} \qquad (3.64)$$

$$\frac{\tau_{j+1/2}^{i+1} - \tau_{j+1/2}^i}{\Delta t} = \mu_{j+\frac{1}{2}} \frac{\dot{v}_{j+1}^{i+\frac{1}{2}} - \dot{v}_j^{i+\frac{1}{2}}}{\Delta x}.$$

As discussed in Aki and Richards (1980, p. 777), the error in this approximation is four times smaller than in (3.63) because the sampling interval has been halved.

Now let us consider the two-dimensional *P-SV* system. In this case $\mathbf{u} = (u, 0, w)$ and we can write:

$$\rho \frac{\partial^2 u}{\partial t^2} = \partial_j \tau_{xj} = \frac{\partial \tau_{xx}}{\partial x} + \frac{\partial \tau_{xz}}{\partial z} \qquad (3.65)$$

$$\rho \frac{\partial^2 w}{\partial t^2} = \partial_j \tau_{zj} = \frac{\partial \tau_{zx}}{\partial x} + \frac{\partial \tau_{zz}}{\partial z}.$$

Using (3.50) we can obtain expressions for τ_{xx}, τ_{xz}, and τ_{zz}:

$$
\begin{aligned}
\tau_{xx} &= \lambda \left[\frac{\partial u}{\partial x} + \frac{\partial w}{\partial z} \right] + \mu \left[2\frac{\partial u}{\partial x} \right] \\
&= (\lambda + 2\mu)\frac{\partial u}{\partial x} + \lambda \frac{\partial w}{\partial z} \\
\tau_{xz} &= \mu \left[\frac{\partial u}{\partial z} + \frac{\partial w}{\partial x} \right] \qquad\qquad (3.66) \\
\tau_{zz} &= \lambda \left[\frac{\partial u}{\partial x} + \frac{\partial w}{\partial z} \right] + \mu \left[2\frac{\partial w}{\partial z} \right] \\
&= (\lambda + 2\mu)\frac{\partial w}{\partial z} + \lambda \frac{\partial u}{\partial x}.
\end{aligned}
$$

Equations (3.49) and (3.51) are a coupled system of equations for two-dimensional *SH*-wave propagation, while (3.65) and (3.66) are the equations for *P-SV* wave propagation. As in the one-dimensional case, it is often convenient to take time derivatives of the equations for the stress (3.51) and (3.66), so that we can express everything in terms of $(\dot{u}, \dot{v}, \dot{w}) = \partial \mathbf{u}/\partial t$. In this case the *SH* equations become:

$$\rho \frac{\partial \dot{v}}{\partial t} = \frac{\partial \tau_{yx}}{\partial x} + \frac{\partial \tau_{yz}}{\partial z}$$

$$\frac{\partial \tau_{yx}}{\partial t} = \mu \frac{\partial \dot{v}}{\partial x} \tag{3.67}$$

$$\frac{\partial \tau_{yz}}{\partial t} = \mu \frac{\partial \dot{v}}{\partial z}$$

and the *P-SV* equations become:

$$\rho \frac{\partial \dot{u}}{\partial t} = \frac{\partial \tau_{xx}}{\partial x} + \frac{\partial \tau_{xz}}{\partial z}$$

$$\rho \frac{\partial \dot{w}}{\partial t} = \frac{\partial \tau_{zx}}{\partial x} + \frac{\partial \tau_{zz}}{\partial z}$$

$$\frac{\partial \tau_{xx}}{\partial t} = (\lambda + 2\mu) \frac{\partial \dot{u}}{\partial x} + \lambda \frac{\partial \dot{w}}{\partial z} \tag{3.68}$$

$$\frac{\partial \tau_{xz}}{\partial t} = \mu \left[\frac{\partial \dot{u}}{\partial z} + \frac{\partial \dot{w}}{\partial x} \right]$$

$$\frac{\partial \tau_{zz}}{\partial t} = (\lambda + 2\mu) \frac{\partial \dot{w}}{\partial z} + \lambda \frac{\partial \dot{u}}{\partial x}.$$

These are first-order systems of equations in velocity and stress which can be solved numerically. In this case, the elastic properties ρ, λ, and μ are specified at a series of model grid points. With suitable starting conditions, the velocities and stresses are also defined at grid points. The program then calculates the required spatial derivatives of the stresses in order to compute the velocities at time $t + \Delta t$. The spatial derivatives of these velocities then allow the computation of new values for the stresses. This cycle is then repeated.

The global finite-difference calculation plotted in Figure 3.3 was performed using an axi-symmetric *SH*-wave algorithm developed and implemented by Igel and Weber (1995), Thorne *et al.* (2007), and Jahnke *et al.* (2008). It uses a staggered grid, with an eight-point operator to compute the spatial derivatives, and can be run on parallel computers with distributed memory.

3.10 Exercises

1. Period T is to angular frequency ω as wavelength Λ is to: (a) wavenumber k, (b) velocity c, (c) frequency f, (d) time t, (e) none of the above.

2. Figure 3.6 plots a harmonic plane wave at $t = 0$, traveling in the x direction at 5 km/s. (a) Write down an equation for this wave that describes displacement, u, as a function of x and t. (b) What is the maximum strain for this wave?

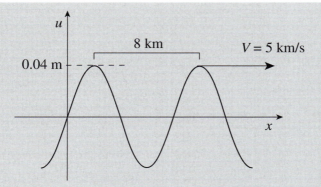

Figure 3.6 Displacement of a harmonic wave at $t = 0$ as a function of distance.

3. Consider two types of monochromatic plane waves propagating in the x direction in a uniform medium: (a) P wave in which $u_x = A \sin(\omega t - kx)$, (b) S wave with displacements in the y direction, i.e., $u_y = A \sin(\omega t - kx)$. For each case, derive expressions for the non-zero components of the stress tensor. Refer to (2.17) to get the components of the strain tensor; then use (2.30) to obtain the stress components.

4. Assume harmonic P waves are traveling through a solid with $\alpha = 10$ km/s. If the maximum strain is 10^{-8}, what is the maximum particle displacement for waves with periods of: (a) 1 s, (b) 10 s, (c) 100 s?

5. Is it possible to have spherical symmetry for S waves propagating away from a point source? Under what conditions could an explosive source generate shear waves?

6. Show that (3.46) satisfies (3.45) for $r \neq 0$.

7. (COMPUTER) In the case of plane-wave propagation in the x direction within a uniform medium, the homogeneous momentum equation (3.9) for shear waves can be expressed as

$$\frac{\partial^2 u}{\partial t^2} = \beta^2 \frac{\partial^2 u}{\partial x^2},$$

where u is the displacement. Write a computer program that uses finite differences to solve this equation for a bar 100 km in length, assuming $\beta = 4$ km/s. Use $dx = 1$ km for the length spacing and $dt = 0.1$ s for the time spacing. Assume a source-time function at u (50 km) of the form

$$u_{50}(t) = \sin^2(\pi t/5), 0 < t < 5 \text{ s}.$$

Apply a stress-free boundary condition at u (0 km) and a fixed boundary condition at u (100 km). Approximate the second derivatives using the finite difference scheme:

$$\frac{\partial^2 u}{\partial x^2} = \frac{u_{i+1} - 2u_i + u_{i-1}}{dx^2}.$$

Plot $u(x)$ at 4 s intervals from 1 to 33 s. Verify that the pulses travel at velocities of 4 km/s. What happens to the reflected pulse at each endpoint? What happens when the pulses cross?

Hint: Here is the key part of a FORTRAN program to solve this problem:

```
(initialize t, dx, dt, tlen, beta and u1,u2,u3 arrays)
10      t=t+dt
        do i=2,100
            rhs=beta**2*(u2(i+1)-2.*u2(i)+u2(i-1))/dx**2
            u3(i)=dt**2*rhs+2.*u2(i)-u1(i)
        enddo
        u3(1)=u3(2)
        u3(101)=0.
        if (t.le.tlen) then
            u3(51)=sin(3.1415927*t/tlen)**2
        end if
        do i=1,101
            u1(i)=u2(i)
            u2(i)=u3(i)
        enddo
(output u2 at desired intervals, stop when t is big
enough)
        go to 10
```

4

Ray theory: Travel times

Seismic ray theory is analogous to optical ray theory and has been applied for over 100 years to interpret seismic data. It continues to be used extensively today, owing to its simplicity and applicability to a wide range of problems. These applications include most earthquake location algorithms, body-wave focal mechanism determinations, and inversions for velocity structure in the crust and mantle. Ray theory is intuitively easy to understand, simple to program, and very efficient. Compared to more complete solutions, it is relatively straightforward to generalize to three-dimensional velocity models. However, ray theory also has several important limitations. It is a high-frequency approximation, which may fail at long periods or within steep velocity gradients, and it does not easily predict any "non-geometrical" effects, such as head waves or diffracted waves. The ray geometries must be completely specified, making it difficult to study the effects of reverberation and resonance due to multiple reflections within a layer.

In this chapter, we will be concerned only with the timing of seismic arrivals, deferring the consideration of amplitudes and other details to later. This narrow focus is nonetheless very useful for many problems; a significant fraction of current research in seismology uses only travel time information. The theoretical basis for much of ray theory is derived from the *eikonal equation* (see Appendix C); however, because these results are not required for most applications we do not describe them here.

4.1 Snell's law

Consider a plane wave, propagating in material of uniform velocity v, that intersects a horizontal interface (Fig. 4.1).

The wavefronts at time t and time $t + \Delta t$ are separated by a distance Δs along the ray path. The ray angle from the vertical, θ, is termed the incidence angle. This

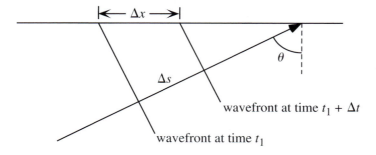

Figure 4.1 A plane wave incident on a horizontal surface. The ray angle from vertical is termed the incidence angle θ.

angle relates Δs to the wavefront separation on the interface, Δx, by

$$\Delta s = \Delta x \, \sin \theta. \tag{4.1}$$

Since $\Delta s = v\Delta t$, we have

$$v\Delta t = \Delta x \, \sin \theta \tag{4.2}$$

or

$$\frac{\Delta t}{\Delta x} = \frac{\sin \theta}{v} = u \sin \theta \equiv p, \tag{4.3}$$

where u is the *slowness* ($u = 1/v$ where v is velocity) and p is termed the *ray parameter*. If the interface represents the free surface, note that by timing the arrival of the wavefront at two different stations, we could directly measure p. The ray parameter p represents the apparent slowness of the wavefront in a horizontal direction, which is why p is sometimes called the *horizontal slowness* of the ray.

Now consider a downgoing plane wave that strikes a horizontal interface between two homogeneous layers of different velocity and the resulting transmitted plane wave in the lower layer (Fig. 4.2). If we draw wavefronts at evenly spaced times along the ray, they will be separated by different distances in the different layers, and we see that the ray angle at the interface must change to preserve the timing of the wavefronts across the interface.

In the case illustrated the top layer has a slower velocity ($v_1 < v_2$) and a correspondingly larger slowness ($u_1 > u_2$). The ray parameter may be expressed in terms of the slowness and ray angle from the vertical within each layer:

$$p = u_1 \sin \theta_1 = u_2 \sin \theta_2. \tag{4.4}$$

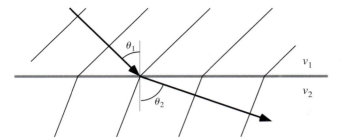

Figure 4.2 A plane wave crossing a horizontal interface between two homogeneous half-spaces. The higher velocity in the bottom layer causes the wavefronts to be spaced further apart.

Notice that this is simply the seismic version of Snell's law in geometrical optics. Equation (4.4) may also be obtained from *Fermat's principle*, which states that the travel time between two points must be stationary (usually, but not always, the minimum time) with respect to small variations in the ray path. Fermat's principle itself can be derived from applying variational calculus to the eikonal equation (e.g., Aki and Richards, 2002, pp. 89–90).

4.2 Ray paths for laterally homogeneous models

In most cases the compressional and shear velocities increase as a function of depth in the Earth. Suppose we examine a ray traveling downward through a series of layers, each of which is faster than the layer above. The ray parameter p remains constant and we have

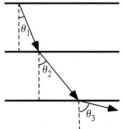

$$p = u_1 \sin \theta_1 = u_2 \sin \theta_2 = u_3 \sin \theta_3. \qquad (4.5)$$

If the velocity continues to increase, θ will eventually equal 90° and the ray will be traveling horizontally.

This is also true for continuous velocity gradients (Fig. 4.3). If we let the slowness at the surface be u_0 and the *takeoff angle* be θ_0, we have

$$u_0 \sin \theta_0 = p = u \sin \theta. \qquad (4.6)$$

When $\theta = 90°$ we say that the ray is at its *turning point* and $p = u_{tp}$, where u_{tp} is the slowness at the turning point. Since velocity generally increases with depth in Earth, the slowness *decreases* with depth. Smaller ray parameters are more steeply dipping at the surface, will turn deeper in Earth, and generally travel farther. In these examples with horizontal layers or vertical velocity gradients, p remains

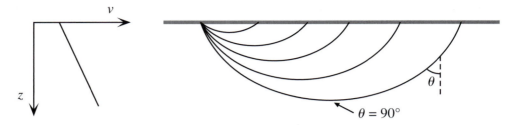

Figure 4.3 Ray paths for a model that has a continuous velocity increase with depth will curve back toward the surface. The ray turning point is defined as the lowermost point on the ray path, where the ray direction is horizontal and the incidence angle is 90°.

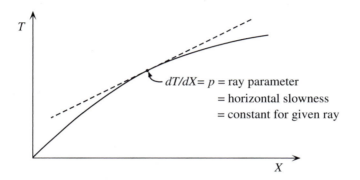

$dT/dX = p$ = ray parameter
= horizontal slowness
= constant for given ray

Figure 4.4 A travel time curve for a model with a continuous velocity increase with depth. Each point on the curve results from a different ray path; the slope of the travel time curve, dT/dX, gives the ray parameter for the ray.

constant along the ray path. However, if lateral velocity gradients or dipping layers are present, then p will change along the ray path.

In a model in which velocity increases with depth,[1] the *travel time curve*, a plot of the first arrival time versus distance, will look like Figure 4.4. Note that p varies along the travel time curve; a different ray is responsible for the "arrival" at each distance X. At any point along a ray, the slowness vector $\mathbf{s}$ can be resolved into its horizontal and vertical components. The length of $\mathbf{s}$ is given by u, the local slowness. The horizontal component, s_x, of the

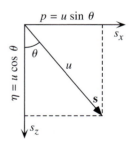

---

[1] Observant readers may notice that our coordinate system has now flipped so that z is no longer pointing upward as it did in Chapters 2 and 3, but rather points downward so that depths from the surface are positive. Both conventions are often used in seismology, depending upon which is more convenient.

slowness is the ray parameter p. In an analogous way, we may define the vertical slowness η by

$$\eta = u \cos \theta = (u^2 - p^2)^{1/2}. \tag{4.7}$$

At the turning point, $p = u$ and $\eta = 0$.

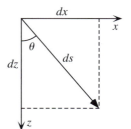

Let us now develop integral expressions to compute travel time and distance along a particular ray. Consider a segment of length ds along a ray path. From the geometry we have

$$\frac{dx}{ds} = \sin \theta, \quad \frac{dz}{ds} = \cos \theta = (1 - \sin^2 \theta)^{1/2}. \tag{4.8}$$

Since $p = u \sin \theta$, we may write

$$\frac{dx}{ds} = \frac{p}{u}, \quad \frac{dz}{ds} = (1 - p^2/u^2)^{1/2} = u^{-1}(u^2 - p^2)^{1/2}. \tag{4.9}$$

From the chain rule

$$\frac{dx}{dz} = \frac{dx}{ds}\frac{ds}{dz} = \frac{dx/ds}{dz/ds} = \frac{p}{u}\frac{u}{(u^2 - p^2)^{1/2}} = \frac{p}{(u^2 - p^2)^{1/2}}. \tag{4.10}$$

This can be integrated to obtain x:

$$x(z_1, z_2, p) = p \int_{z_1}^{z_2} \frac{dz}{(u^2(z) - p^2)^{1/2}}. \tag{4.11}$$

If we let z_1 be the free surface ($z_1 = 0$) and z_2 be the turning point z_p, the distance x from the surface source to the surface point over the turning point is

$$x(p) = p \int_0^{z_p} \frac{dz}{(u^2(z) - p^2)^{1/2}}. \tag{4.12}$$

Because the ray is symmetric about the turning point, the total distance $X(p)$ from surface source to surface receiver is just twice this expression, that is,

$$X(p) = 2p \int_0^{z_p} \frac{dz}{\left(u^2(z) - p^2\right)^{1/2}}. \tag{4.13}$$

In a similar way, we can derive an expression for the travel time $t(p)$:

$$dt = u\, ds, \quad dt/ds = u,$$ (4.14)

$$\frac{dt}{dz} = \frac{dt}{ds}\frac{ds}{dz} = \frac{dt/ds}{dz/ds} = \frac{u^2}{\left(u^2(z) - p^2\right)^{1/2}},$$ (4.15)

and we obtain

$$t(p) = \int_0^{z_p} \frac{u^2(z)}{\left(u^2(z) - p^2\right)^{1/2}}\, dz.$$ (4.16)

This gives the travel time from the surface to the turning point, z_p. The total surface-to-surface travel time, $T(p)$, is given by

$$T(p) = 2\int_0^{z_p} \frac{u^2(z)}{\left(u^2(z) - p^2\right)^{1/2}}\, dz.$$ (4.17)

Equations (4.13) and (4.17) are suitable for a model in which $u(z)$ is a continuous function of depth.

The simplest velocity models are specified as a stack of homogeneous layers. In this case the integrals for X and T become summations:

$$X(p) = 2p \sum_i \frac{\Delta z_i}{\left(u_i^2 - p^2\right)^{1/2}}, \quad u_i > p,$$ (4.18)

and

$$T(p) = 2 \sum_i \frac{u_i^2 \Delta z_i}{\left(u_i^2 - p^2\right)^{1/2}}, \quad u_i > p.$$ (4.19)

Note that we sum over the layers from the top downward until the layer slowness is less than the ray parameter; we don't want $(u^2(z) - p^2)^{1/2}$ to become imaginary.

4.2.1 Example: Computing X(p) and T(p)

Consider a homogeneous three-layer model with 3 km layer thicknesses and velocities 4, 6 and 8 km/s for the top, middle and bottom layers, respectively. What is the surface-to-surface distance and travel time for a ray with $p = 0.15$ s/km? We first convert the velocities to slownesses and obtain $u_1 = 0.25$, $u_2 = 0.167$, and $u_3 = 0.125$ s/km. We also have $\Delta z_1 = \Delta z_2 = \Delta z_3 = 3$ km.

In equations (4.18) and (4.19), note that u is only greater than p for layers 1 and 2. This means that the ray will pass through these layers but will be reflected off the top of layer 3. We thus have:

$$
\begin{aligned}
X(p) &= 2p\frac{\Delta z_1}{\left(u_1^2 - p^2\right)^{1/2}} + 2p\frac{\Delta z_2}{\left(u_2^2 - p^2\right)^{1/2}} \\
&= \frac{2 \cdot 3 \cdot 0.15}{\left(0.25^2 - 0.15^2\right)^{1/2}} + \frac{2 \cdot 3 \cdot 0.15}{\left(0.1677^2 - 0.15^2\right)^{1/2}} \\
&= 16.9\,\mathrm{km}
\end{aligned}
$$

and

$$
\begin{aligned}
T(p) &= 2\frac{u_1^2 \Delta z_1}{\left(u_1^2 - p^2\right)^{1/2}} + 2\frac{u_2^2 \Delta z_2}{\left(u_2^2 - p^2\right)^{1/2}} \\
&= \frac{2 \cdot 3 \cdot 0.25^2}{\left(0.25^2 - 0.15^2\right)^{1/2}} + \frac{2 \cdot 3 \cdot 0.167^2}{\left(0.1677^2 - 0.15^2\right)^{1/2}} \\
&= 4.17\,\mathrm{s}
\end{aligned}
$$

The ray travels for 4.17 s and hits the surface 16.9 km from the source.

4.2.2 Ray tracing through velocity gradients

When velocity gradients are present, (4.18) and (4.19) are not very convenient since a "staircase" model with a large number of homogeneous layers must be evaluated to give accurate results. A better strategy is to parameterize the velocity model at a number of discrete points in depth and evaluate the integrals (4.12) and (4.16) by assuming an appropriate interpolation function between the model points. For a linear velocity gradient between model points of the form $v(z) = a + bz$, the slope of the gradient, b, between $v_1(z_1)$ and $v_2(z_2)$ is given by

$$
b = \frac{v_2 - v_1}{z_2 - z_1}. \tag{4.20}
$$

Evaluating the integrals for $t(p)$ and $x(p)$, one can then obtain (e.g., Chapman *et al.*, 1988)

$$
x(p) = \left. \frac{\eta}{bup} \right|_{u_2}^{u_1} \tag{4.21}
$$

and

$$t(p) = \frac{1}{b}\left[\ln\left(\frac{u+\eta}{p}\right) - \frac{\eta}{u}\right]\Bigg|_{u_2}^{|u_1|} + px(p), \qquad (4.22)$$

where the vertical slowness $\eta = (u^2 - p^2)^{1/2}$. If the ray turns within the layer, then there is no contribution to these integrals from the lower point. A computer subroutine that uses these expressions to compute $x(p)$ and $t(p)$ for a layer with a linear velocity gradient is provided in Appendix D; this is needed in some of the Exercises.

4.3 Travel time curves and delay times

Generally in the Earth, $X(p)$ will increase as p decreases; that is, as the takeoff angle decreases, the range increases:

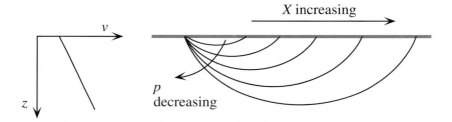

In this case the derivative dX/dp is negative. When $dX/dp < 0$, we say that this *branch* of the travel time curve is *prograde*. Occasionally, because of a rapid velocity transition in the Earth, $dX/dp > 0$, and the rays turn back on themselves:

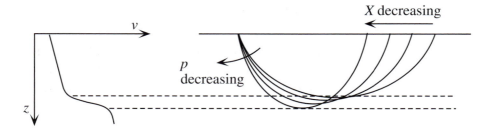

When $dX/dp > 0$ the travel time curve is termed *retrograde*. The transition from prograde to retrograde and back to prograde generates a *triplication* in the travel time curve (Fig. 4.5). The endpoints on the triplication are termed *caustics*. These

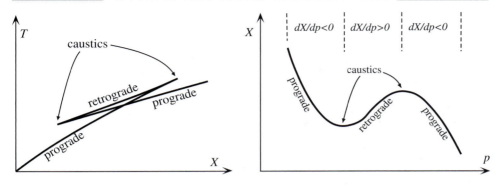

Figure 4.5 A triplication in the travel time curve and the corresponding $X(p)$ curve resulting from a steep velocity increase.

are the points where $dX/dp = 0$. Energy is focused to these points since rays at different takeoff angles arrive at the same range (as we shall see later, geometrical ray theory predicts infinite amplitude at these points!). The triplication may be "unraveled" by considering the $X(p)$ function.

At large values of p the rays turn at shallow depths and travel only short distances. As the ray parameter decreases, the turning point depth increases and the range, X, increases. When the turning points enter the steep velocity gradient, X begins decreasing with decreasing p. Once the rays break through the steep velocity gradient and turn in the more shallow gradient below, X once again increases with decreasing p. The caustics are the stationary points on the $X(p)$ curve. However, it should be noted that sharp changes in the velocity gradient (i.e., discontinuities in dv/dz) will produce sharp bends in the $X(p)$ curve so that it will not always have the smooth appearance shown in Figure 4.5. In particular, local maxima or minima in $X(p)$ may be present where dX/dp is discontinuous and a caustic does not occur.

4.3.1 Reduced velocity

Travel time curves can often be seen in more detail if they are plotted using a *reduction velocity* that is subtracted from the travel times (Figure 4.6). In this case the time scale is shifted an amount equal to the range divided by the reducing velocity. Velocities that are equal to the reduction velocity will plot as horizontal lines, and a greatly expanded time scale becomes possible.

4.3.2 The $\tau(p)$ function

The function $X(p)$ is more nicely behaved than $T(X)$ since it does not cross itself (there is a single value of X for each value of p), but the inverse function $p(X)$ is

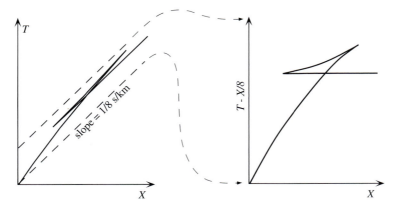

Figure 4.6 A reduction velocity can be used to expand the time scale to show more detail in travel time curves.

multivalued. An even nicer function is the combination

$$\tau(p) = T(p) - pX(p), \tag{4.23}$$

where τ is called the *delay time*. It can be calculated very simply from (4.13) and (4.17):

$$
\begin{aligned}
\tau(p) &= 2\int_0^{z_p} \left[\frac{u^2}{(u^2 - p^2)^{1/2}} - \frac{p^2}{(u^2 - p^2)^{1/2}} \right] dz \\
&= 2\int_0^{z_p} \left(u^2(z) - p^2 \right)^{1/2} dz \\
&= 2\int_0^{z_p} \eta(z)\, dz.
\end{aligned} \tag{4.24}
$$

For our simple layered medium, we have

$$\tau(p) = 2\sum_i \left(u_i^2 - p^2 \right)^{1/2} \Delta z_i = 2\sum_i \eta_i \Delta z_i, \quad u_i > p. \tag{4.25}$$

Consider a point on a travel time curve $t(x)$ at distance X and time T (Fig. 4.7). The equation of the straight line tangent to the travel time curve is $t = T + p(x - X)$. At $x = 0$, $t = T - pX = \tau(p)$, so the intercept of the line is $\tau(p)$ while the slope is p. The slope of the τ versus p curve is

$$\frac{d\tau}{dp} = \frac{d}{dp} 2\int_0^{z_p} \left(u^2 - p^2 \right)^{1/2} dz = -2p \int_0^{z_p} \frac{dz}{\left(u^2 - p^2 \right)^{1/2}} \tag{4.26}$$

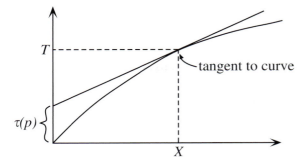

Figure 4.7 The delay time, $\tau(p) = T - pX$, is given by the intercept of the tangent to the travel time curve.

and thus

$$\frac{d\tau}{dp} = -X(p). \tag{4.27}$$

The slope of the $\tau(p)$ curve is $-X$. Because $X \geq 0$, the $\tau(p)$ curve is always decreasing, or *monotonically decreasing*. The $\tau(p)$ curve remains monotonically decreasing $(d\tau/dp < 0)$ even in the presence of a triplication in the $T(X)$ curve. The second derivative of τ is simply

$$\frac{d^2\tau}{dp^2} = \frac{d}{dp}(-X) = -\frac{dX}{dp}. \tag{4.28}$$

The $\tau(p)$ curve is concave upward for a prograde branch and concave downward for a retrograde branch (Fig. 4.8). The functions $X(p)$, $T(p)$, and $\tau(p)$ are "proper" in that they are single valued. $T(X)$, on the other hand, is not a proper function since

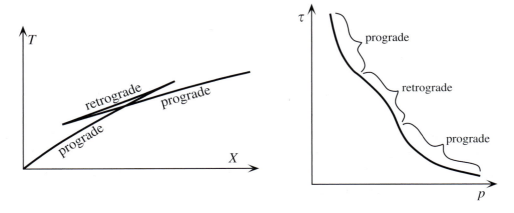

Figure 4.8 The $\tau(p)$ function "unravels" triplications in travel time curves. Prograde branches have concave upward $\tau(p)$ curves; retrograde branches have concave downward $\tau(p)$ curves.

it will generally be multivalued. This difference means that $\tau(p)$ data, if available, are generally easier to interpret than $T(X)$ data. We shall see in the next chapter that working in the $\tau(p)$ domain has advantages in computing velocity versus depth functions from travel time data.

4.4 Low-velocity zones

In all of the examples plotted above, we have assumed that velocity always increases with depth. However, occasionally we will encounter the case where velocity decreases with depth, creating a *low-velocity zone* (LVZ). The outstanding example in Earth is the core, where the P velocity drops from about 14 km/s in the lowermost mantle to 8 km/s in the outermost core. There is also considerable evidence for a LVZ in the upper mantle, at least in shear velocity, within the asthenosphere (about 80 to 200 km depth). Within the negative velocity gradient at the top of the LVZ, the rays are bent downward (Fig. 4.9). Note that no rays originating at the surface

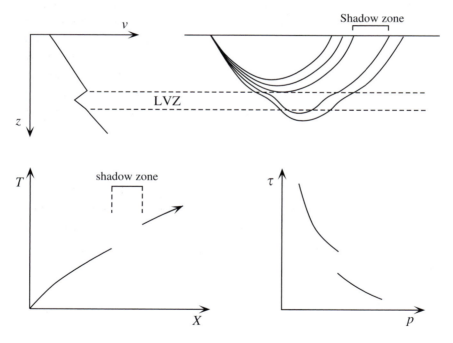

Figure 4.9 A low velocity zone (LVZ) results from a velocity decrease with depth. Rays curve downward as the velocity decreases, creating a shadow zone on the surface and gaps in the $T(X)$ and $\tau(p)$ curves.

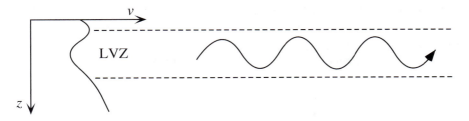

Figure 4.10 A low-velocity zone (LVZ) can trap waves, creating a wave guide.

can turn within the LVZ itself. Rays with horizontal slownesses corresponding to values within the LVZ turn above the zone. Only when the ray parameter becomes small enough do rays enter the LVZ. These rays then pass through the low velocity zone and turn in the region below the LVZ in which velocities are once again higher than any velocities in the overlying material.

The existence of a low velocity zone creates a gap, termed a *shadow zone*, in both the $T(X)$ and $\tau(p)$ curves. The absence of rays turning within the LVZ often makes the velocity structure within low velocity zones difficult to determine. A related phenomenon arises when rays originate within a low velocity zone. In this case, some rays are trapped within the LVZ and are forever curving back in toward the velocity minimum (Fig. 4.10). The LVZ acts as a *wave guide* and, in cases of low attenuation, seismic energy can propagate very long distances. The most notable example of a wave guide in Earth is the sound channel in the ocean. Acoustic waves trapped within the sound channel can be detected over distances of thousands of kilometers.

4.5 Summary of 1-D ray tracing equations

Let us now review the important equations for ray tracing in laterally homogeneous Earth models. First, the ray parameter or horizontal slowness p is defined by several expressions:

$$p = u(z) \sin \theta = \frac{dT}{dX} = u_{\text{tp}} = \text{constant for given ray}, \qquad (4.29)$$

where $u = 1/v$ is the slowness, θ is the ray incidence angle (from vertical), T is the travel time, X is the horizontal range, and u_{tp} is the slowness at the ray turning point. We also defined the vertical slowness

$$\eta = \left(u^2 - p^2\right)^{1/2} \tag{4.30}$$

and integral expressions for the surface-to-surface travel time $T(p)$, range $X(p)$, and delay time $\tau(p)$:

$$T(p) = 2 \int_0^{z_p} \frac{u^2(z)}{\left(u^2(z) - p^2\right)^{1/2}} \, dz = 2 \int_0^{z_p} \frac{u^2(z)}{\eta} \, dz, \tag{4.31}$$

$$X(p) = 2p \int_0^{z_p} \frac{dz}{\left(u^2(z) - p^2\right)^{1/2}} = 2p \int_0^{z_p} \frac{dz}{\eta}, \tag{4.32}$$

and

$$\tau(p) = 2 \int_0^{z_p} \left(u^2(z) - p^2\right)^{1/2} \, dz = 2 \int_0^{z_p} \eta(z) \, dz, \tag{4.33}$$

where z_p is the turning point depth ($p = u$ at the turning point). We also have

$$\tau(p) = T(p) - pX(p) \tag{4.34}$$

and

$$X(p) = -\frac{d\tau}{dp}. \tag{4.35}$$

The $T(X)$, $X(p)$, $\tau(p)$ functions can be quite sensitive to even small changes in the velocity-depth model. Figure 4.11 illustrates the relationships between these functions for some simple models. These examples demonstrate that even slight changes in seismic velocity gradients can have large effects on the resulting ray paths. It is also apparent that many if not most triplications will not have the simple form illustrated in Figure 4.5 where the $X(p)$ curve has smooth maxima and minima. Discontinuities in the vertical velocity gradient produce sharp bends in the $X(p)$ function, and triplications can exist without formal caustics at their endpoints where $dX/dp = 0$. Staircase models are interesting because all of their ray paths involve reflections off the top of layers and thus are on retrograde branches, connected by straight lines on the $T(X)$ curve that define the first arrivals. These first arrivals represent head waves traveling along the top of each layer. They have zero energy in geometrical ray theory but more complete synthetic seismogram calculations show that they do produce observable arrivals, although generally of low amplitude.

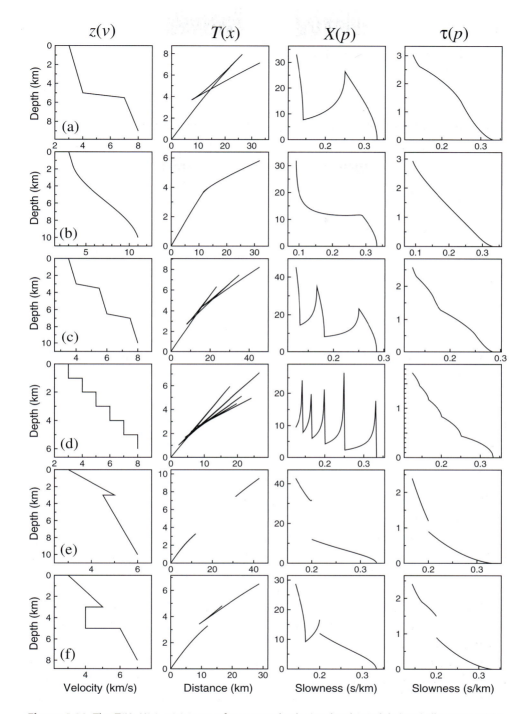

Figure 4.11 The $T(X), X(p), \tau(p)$ curves for assorted velocity-depth models ($z(v)$), illustrating (a) a simple triplication, (b) a velocity model that focuses many of the rays to a single $T(X)$ point, (c) a double triplication, (d) a staircase model, (e) a low-velocity zone, and (f) a low-velocity zone with a sharp velocity jump. Note: T and τ are in s, x and z are in km.

4.6 Spherical-Earth ray tracing

The ray tracing equations described above are for a horizontally layered Earth. They are adequate for modeling crustal arrivals in the upper 30 km or so. However, for deeper rays, it is necessary to take into account Earth's sphericity. There are two ways in which this can be done: (1) changing the definition of the ray parameter to account for the spherical geometry and (2) applying a transformation (the *Earth-flattening transformation*) to the spherical model to permit direct use of the flat-Earth ray tracing equations.

In the first method we modify the definition of the ray parameter to account for the fact that the ray angle from the radius (the local vertical) changes along the ray path, even within homogeneous shells. Consider two such spherical shells in a spherically symmetric Earth (Fig. 4.12). At the interface between shell 1 and shell 2, we have from Snell's law

$$u_1 \sin \theta_1 (r_1) = u_2 \sin \theta_2 (r_1). \qquad (4.36)$$

As the ray travels through shell 2, note that the incidence angle changes ($\theta_2(r_1) \neq \theta_2(r_2)$). If we project the ray to its "turning point" as if layer 2 continued down, we see from the geometry of the triangles that we can express the incidence angle within layer 2 as a function of radius:

$$\sin \theta_2 (r) = r_{min}/r. \qquad (4.37)$$

Thus $\theta_2(r_1)$ in (4.36) is related to $\theta_2(r_2)$ by the expression

$$r_1 \sin \theta_2 (r_1) = r_2 \sin \theta_2 (r_2)$$

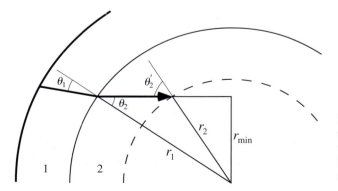

Figure 4.12 The ray geometry for spherical shells of constant velocity. Note that $\theta_2 \neq \theta_2'$ because of the changing angle of the radius vector.

or   (4.38)

$$\sin \theta_2(r_1) = (r_2/r_1) \sin \theta_2(r_2).$$

Substituting into (4.36), we obtain the generalization of Snell's law for spherically symmetric media:

$$r_1 u_1 \sin \theta_1 = r_2 u_2 \sin \theta_2.$$ (4.39)

In this case, the ray parameter p becomes

$$p_{\text{sph}} = ru \sin \theta.$$ (4.40)

Recall, in the case of the flat-Earth ray parameter, that p is a measure of the horizontal slowness

$$p_{\text{f}} = u \sin \theta = \frac{dT}{dX}.$$ (4.41)

In the spherical Earth, $dX = d\Delta\, r$, where Δ is the angle in radians. Thus

$$p_{\text{sph}} = \frac{r dT}{dX} = \frac{dT}{d\Delta}.$$ (4.42)

Note that the spherical-Earth ray parameter p_{sph} has units of time (s/radian), whereas the flat-Earth ray parameter has units of slowness (time/distance). Expressions for travel time and range as a function of p_{sph} are very similar to those we derived earlier:

$$T(p_{\text{sph}}) = 2 \int_{r_{\text{tp}}}^{r_{\text{e}}} \frac{(ur)^2}{\left[(ur)^2 - p_{\text{sph}}^2\right]^{1/2}} \frac{dr}{r}$$ (4.43)

and

$$\Delta(p_{\text{sph}}) = 2 p_{\text{sph}} \int_{r_{\text{tp}}}^{r_{\text{e}}} \frac{1}{\left[(ur)^2 - p_{\text{sph}}^2\right]^{1/2}} \frac{dr}{r},$$ (4.44)

where r_{e} is the Earth radius. The distinction that we make here between p_{f} and p_{sph} is generally not made in the literature – you have to look at the equations to tell the difference. If there is an r present, then the authors are using p_{sph}.

4.7 The Earth-flattening transformation

Our derivations for $T(p)$, $X(p)$, and $\tau(p)$ all assumed that the rays were traveling in a "flat" Earth. For a flat, homogeneous half-space, the travel time curve is a straight line and none of the rays leaving the source at incidence angles other than 90° return to the surface since they cannot turn at depth. However, in a homogeneous spherical Earth, all rays will eventually return to the surface and the travel time curve is not straight. The curvature of the travel time curve in a spherical Earth can be simulated in a flat-Earth model if a special velocity gradient is introduced in the half-space (e.g., Müller, 1971). A new depth variable, z_f, is defined:

$$e^{-z_f/a} = r/a \quad \text{or} \quad z_f = -a \ln(r/a), \tag{4.45}$$

where r is the distance from the center of the Earth and a is the radius of the Earth, 6371 km. Note that $r = a - z_s$, where z_s is the depth in a spherical Earth. The radius $r = a$ (the free surface) corresponds to a flat-Earth depth of $z = -a \ln(a/a) = 0$, while a radius of $r = 0$ corresponds to an infinite depth. The velocities transform as

$$v_f(z_f) = (a/r)v_s(r) \tag{4.46}$$

with separate transformations for P-wave and S-wave velocities. For a given spherical-Earth model, we can use the above transformations to obtain a corresponding flat-Earth model that will predict identical travel time behavior. Thus, all of our ray tracing equations can be used without modification. The ranges, X_{km}, calculated for a flat Earth can be converted into degrees, Δ_{deg}, on a spherical Earth by using $\Delta_{deg} = X_{km}[360/(2\pi a)]$ since $2\pi a$ is the circumference of the Earth in kilometers.

At a depth of 30 km below the Earth's surface, $a/r = 6371/(6371 - 30) = 1.005$ and the Earth-flattening transformation is probably unnecessary. However, at a depth of 150 km in the Earth, a spherical Earth velocity, v_s, of 8.6 km s^{-1} becomes $v_f = 8.6[6371/(6371 - 150)] = 8.81$ km s^{-1}. In this case, the Earth-flattening transformation starts to assume significance, and of course it becomes increasingly important for deeper rays. Figure 4.13 shows the Earth-flattening transformation applied to the PREM P-velocity model. Notice that both the depth and velocity increase for features at depth. For example, the core–mantle boundary (CMB) depth increases from 2891 km to 3853 km, while the velocity at the base of the mantle increases from 13.7 km/s to 25.1 km/s. The change for the inner-core boundary (ICB) is even more dramatic, with its depth increasing from 5149 km to 10 523 km and the velocity at the base of the outer core increasing from 10.4 km/s to 54 km/s.

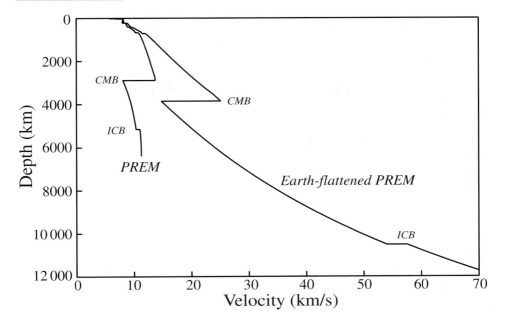

Figure 4.13 The Earth-flattening transformation applied to the PREM P-velocity model.

There is a singularity in the Earth-flattening transformation at the Earth's center where $r = 0$.

4.8 Three-dimensional ray tracing[†]

In the preceding sections, we have solved for ray paths by assuming that seismic velocity varies only with depth or radius. However, ray tracing often must be performed for models with general 3-D velocity variations, which makes the equations considerably more complicated. Detailed descriptions of ray tracing theory in seismology may be found in Cerveny (2001) and Chapman (2004). Here we adapt results presented by Müller (2007).

Solutions in this case usually begin with the eikonal equation (see Appendix C)

$$|\nabla T|^2 = \left(\frac{\partial T}{\partial x}\right)^2 + \left(\frac{\partial T}{\partial y}\right)^2 + \left(\frac{\partial T}{\partial z}\right)^2 = u^2, \qquad (4.47)$$

which gives the relationship between the phase factor T and the local slowness u. The wavefronts are defined by the surfaces $T(\mathbf{x}) = constant$ and lines perpendicular to $T(\mathbf{x})$ (parallel to $\nabla T(\mathbf{x})$) define the rays. The ray direction is given by the gradient of T

$$\nabla T = u\hat{\mathbf{s}} = \mathbf{s} \tag{4.48}$$

where $\hat{\mathbf{s}}$ is a unit vector in the local ray direction and $\mathbf{s}$ is the slowness vector. If we consider a position vector $\mathbf{x} = (x, y, z)$ along the ray path, then

$$\mathbf{s} = u\frac{d\mathbf{x}}{ds} = \nabla T \tag{4.49}$$

where ds is the incremental length along the ray path. For ray tracing purposes, we are interested in how things change along the ray path. Thus let us evaluate the expression

$$\frac{d}{ds}(\nabla T) = \frac{d}{ds}\left(\frac{\partial T}{\partial x}, \frac{\partial T}{\partial y}, \frac{\partial T}{\partial z}\right). \tag{4.50}$$

Considering for now only the $\partial/\partial x$ term, we have (e.g., Müller, 2007)

$$\frac{d}{ds}\left(\frac{\partial T}{\partial x}\right) = \frac{\partial^2 T}{\partial x^2}\frac{dx}{ds} + \frac{\partial^2 T}{\partial x\partial y}\frac{dy}{ds} + \frac{\partial^2 T}{\partial x\partial z}\frac{dz}{ds} \tag{4.51}$$

$$= \frac{1}{u}\left(\frac{\partial^2 T}{\partial x^2}\frac{\partial T}{\partial x} + \frac{\partial^2 T}{\partial x\partial y}\frac{\partial T}{\partial y} + \frac{\partial^2 T}{\partial x\partial z}\frac{\partial T}{\partial z}\right) \tag{4.52}$$

$$= \frac{1}{2u}\frac{\partial}{\partial x}\left[\left(\frac{\partial T}{\partial x}\right)^2 + \left(\frac{\partial T}{\partial y}\right)^2 + \left(\frac{\partial T}{\partial z}\right)^2\right] \tag{4.53}$$

$$= \frac{1}{2u}\frac{\partial}{\partial x}\left(u^2\right) \tag{4.54}$$

$$= \frac{1}{2u}2u\left(\frac{\partial u}{\partial x}\right) = \frac{\partial u}{\partial x} \tag{4.55}$$

where we used (4.49) and (4.47). In a similar fashion we may obtain:

$$\frac{d}{ds}\left(\frac{\partial T}{\partial y}\right) = \frac{\partial u}{\partial y} \tag{4.56}$$

$$\frac{d}{ds}\left(\frac{\partial T}{\partial z}\right) = \frac{\partial u}{\partial z} \tag{4.57}$$

and thus

$$\frac{d}{ds}(\nabla T) = \left(\frac{\partial u}{\partial x}, \frac{\partial u}{\partial y}, \frac{\partial u}{\partial z}\right) = \nabla u. \tag{4.58}$$

From (4.49), we then have

$$\frac{d}{ds}\left(u\frac{d\mathbf{x}}{ds}\right) = \nabla u. \tag{4.59}$$

This second-order equation and (4.49) can be replaced with two coupled first-order equations:

$$\frac{d\mathbf{s}}{ds} = \nabla u \tag{4.60}$$

and

$$\frac{d\mathbf{x}}{ds} = \frac{\mathbf{s}}{u}, \tag{4.61}$$

which can be solved numerically for the position, $\mathbf{x}$, and the slowness vector, $\mathbf{s}$, along the ray path, assuming initial values for $\mathbf{x}$ and $\mathbf{s}$ and specified 3-D variations in the local medium slowness u. Once the ray path is known, it is straightforward to also solve for the travel time along the ray using (4.48).

For models where slowness varies only with depth, these equations reduce to those of the previous sections. In this case, $\nabla u = (0,\ 0,\ du/dz)$ and, from (4.60), s_x and s_y are constant and s_z is the vertical slowness η. With no loss of generality, we may rotate the coordinates such that $s_y = 0$ and $s_x = p$, the ray parameter (horizontal slowness). Equation (4.61) is thus

$$\frac{dx}{ds} = \frac{p}{u}, \tag{4.62}$$

which is seen to match equation (4.9). Equation (4.60) becomes

$$\frac{d\eta}{ds} = \frac{du}{dz} \quad \text{or} \quad \frac{d\eta}{du} = \frac{ds}{dz}, \tag{4.63}$$

which is consistent with $\eta = u\cos\theta = (u^2 - p^2)^{1/2}$ of equation (4.7).

Equations (4.60) and (4.61) are easiest to solve numerically if the slowness variations are smooth so that the spatial derivatives are all defined. However, slowness discontinuities can be handled through a generalization of Snell's law to material interfaces of any orientation. Numerical ray tracing can solve for ray theoretical amplitudes as well as travel times, but we defer discussion of amplitude variations until Chapter 6. Finally, we note that these equations sometimes have difficulty in solving the problem of finding the travel time between two specified points (the two-point ray tracing problem, discussed in more detail in Chapter 5), because ray paths can bend wildly in models with strong velocity variations and it can be hard to find an exact ray path that connects the points. In this case, other numerical schemes, such as finite-difference ray tracing (e.g., Vidale, 1988) and graph theory (e.g., Moser, 1991) have often proven more stable. These methods also have the convenience of requiring model slowness values only at fixed grid points (usually evenly spaced) within the model.

4.9 Ray nomenclature

The different layers in the Earth (e.g., crust, mantle, outer core, and inner core), combined with the two different body-wave types (P, S), result in a large number of possible ray geometries, termed *seismic phases*. The following naming scheme has achieved general acceptance in seismology:

4.9.1 Crustal phases

Earth's crust is typically about 6 km thick under the oceans and 30 to 50 km thick beneath the continents. Seismic velocities increase sharply at the Moho discontinuity between the crust and upper mantle. A P wave turning within the crust is called Pg, whereas a ray turning in or reflecting off the Moho is called PmP (Fig. 4.14). The m in PmP denotes a reflection off the Moho and presumes that the Moho is a first-order discontinuity. However, the Moho might also be simply a strong velocity gradient, which causes a triplication that mimics the more simple case of a reflection. Finally, Pn is a ray traveling in the uppermost mantle below the Moho. The *crossover point* is where the first arrivals change abruptly from Pg to Pn. The crossover point is a strong function of crustal thickness and occurs at about $X = 30$ km for oceanic crust and at about $X = 150$ km for continental crust. There are, of course, similar names for the S-wave phases (SmS, Sn, etc.) and converted phases such as SmP.

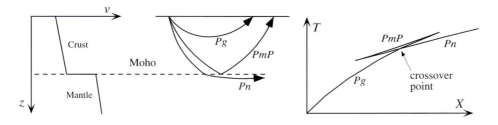

Figure 4.14 Ray geometries and names for crustal P phases. The sharp velocity increase at the Moho causes a triplication in the travel time curve.

4.9.2 Whole Earth phases

Here the main layers are the mantle, the fluid outer core, and the solid inner core. *P*- and *S*-wave legs in the mantle and core are labeled as follows:

P – *P* wave in the mantle
K – *P* wave in the outer core
I – *P* wave in the inner core
S – *S* wave in the mantle
J – *S* wave in the inner core
c – reflection off the core–mantle boundary (CMB)
i – reflection off the inner-core boundary (ICB)

For *P* and *S* waves in the whole earth, the above abbreviations apply and stand for successive segments of the ray path from source to receiver. Some examples of these ray paths and their names are shown in Figure 4.15. Notice that surface multiple

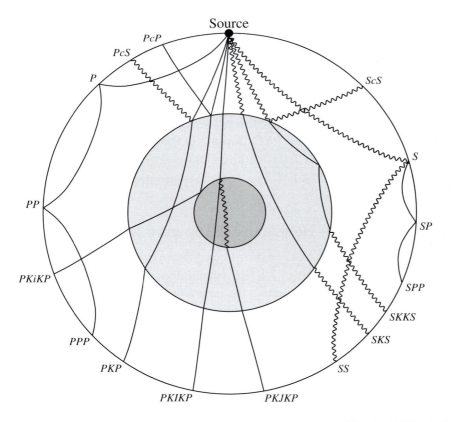

Figure 4.15 Global seismic ray paths and phase names, computed for the PREM velocity model. *P* waves are shown as solid lines, *S* waves as wiggly lines. The different shades indicate the inner core, the outer core, and the mantle.

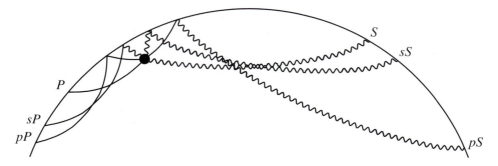

Figure 4.16 Deep earthquakes generate surface-reflected arrivals, termed depth phases, with the upgoing leg from the source labeled with a lower-case *p* or *s*. Ray paths plotted here are for an earthquake at 650 km depth, using the PREM velocity model.

phases are denoted by *PP*, *PPP*, *SS*, *SP*, and so on. For deep focus earthquakes, the upgoing branch in surface reflections is denoted by a lowercase *p* or *s*; this defines *pP*, *sS*, *sP*, etc. (see Fig. 4.16). These are termed *depth phases*, and the time separation between a direct arrival and a depth phase is one of the best ways to constrain the depth of distant earthquakes. *P*-to-*S* conversions can also occur at the CMB; this provides for phases such as *PcS* and *SKS*. Ray paths for the core phase *PKP* are complicated by the Earth's spherical geometry, leading to several triplications in the travel time curve for this phase. Often the inner-core *P* phase *PKIKP* is labeled as the *df* branch of *PKP*. Because of the sharp drop in *P* velocity at the CMB, *PKP* does not turn in the outer third of the outer core. However, *S*-to-*P* converted phases, such as *SKS* and *SKKS*, can be used to sample this region.

4.9.3 *PKJKP*: The Holy Grail of body wave seismology

The phase *PKJKP* is a *P* wave that converts to an *S* wave during passage through the solid inner core. Observations of *PKJKP* potentially could provide improved estimates of the shear-velocity structure of the inner core, which is otherwise only constrained by normal mode observations. However, the predicted amplitude of *PKJKP* is well below typical noise levels (Doornbos, 1974), owing to the small *P*-to-*S* and *S*-to-*P* transmission coefficients at the ICB and strong inner-core attenuation. To meet the challenge of its expected weak amplitude, seismologists have deployed stacking and other methods to enhance the visibility of *PKJKP*, and recently a number of observations have been published (Okal and Cansi, 1998; Deuss *et al.*, 2000; Cao *et al.*, 2005; Wookey and Helffrich, 2008). These results generally suggest an average inner-core shear velocity close to that derived from normal mode data. However, *PKJKP* cannot yet be observed routinely and it remains puzzling why the amplitudes of the existing *PKJKP* observations are much larger than those predicted by standard velocity and attenuation models of the inner core. One possibility is that inner-core anisotropy (see Chapter 11) produces larger *PKJKP* amplitudes along certain ray paths (Wookey and Helffrich, 2008).

4.10 Global body-wave observations

The visibility of different body-wave phases depends upon their amplitude, polarization, and frequency content. Modern seismographs record all three components of ground motion (using a vertical and two orthogonal horizontal sensors) over a wide frequency range. The horizontal records are normally rotated into the *radial* component parallel to the azimuth to the source and the *transverse* component perpendicular to this azimuth. Figure 4.17 plots the vertical, radial, and transverse

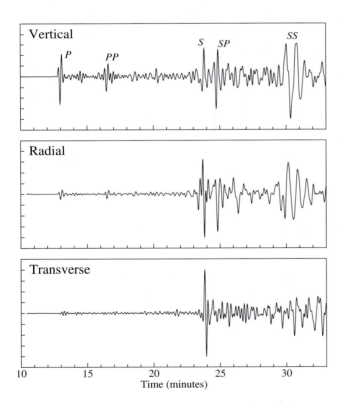

Figure 4.17 The vertical, radial, and transverse components of ground motion (velocity) from the January 17, 1994 Northridge earthquake recorded at the IRIS/IDA station OBN at 88.5° range. The original broadband records have been filtered to between 15 and 100 s period. Time is in minutes relative to the earthquake origin time; amplitudes are self-scaled.

component records for the 1994 Northridge earthquake in southern California, recorded at station OBN in Russia, 88.5 degrees away, and identifies some of the major body-wave phases. Note that the *P* waves are most visible on the vertical component, with little *P* energy arriving on the transverse component.

The time of the first discernible motion of a seismic phase is called the *arrival time*, and the process of making this measurement is termed *picking* the arrival. In the past, particularly before digital data became available, arrival time picking was a major part of the operation of seismographic stations. Even today, many seismic records are still picked by hand, since devising automatic picking schemes that are reliable in the presence of noise or multiple events has proven difficult. By measuring the arrival times of seismic phases at a variety of source–receiver ranges, seismologists are able to construct travel time curves for the major phases and use these to infer Earth's average radial velocity structure. This was largely accomplished in the early part of the twentieth century, and the JB travel time tables, completed by Jeffreys and Bullen in 1940, are still widely used, differing by no more than a few seconds from the best current models.

Figure 4.18 plots over five million travel time picks archived by the International Seismological Centre (ISC) from 1964 to 1987. The major body-wave phases are easily seen and may be identified from the travel time curves plotted in Figures 4.20 and 4.24. The ISC data have proven to be an invaluable resource in seismology, and they are used extensively both to locate earthquakes and to perform three-dimensional velocity inversions (see Chapter 5). They have the advantage of covering over three decades in time and providing data from many more stations (some still analog) than are currently available in digital form.

However, the ISC data only provide travel times, and many of the later arriving phases are sparsely picked. A more complete picture of the entire seismic wavefield may be obtained by stacking data from the modern digital seismic networks. In this procedure, records at common source–receiver ranges are averaged to produce a composite seismogram. Figures 4.19 and 4.21–4.23 plot stacks of almost 100 000 seismograms from the global networks for all earthquakes larger than magnitude 5.7 between 1988 and 1994 (figures from Astiz *et al.*, 1996). At higher frequencies, the arrivals appear sharper, but fewer phases can be distinguished. Figures 4.20 and 4.24 show theoretical travel time curves, calculated from the reference velocity model IASP91 (Kennett and Engdahl, 1991), for the phases visible in the stacked images.

The shape of the observed travel time curves is related to Earth's radial velocity structure, plotted in Figure 1.1 and also tabulated in Appendix A. The ray tracing equations derived earlier in this chapter can be used to compute theoretical travel time curves for any radial Earth model. In the next chapter we will discuss the inverse problem – how velocity models can be derived from travel time data.

ISC travel times

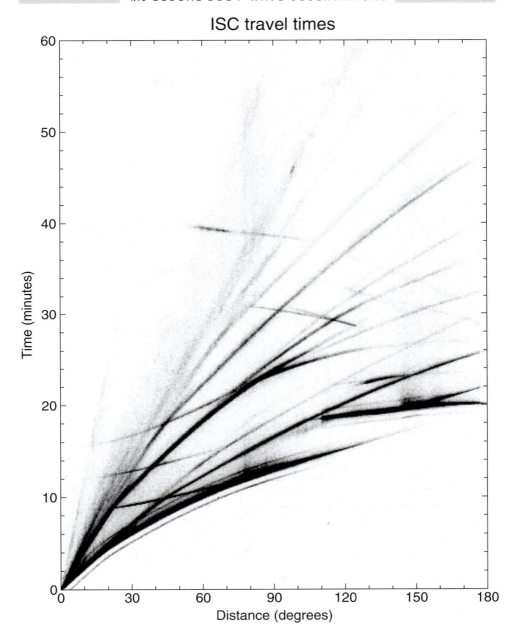

Figure 4.18 Travel time picks collected by the ISC between 1964 and 1987 for events shallower than 50 km. Over five million individual picks are plotted, the bulk of which are *P*, *PKP*, and *S* arrivals. However, several later arriving branches can also be seen, including *PP*, *PKS*, *PcP*, *PcS*, *ScS*, *PKKP*, and *PKPPKP*. See Figures 4.20 and 4.24 for a key to the phase names. The phases visible at ±1 minute from the *P* wave are due to errors in assigning times.

Short-period (vertical)

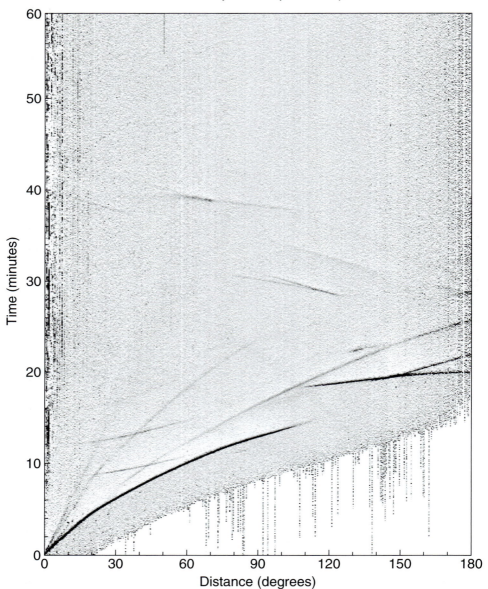

Figure 4.19 A stack of short-period (< 2 s), vertical component data from
the global networks between 1988 to 1994. See Figure 4.20 for a key to the phase names.
(From Astiz *et al.*, 1996.)

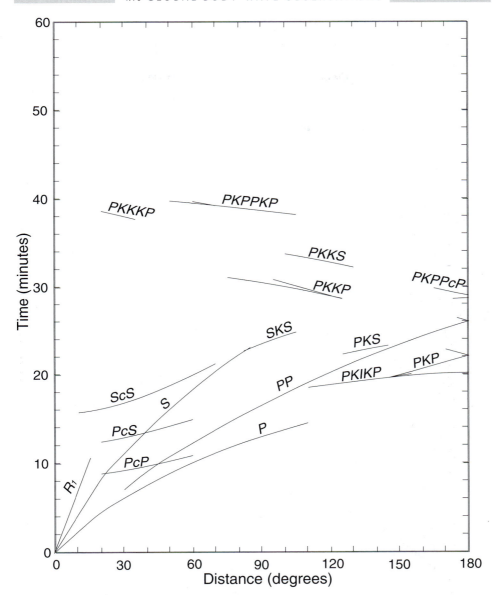

Figure 4.20 A key to the phases visible in the short-period stack plotted in Figure 4.19. Travel time curves are calculated using the IASP91 velocity model (Kennett and Engdahl, 1991). (From Astiz *et al.*, 1996.)

Long-period (vertical)

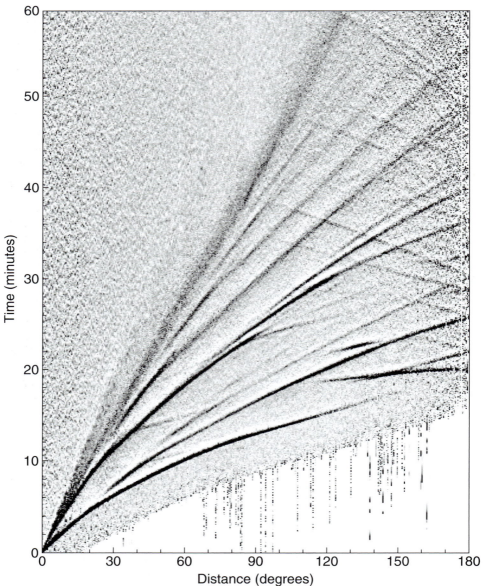

Figure 4.21 A stack of long-period (> 10 s), vertical component data from the global networks between 1988 to 1994. See Figure 4.24 for a key to the phase names.
(From Astiz *et al.*, 1996.)

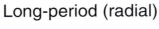

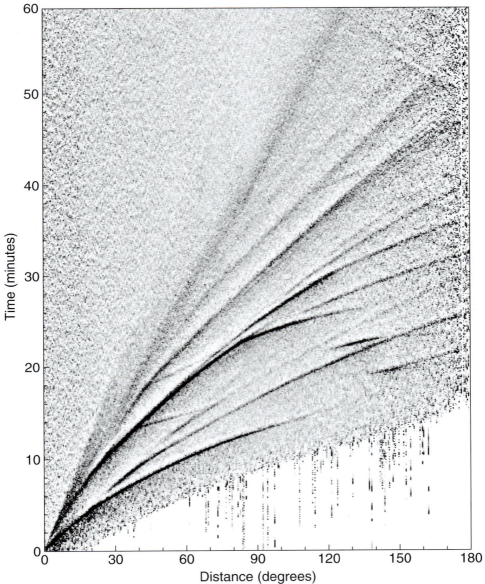

Figure 4.22 A stack of long-period (> 10 s), radial component data from the global networks between 1988 to 1994. See Figure 4.24 for a key to the phase names. (From Astiz *et al.*, 1996.)

Long-period (transverse)

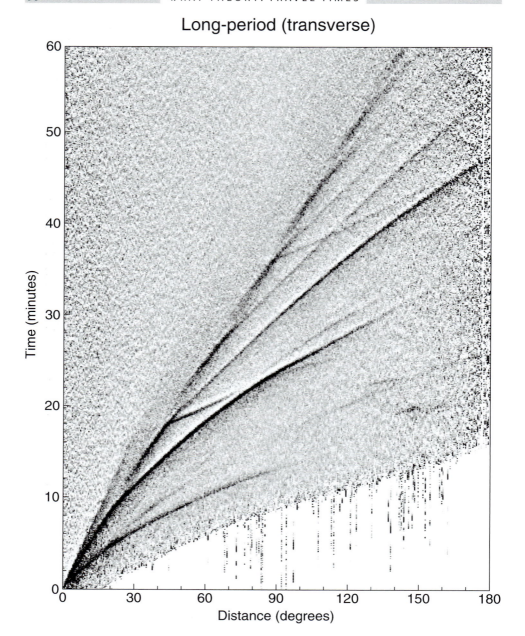

Figure 4.23 A stack of long-period (> 10 s), transverse component data from the global networks between 1988 to 1994. See Figure 4.24 for a key to the phase names. (From Astiz *et al.*, 1996.)

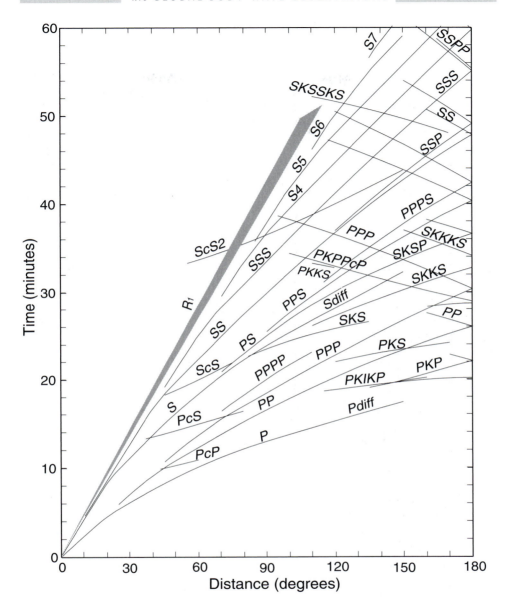

Figure 4.24 The phases visible in the long-period stacks shown in Figures 4.21–4.23. Travel time curves are calculated using the IASP91 velocity model (Kennett and Engdahl, 1991). (From Astiz *et al.*, 1996.)

4.11 Exercises

1. Show that the minimum time path between points A and B for the ray geometry in Figure 4.25 gives the same result as Snell's law. Hint: Express the total travel time as a function of the position of the ray bending point on the interface.

2. A downgoing P wave in a medium with a P velocity of 6 km/s travels through this "corner" shaped structure. If the incident ray is at an angle of 60° from the horizontal and the final ray is at an angle of 75° from the vertical, what is the P velocity within the corner-shaped medium?

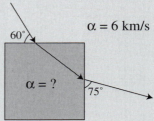

3. As shown in Figure 4.26, a vertically propagating upgoing plane wave is incident on a sediment-basement interface where the seismic velocity drops from 4 km/s to 2 km/s. A cylindrically shaped depression (with radius of curvature r = 1 km) focuses (approximately) the ray paths to a focus point within the sediments. Solve for the height of the focus point above the lowest point of the interface. Hint: Assume that the maximum slope of the depression is sufficiently small that $\sin \theta \approx \theta$ when considering the ray angles.

4. Show that rays in a linear velocity gradient of the form $v = a + bz$ will have ray paths that are circular. Derive an expression for the radius of the circle in terms of b and the ray parameter p. (It is also possible to derive a more general constraint, applying to ray geometries for linear velocity gradients in anisotropic media (see Shearer and Chapman, 1988); the circular result for isotropic media follows as a special case.)

5. Assume that velocity varies nicely with depth in such a way that the $T(X)$ curve for P (surface to surface) has the simple analytical form $T = 2(X + 1)^{1/2} - 2$.

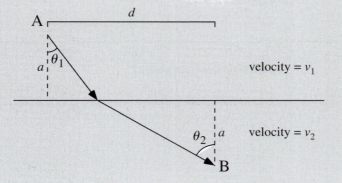

Figure 4.25 The bending of a ray path between two homogeneous layers.

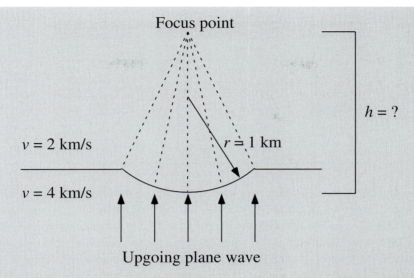

Figure 4.26 The ray geometry for an upcoming plane wave incident on a cylindrical basin.

Now consider the phase PP for this velocity structure, containing two P legs with a surface bounce point between source and receiver. Using the above $T(X)$ function, show that for the travel time to be stationary with respect to changes in the bouncepoint position, the bouncepoint must occur at the midpoint between source and receiver. Show that this is a *maximum* time point. (Note: Consider only bounce-points within the vertical plane connecting source and receiver. The midpoint is a *minimum* time point for perturbations perpendicular to the source–receiver plane, and thus the PP bouncepoint on a two-dimensional surface is actually a minimax or saddle point.)

6. Give a graphical argument using the $T(X)$ curve that shows that the PP midpoint must be a maximum time point along any prograde branch of the travel time curve.

7. Transform the variables in equation (4.32) for $X(p)$ to the spherical-Earth case by applying the flat-Earth transformation (4.45)–(4.46) and changing the equation from a dz integral to a dr integral. That is, replace z with $z = -a \ln(r/a)$, $u = u_f$ with the appropriate function of u_s, etc. Show that the result is the spherical-Earth equation for $\Delta(p)$ (equation 4.44). Hint: Remember that $x \neq \Delta$ and to transform the u on the right side of equation (4.40).

8. (COMPUTER) Consider MARMOD, a velocity-versus-depth model, which is typical of much of the oceanic crust (Table 4.1). Linear velocity gradients are assumed to exist at intermediate depths in the model; for example, the P velocity at 3.75 km is 6.9 km/s. Write a computer program to trace rays through this model

Table 4.1: MARMOD, a generic marine seismic model			
Depth (km)	α (km/s)	β (km/s)	ρ (g/cc)
0.0	4.50	2.40	2.0
1.5	6.80	3.75	2.8
6.0	7.00	3.85	2.9
6.5	8.00	4.60	3.1
10.0	8.10	4.70	3.1

and produce a P-wave $T(X)$ curve, using 100 values of the ray parameter p equally spaced between 0.1236 and 0.2217 s/km. You will find it helpful to use subroutine LAYERXT (provided in Fortran in Appendix D and in the supplemental web material as a Matlab script), which gives dx and dt as a function of p for layers with linear velocity gradients. Your program will involve an outer loop over ray parameter and an inner loop over depth in the model. For each ray, set x and t to zero and then, starting with the surface layer and proceeding downward, sum the contributions, dx and dt, from LAYERXT for each layer until the ray turns. This will give x and t for the ray from the surface to the turning point. Multiply by two to obtain the total surface-to-surface values of $X(p)$ and $T(p)$. Now produce plots of: (a) $T(X)$ plotted with a reduction velocity of 8 km/s, (b) $X(p)$, and (c) $\tau(p)$. On each plot, label the prograde and retrograde branches. Where might one anticipate that the largest amplitudes will occur?

9. (COMPUTER) Construct a P-wave travel time curve for Earth using the PREM model (see Appendix A). Your program should first read in the depth and P velocity for the different layers in the model. Next, apply the Earth-flattening transformation (4.45)–(4.46) to convert these depths and velocities to their flat-Earth equivalent values. Then, use the subroutine LAYERXT (provided in Fortran in Appendix D and in the supplemental web material as a Matlab script) to trace rays through this model and produce a P-wave $T(X)$ curve, using 201 values of the ray parameter p equally spaced between 0.0017 and 0.1128 s/km. Your program can be structured as described in Exercise 4.8; however, you should convert the X values returned by LAYERXT from kilometers to degrees along the Earth's surface. Now produce plots of:

    (a) $T(X)$ with $X = 0$ to $180°$, $T = 0$ to 25 minutes, and no reduction velocity. If you connect the individual $T(X)$ points with a line, be careful to avoid filling in the shadow zone between P and PKP. Compare your result with Figure 4.20.

    (b) $T(X)$ with $X = 10$ to $35°$, $T = 50$ to 100 s, and a reduction velocity of 0.1 degree/s. This should produce an enlarged view of the triplications associated with the upper mantle discontinuities at 400 and 670 km depth in the PREM model. On the plot, label the travel time branch that represents: (1) rays that turn above 400 km, (2) rays that turn at 400 km, (3) rays that turn between 400 and 670 km, (4) rays that turn at 670 km, and (5) rays that turn below 670 km.

Note: The flat-Earth transformation blows up at the center of the Earth and your program may produce strange results at small r values; thus do not attempt to transform the $(r = 0, z = 6{,}371)$ level in PREM. As a kluge, simply change the final depth in the model to 6360 km. This means that you will not be able to include the vertical ray that goes straight through the center of the inner core; this is why you are asked to use $p = 0.0017$ as a minimum ray parameter. Warning: Your computed P travel times are only approximate, owing to the relatively coarse sampling of PREM in Appendix 1. The true PREM model does not contain linear velocity gradients between depth points, as the LAYERXT subroutine assumes.

10. For spherically symmetric Earth models, P-SV wave motion separates completely from SH motion. Despite this, P waves are often observed (weakly) on the transverse component (e.g., Figures 4.17 and 4.23). Give several reasons why this might occur.

11. The stacked images in Figures 4.21–4.23 appear "grainier" at source–receiver distances near 0° and 180° than at 90°, owing to a smaller number of seismograms available at these distances. Why might one expect the number of seismograms to diminish at small and long ranges?

12. Some of the seismic phases in Figures 4.18–4.24 are plotted with negative slopes, that is, they arrive sooner at longer distances. How can this be?

13. Using Figures 4.18–4.24, identify the period (short or long), component (vertical, radial, or transverse), and the source–receiver distance at which one can most easily observe the following seismic phases:

   (a) *SKS*,

   (b) *PKKP*,

   (c) *Sdiff* (*S* wave diffracted along the CMB),

   (d) *PKPPcP*,

   (e) *PcP*,

   (f) *ScSScS* (labeled *ScS2* in Figure 4.24),

   (g) *PKPPKP* (often called *P'P'*).

5

Inversion of travel time data

In the preceding chapter we examined the problem of tracing rays and calculating travel time curves from a known velocity structure. We derived expressions for ray tracing in a one-dimensional (1-D) velocity model in which velocity varies only with depth; ray tracing in general three-dimensional (3-D) structures is more complex but follows similar principles. We now examine the case where we are given travel times obtained from observations and wish to invert for a velocity structure that can explain the data. As one might imagine, the inversion is much more complicated than the forward problem. The main strategy used by seismologists, both in global and crustal studies, has generally been to divide the problem into two parts:

1. A 1-D "average" velocity model is determined from all the available data. This is generally a non-linear problem but is tractable since we are seeking a single function of depth. Analysis often does not proceed beyond this point.
2. If sufficient 3-D ray coverage is present, the 1-D model is used as a reference model and a travel time residual is computed for each datum by subtracting the predicted time from the observed time. A 3-D model is obtained by inverting the travel time residuals for velocity perturbations relative to the reference model. If the velocity perturbations are fairly small, this problem can be linearized and is computationally feasible even for large data sets. This is the basis for *tomographic* inversion techniques.

We now consider each of these problems in turn. For now we will assume that the source locations are precisely known, deferring discussion of the earthquake location problem to the end of the chapter.

5.1 One-dimensional velocity inversion

Before beginning it is useful to imagine how one might obtain a 1-D velocity structure from travel times. Assume that we are given a simple travel time curve

103

without any triplications or low-velocity zones. Each point on the $T(X)$ curve has a slope, which gives the velocity at the turning point of the ray. Thus, we know that a particular velocity must be present; the problem is to determine where. This is equivalent to assigning a depth to each point along the travel time curve. To do this we need to know the velocity structure above the depth in question, and so it makes sense to start at the surface and work our way down. At the origin we know both the depth (zero) and the velocity (the slope of the $T(X)$ curve). We could then examine a nearby point on the $T(X)$ curve, compute the velocity, and find the depth at which the predicted travel time curve would pass through the observed point. In this way we could continue along the $T(X)$ curve and down in depth.

However, this is hardly a rigorous approach and leaves several questions un-answered. Is it always possible to obtain a velocity model? Could there be more than one velocity model that predicts the same travel time curve? We now explore these issues, generally following the treatment in Aki and Richards (2002, pp. 414–22) to which the reader is referred for more details.

Recall from Chapter 4 the formulas for the surface-to-surface travel time and distance for a 1-D velocity model:

$$T(p) = 2 \int_0^{z_p} \frac{u^2(z)}{\left(u^2(z) - p^2\right)^{1/2}} \, dz, \quad X(p) = 2p \int_0^{z_p} \frac{dz}{\left(u^2(z) - p^2\right)^{1/2}}. \quad (5.1)$$

Assume that we are given a complete $T(X)$ curve. By measuring the slope of the $T(X)$ curve, we can obtain $p = dT/dX$ and thus both $X(p)$ and $T(p)$. Our goal is to invert for $u(z)$.

It turns out that this inversion problem is analo-gous to a very old problem that Abel (pronounced "ah-buhl") solved in 1826. Abel's problem was to find the shape of a hill, given measurements of how long it takes a ball to roll up and back down the hill, as a function of the ball's initial velocity (the unrealistic assumptions typical of first-year physics problems ap-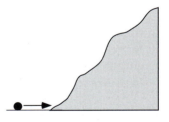ply: the ball is assumed to be frictionless, to have no rotational inertia, and to be "stuck" to the hill such that it never becomes airborne). The highest point the ball reaches can be computed from the initial velocity by equating kinetic and potential energy.

Abel showed that the solution can be obtained from the integral transform pair:

$$t(x) = \int_x^a \frac{f(\xi)}{\sqrt{\xi - x}} \, d\xi, \quad (5.2)$$

$$f(\xi) = -\frac{1}{\pi} \frac{d}{d\xi} \int_{\xi}^{a} \frac{t(x)}{\sqrt{x - \xi}} \, dx, \tag{5.3}$$

where x is the highest point of the ball and t is the travel time. The $X(p)$ equation (5.1) can be put into an analogous form by using u^2 as the integration variable:

$$\frac{X(p)}{2p} = \int_{u_0^2}^{p^2} \frac{dz/d(u^2)}{(u^2 - p^2)^{1/2}} \, d(u^2), \tag{5.4}$$

where u_0 is the slowness at $z = 0$. Now compare with (5.2) and (5.3) and let $t(x) = X(p)/2p$, $x = p^2$, $\xi = [u(z)]^2$, and $f(\xi) = dz/d(u^2)$ to obtain

$$dz/d(u^2) = -\frac{1}{\pi} \frac{d}{d(u^2)} \int_{u_0^2}^{u^2} \frac{X(p)/2p}{\sqrt{p^2 - u^2}} \, d(p^2),$$

$$z(u) = -\frac{1}{\pi} \int_{u_0^2}^{u^2} \frac{X(p)/2p}{\sqrt{p^2 - u^2}} \, d(p^2)$$

$$= -\frac{1}{\pi} \int_{u_0}^{u} \frac{X(p)}{\sqrt{p^2 - u^2}} \, d(p). \tag{5.5}$$

Integrating by parts, we can obtain

$$z(u) = \frac{1}{\pi} \int_{0}^{X(u)} \cosh^{-1}(p/u) \, dX. \tag{5.6}$$

Equations (5.5) and (5.6) were derived in seismology in the period 1903 to 1910 by three independent investigators and are referred to as the *Herglotz–Wiechert–Bateman formulas* (often just *Herglotz–Wiechert*). Similar formulas can be derived for the spherical Earth case (see Aki and Richards, 2002, p. 419). In order to use this equation to obtain a velocity depth function, we select a value for the slowness u. The upper limit of integration in (5.6), $X(u)$, represents the range for a ray with ray parameter $p = u$ and is obtained from the $X(p)$ curve. The integral is then computed for values of X ranging from 0 to $X(u)$ (note that p in (5.6) is a function of X). This gives $z(u)$, the depth to the slowness u. By repeating this calculation for different values of u we can obtain $u(z)$ and thus the desired velocity profile.

These formulas are invalid when a low-velocity zone is present, in which case $X(p)$ is discontinuous and there is no unique solution. A simple illustration of this fact is to imagine a number of homogeneous layers of varying velocity within the LVZ. Since no rays turn within these layers, they can be shuffled arbitrarily and the integrated travel time and distance for rays passing through the LVZ will be

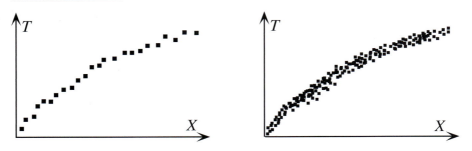

Figure 5.1 Travel time observations often exhibit scatter, complicating inversions for velocity profiles.

unchanged. The thickness of a LVZ also cannot be uniquely determined, although limits can be placed on its maximum thickness.

Despite its analytical elegance, the Herglotz–Wiechert (HW) formula is seldom used in modern seismology. There are at least two reasons for this. First, HW assumes that we are given a *continuous* $T(X)$ curve. In practice, we always have only a finite number of travel time points. This means that the $T(X)$ curve will need to be interpolated between data points, and differences in interpolation schemes will lead to different velocity profiles. Indeed, there are an infinite number of slightly different velocity models that are compatible with a finite number of $T(X)$ points. However, a more serious problem is that real seismic data are generally somewhat noisy and self-contradictory. Typical examples of real data are shown in Figure 5.1. In the example on the left, small timing shifts result in "jitter" in the $T(X)$ points. It is impossible to connect these points with the smooth, physically realizable $T(X)$ curve that is expected for a 1-D velocity model. In the example on the right, data from a number of different earthquakes have been combined and we are interested in determining an "average" velocity profile for these data.

In both of these cases, the HW formula cannot be applied in a straightforward manner. We will now go on to explore some of the ways that seismologists invert these imperfect data sets. The main usefulness of the HW formula is the demonstration that a precisely specified $T(X)$ curve does produce a unique solution for the velocity profile. Thus, in many inversion strategies the problem of finding the "best" velocity model is reduced to the problem of finding the "best" $T(X)$ curve through the data.

5.2 Straight-line fitting

One of the simplest approaches to velocity inversion is to fit the travel time data with a series of straight lines. This was used extensively by seismologists in the

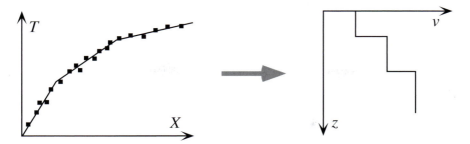

Figure 5.2 Straight lines fit to $T(X)$ data can be inverted for a "layer-cake" velocity model.

1950s and early 1960s to interpret results of their marine refraction experiments. Lines were generally drawn in by hand at slopes and positions estimated by eye (this was before computers became commonly available). Because the slope of each line determines a seismic velocity, it is straightforward to invert the data for a simple model consisting of a small number of homogeneous layers (Fig. 5.2). The model can be obtained most easily by converting each line segment to a point in $\tau(p)$. Recall (4.25) for $\tau(p)$ in the case of a series of homogeneous layers:

$$\tau(p) = 2 \sum_i \left(u_i^2 - p^2\right)^{1/2} \Delta z_i, \qquad u_i > p. \tag{5.7}$$

Note that $\tau_1 = 0$ and that the slowness of the top layer is determined by the slope of the first line segment ($u_1 = p_1$), the slowness of the second layer is set by the slope of the second line segment ($u_2 = p_2$), etc. The thickness of the first layer is determined by the slopes and delay times of the first two line segments, that is, given τ_2 and $u_2 = p_2$, we can solve this equation for z_1 (also using $\tau_1 = 0$, $u_1 = p_1$). We then use τ_3 and $u_3 = p_3$ to solve for z_2, etc.

Travel time data from the oceanic crust were generally fit with three line segments, leading to models containing three homogeneous layers: layer 1 for the sediments, layer 2 for the upper crust ($\sim$2 km thick), and layer 3 for the lower crust ($\sim$4 km thick). These layer designations have become a standard terminology for describing the oceanic crust, although we now know that the upper crust (layer 2) generally consists of several structures and is characterized by a steep velocity gradient rather than a constant velocity.

Note that layer-cake models such as this predict triplications and secondary arrivals around each "corner" in the travel time curve. There are a large number of alternative models that will produce identical first-arrival times, differing only in the shape of the secondary branches (Fig. 5.3). For the case shown on the bottom of Figure 5.3, the model focuses all the intermediate rays to land at the same range (the velocity profile that accomplishes this was first obtained by L. B. Slichter in

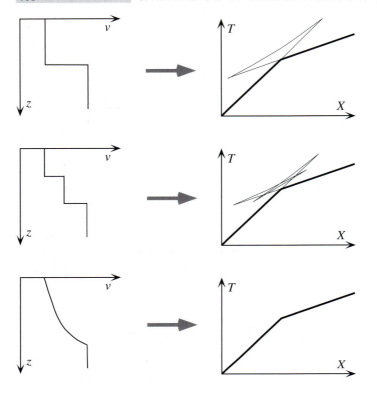

Figure 5.3 Each of the velocity models on the left produces identical first arrivals; the differences only appear in the secondary branches of the travel time curve.

1932). Without secondary branch data, there is no way to discriminate between these models. The homogeneous layer approach only makes sense in cases where we have some a priori reason to expect the velocity model to be characterized by a small number of layers of nearly constant velocity.

5.2.1 Example: Solving for a layer-cake model

Given the travel time curve to the right, solve for a homogeneous layer model. The curve has three straight line segments, so the first step is to solve for the ray parameter, p, and delay time, τ, for each line. This is straightforward, but note that the plotted times are reduced by 6 km/s. Thus,

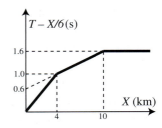

$p_1 = \Delta T/\Delta X = (1 + 4/6)/4 = 0.417$ s/km,

$p_2 = ((1.6 + 10/6) - (1 + 4/6))/(10 - 4) = 0.267$ s/km, and $p_3 = 1/6 = 0.167$ s/km (because horizontal lines occur at the reduction velocity).

The τ values are obtained by the y-intercepts of each line, i.e., $\tau_1 = 0$ s, $\tau_2 = 0.6$ s, $\tau_3 = 1.6$ s. For a layer cake model, we set $u_1 = p_1$, $u_2 = p_2$, and $u_3 = p_3$, corresponding to velocities $v_1 = 1/u_1 = 2.4$, $v_2 = 1/u_2 = 3.75$, and $v_3 = 6$ km/s. We then use (5.7) to solve for Δz_1 and Δz_2, i.e.,

$$\tau_2(p_2) = 0.6 = 2(u_1^2 - p_2^2)^{1/2}\Delta z_1 = 2(0.417^2 - 0.267^2)^{1/2}\Delta z_1$$

from which we obtain $\Delta z_1 = 0.937$ km. We then have

$$\tau_3(p_3) = 1.6 = 2(u_1^2 - p_3^2)^{1/2}\Delta z_1 + 2(u_2^2 - p_3^2)^{1/2}\Delta z_2$$
$$= 2(0.417^2 - 0.167^2)^{1/2}0.937 + 2(0.267^2 - 0.167^2)^{1/2}\Delta z_2$$

from which we obtain $\Delta z_2 = 2.12$ km. The model consists of a 0.937-km-thick surface layer of velocity 2.40 km/s, above a 2.12-km-thick layer of velocity 3.75 km/s. Note that the thickness of the bottom layer with velocity of 6 km/s is unconstrained.

5.2.2 Other ways to fit the *T(X)* curve

Suppose we abandon the crude straight-line fitting approach discussed above and attempt to determine directly the $T(X)$ curve that in some sense (e.g., least squares) comes closest to our observed discrete $T(X)$ points. Any simple discrete representation of the $T(X)$ curve will often lead to $T(X)$ curves with patches that are concave up and thus physically unrealizable for first-arriving branches. We can avoid this problem if we add a positivity constraint to the change in slope of the $T(X)$ curve (for example, applying a non-negative least-squares algorithm). However, the positivity constraint will lead to $T(X)$ curves characterized by a series of straight-line segments. This result, which may seem surprising at first, is typical of positivity constraints in inverse problems. The $T(X)$ curve is "banging against the stops" in places where it really would like to be concave up. This $T(X)$ curve is best fitting in the sense that it is the closest physically realizable curve to the data points. However, as we saw above, parameterizing the $T(X)$ curve with a series of line segments hardly leads to a unique solution for the velocity versus depth profile. Furthermore, a segmented $T(X)$ curve results from this approach even if the underlying velocity distribution is smooth.

In cases where we expect a smooth velocity profile and no triplications, better results may be obtained by fitting a smooth curve to the data. This can be done using a polynomial or spline representation. The travel time data must be sufficiently curved that concave upward segments are not a problem. This approach has been used with some success in the lower mantle where triplications are generally not thought to be present. However, uniqueness is still a significant problem since different curve-fitting procedures will lead to different $T(X)$ curves (and different velocity models).

5.3 $\tau(p)$ Inversion

The Herglotz–Wiechert formula is awkward for performing inversions because velocity is a non-linear function of $X(p)$. Working in the $\tau(p)$ domain can be preferable, since a linear formulation is possible (e.g., Garmany *et al.*, 1979). Recall the $\tau(p)$ formula for the surface-to-surface ray geometry

$$\tau(p) = 2 \int_0^{z_p} \left(u^2(z) - p^2 \right)^{1/2} dz. \tag{5.8}$$

Now change the integration variable from z to u. If $u(z)$ is monotonically decreasing (no low-velocity zones), then switching the limits on the integral is straightforward:

$$\tau(p) = 2 \int_{u_0}^{u_{tp}=p} \left(u^2(z) - p^2 \right)^{1/2} \frac{dz}{du} du, \tag{5.9}$$

where u_0 is the slowness at the surface (the maximum slowness) and the slowness at the turning point, u_{tp}, is equal to the ray parameter p. If we integrate this equation by parts, we obtain

$$\tau(p) = \left[2z(u)(u^2 - p^2)^{1/2} \right]_{u_0}^{p} - 2 \int_{u_0}^{p} u z(u) \left(u^2 - p^2 \right)^{-1/2} du. \tag{5.10}$$

The left-hand term vanishes at both limits, and so we are left with

$$\tau(p) = 2 \int_p^{u_0} \frac{u}{\left(u^2(z) - p^2 \right)^{1/2}} z(u) \, du. \tag{5.11}$$

Note that this expression is linear with respect to changes in $z(u)$. If the $z(u)$ function is multiplied by 2 then the $\tau(p)$ function is also doubled. This linearity permits the application of many of the techniques of linear inverse theory to the problem of finding the velocity–depth profile.

As a simple example, consider the case where we know $\tau(p)$ at a series of discrete values of ray parameter p_j, $(j = 1, 2, \ldots, m)$ and we decide to parameterize our model as a series of homogeneous layers $(u_i, \ i = 1, 2, \ldots, n)$. Our integral for $\tau(p)$, equation (5.8), becomes a summation:

$$\tau(p_j) = 2 \sum_i^n h_i \left(u_i^2 - p_j^2 \right)^{1/2}, u_i > p_j, \tag{5.12}$$

where h_i is the thickness of the ith layer. We might write this out as

$$\tau(p_1) = 2h_1 \left(u_1^2 - p_1^2\right)^{1/2},$$

$$\tau(p_2) = 2h_1 \left(u_1^2 - p_2^2\right)^{1/2} + 2h_2 \left(u_2^2 - p_2^2\right)^{1/2},$$

$$\tau(p_3) = 2h_1 \left(u_1^2 - p_3^2\right)^{1/2} + 2h_2 \left(u_2^2 - p_3^2\right)^{1/2} + 2h_3 \left(u_3^2 - p_3^2\right)^{1/2},$$

etc. In matrix form, this becomes

$$
\begin{bmatrix}
\tau(p_1) \\
\tau(p_2) \\
\tau(p_3) \\
\cdot \\
\cdot \\
\cdot \\
\tau(p_m)
\end{bmatrix}
=
\begin{bmatrix}
2\left(u_1^2 - p_1^2\right)^{1/2} & 0 & 0 & \cdots \\
2\left(u_1^2 - p_2^2\right)^{1/2} & 2\left(u_2^2 - p_2^2\right)^{1/2} & 0 & \cdots \\
& \cdot & & \\
& \cdot & & \\
& \cdot & & \\
2\left(u_1^2 - p_m^2\right)^{1/2} & & \cdots &
\end{bmatrix}
\begin{bmatrix}
h_1 \\
h_2 \\
h_3 \\
\cdot \\
\cdot \\
\cdot \\
h_n
\end{bmatrix}
\tag{5.13}
$$

or, in shorthand,

$$\boldsymbol{\tau} = \mathbf{G}\mathbf{h}, \tag{5.14}$$

where $\mathbf{G}$ is the matrix defined above. Note that all the τ, p, and u values are known; the only unknowns are the layer thicknesses, which are contained in the $\mathbf{h}$ vector. Because this is a linear system, we can use standard techniques to solve for h. If the number of layers is less than the number of τ values ($n < m$), this will generally be an overdetermined problem for which least squares methods can be used. If the number of layers is greater than the number of τ values ($n > m$), then the problem is underdetermined, and some form of regularization will be required to obtain a solution. If $n = m$ and we set $u_1 = p_1$, $u_2 = p_2$, etc., then we are solving for the layer-cake model that fits a $T(X)$ curve consisting of a series of straight line segments. In this case, the diagonal terms of $\mathbf{G}$ in (5.13) are all zero, the first term of $\boldsymbol{\tau}$ and the last value of $\mathbf{h}$ can be deleted (because $\tau(p_1) = 0$ and h_n is unconstrained) and a new square $\mathbf{G}$ matrix formed with non-zero diagonal, analogous to equation (5.7).

5.3.1 Example: The layer-cake model revisited

In 5.2.1, we solved for a three-layer model using equation (5.7). We can perform the same calculation using equation (5.13). Again, we assume $u_1 = p_1 = 0.417$ s/km, $u_2 = p_2 = 0.267$ s/km, and $u_3 = p_3 = 0.167$ s/km. We also have $\tau_1 = 0$ s, $\tau_2 = 0.6$ s and $\tau_3 = 1.6$ s. The diagonal terms of the matrix in (5.13) are zero and we have

$$\begin{bmatrix} \tau(p_2) \\ \tau(p_3) \end{bmatrix} = \begin{bmatrix} 2\left(u_1^2 - p_2^2\right)^{1/2} & 0 \\ 2\left(u_1^2 - p_3^2\right)^{1/2} & 2\left(u_2^2 - p_3^2\right)^{1/2} \end{bmatrix} \begin{bmatrix} h_1 \\ h_2 \end{bmatrix}$$

These are the same equations as before and substituting for the u, p and τ values, we obtain $h_1 = 0.937$ km and $h_2 = 2.12$ km.

5.3.2 Obtaining $\tau(p)$ constraints

It is reasonably straightforward to generalize this approach to solve for a smoother velocity model than that resulting from homogeneous layers. For example, the velocities could be parameterized with linear slowness gradients connecting each (u, z) point in the model. In this case a much smaller number of model points are required to achieve a smooth velocity gradient. The single-valued nature of the $\tau(p)$ function (triplications in $T(X)$ are "unraveled") and the linearity of the equations relating $\tau(p)$ to velocity make it much easier to work in the $\tau(p)$ domain than in the $T(X)$ domain. Unfortunately, seismic data typically are in $T(X)$, making it necessary to convert to $\tau(p)$ before analysis can begin. In principle, given noise-free continuous data, it is trivial to compute $\tau(p)$ from $T(X)$ and vice versa, using

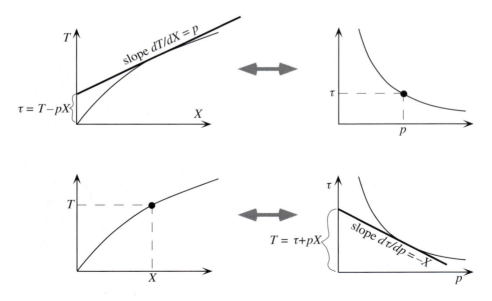

Figure 5.4 Lines in the $T(X)$ domain correspond to points in the $\tau(p)$ domain, and vice versa.

the relationships

$$\tau = T - pX, \quad T = \tau + pX, \quad p = dT/dX, \quad X = -\frac{d\tau}{dp}. \tag{5.15}$$

Note that a line in $T(X)$ becomes a point in $\tau(p)$, and that a line in $\tau(p)$ becomes a point in $T(X)$ (Fig. 5.4).

In practice, we normally have $T(X)$ data as a series of points. One way to construct a $\tau(p)$ curve from a series of $T(X)$ points is to convert each of the $T(X)$ points into a line in $\tau(p)$. The desired $\tau(p)$ curve is then given graphically by the envelope formed by the intersection of these lines (Bessonova et al., 1974, 1976). This technique does not give an exact solution for the $\tau(p)$ curve; some form of fitting procedure is still required to get a single curve from the envelope of lines.

It is also possible to convert waveform data directly into the $\tau(p)$ domain by performing a *slant stack* (also called a τ-p or *Radon transform*), in which each point in the $\tau(p)$ image is generated by summing the data points (this is called *stacking*) along the corresponding line in $T(X)$. To see how this works, consider the record section plot of seismograms versus distance in Figure 5.5. The slant stack is built by summing the data along lines defined by a ray parameter and delay time. For $p = 0.128$ s/km and $\tau = 3$ s, the line does not align with any arrivals in the record section and thus there is little contribution to the

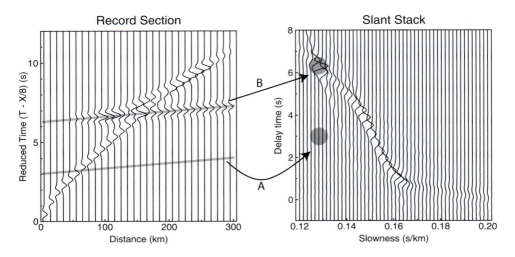

Figure 5.5 Stacking seismograms along lines defined by a slope and a delay time can be used to map a $T(X)$ image into a $\tau(p)$ (slant stack) image.

slant stack ('A' in Fig. 5.5). However, the line with $\tau = 6.4$ s and the same slope aligns with many peaks in the seismograms and the slant stack will have a large amplitude at the $p = 0.128$ s/km and $\tau = 6.6$ s point ('B' in Fig. 5.5). This method can work well if the data are evenly distributed with distance and the arrivals are reasonably coherent, and it has the advantage of not requiring any picking of the seismograms and naturally including the secondary branches of triplications. Slant stacking is commonly used in seismology to process data from arrays of seismic stations and to enhance the visibility of phases with poor signal-to-noise in individual records. However, irregularly spaced data and edge effects often produce artifacts in the slant stack image. In addition, to invert for a $v(z)$ model using the techniques discussed here, the $\tau(p)$ curve must still be parameterized from the slant-stack image.

Sometimes this type of analysis is used to estimate upper and lower *bounds* on the $\tau(p)$ curve. These can be measured, somewhat subjectively, from the graphical envelope shown above or the slant stack image. Alternatively, bounds on τ can be obtained for specific values of p by finding the lines of appropriate slope that limit the $T(X)$ curve (Fig. 5.6). Even this procedure is not entirely satisfactory since there is some subjectivity in defining the lower bounds, and undetected secondary arrivals from triplicated branches could, in some circumstances, lie outside the assumed $T(X)$ bounds. The problem of unraveling the triplications is still present, and the uniqueness of the velocity inversion is severely limited if only first-arrival data are available.

However, let us assume reliable upper and lower bounds on the $\tau(p)$ curve are available. There are an infinite number of possible $\tau(p)$ curves within these limits, corresponding to an infinite number of possible velocity models. How do we choose between these models? We will discuss two approaches: (1) extremal inversions to obtain limits on $v(z)$ and (2) smoothness constraints on $v(z)$.

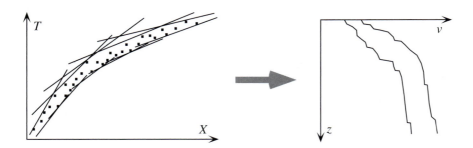

Figure 5.6 Upper and lower bounds obtained from travel time data can be used to place limits on $\tau(p)$ and velocity vs. depth profiles. The two lines in $z(v)$ show the minimum and maximum depths for which a particular velocity can be present and still satisfy the travel time constraints.

5.4 Linear programming and regularization methods

Perhaps the most conservative approach we could take is to find the limits on $v(z)$ that correspond to the limits on $\tau(p)$. "Limits" on $v(z)$ are generally defined by asking the question: What is the minimum and maximum depth possible to a given velocity? Once again, we must exclude low-velocity zones to obtain a well-defined answer. This problem can be formulated using linear equations such as the example shown in equation (5.13), and by applying linear programming theory to obtain the maximum or minimum values of $z(v)$ (e.g., Garmany *et al.*, 1979; Orcutt, 1980; Stark *et al.*, 1986). The result is a corridor of permitted $z(v)$ functions (Fig. 5.6).

These results are sometimes misinterpreted to indicate that any velocity model within the boundaries is permitted. This is incorrect. Most of the velocity models that one could draw between the boundaries will produce $\tau(p)$ curves that exceed the limits on $\tau(p)$. The boundaries shown represent the minimum and maximum depths permitted *for a specific velocity*; the velocity profile that produces a maximum or minimum depth at that velocity will never lie along the $z(v)$ boundary.

This extremal inversion approach is appealing in that it can produce rigorous limits on the velocity profile. However, in practice these limits often prove to be so wide as to suggest that only very crude resolution in $v(z)$ is possible (naturally, this has discouraged their use!). The temptation exists to use other methods that, however naively, appear to give finer resolution. Part of the problem is discussed above; the simple maximum and minimum depth bounds don't necessarily tell us about the fine structure of the $v(z)$ curve that we might be interested in. The linear programming approach can return more than these bounds, but it is difficult to plot the suite of permitted $v(z)$ models in an understandable way. Another difficulty with extremal inversions is that the $T(X)$ "outliers" are the points that really limit the models. The bulk of the data might appear quite consistent, but a few outlying points ensure that the $\tau(p)$ bounds and $z(v)$ bounds will be broad. It is always tempting to throw out these points, but where does this process end? A less subjective approach involves assuming some model for the errors in the T and τ (e.g., Gaussian) and then solving for a best-fitting $v(z)$ model that includes statistical confidence limits (e.g., Dorman and Jacobson, 1981). However, it remains difficult to visualize the permitted $v(z)$ models that lie within the computed confidence limits.

An alternative to the extremal bound approach for dealing with non-uniqueness in inverse problems is provided by *regularization*, which involves finding the single model among the infinity of permitted models that maximizes some property of the

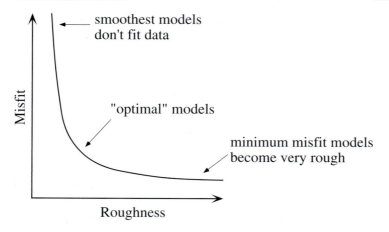

Figure 5.7 Solutions to geophysical inverse problems often involve a tradeoff between model roughness and the misfit to the data.

model. For example, one could search for the "smoothest" model that is consistent with the data. The advantage of this approach is that any structure (e.g., "rough-ness") present in the smoothest model must be real since the method has tried to remove it.

In the case of seismic velocity inversion, the second derivative of the $v(z)$ function can be used as a measure of the roughness of the model. A linear formulation of the problem is possible if $\tau(p)$ data are available (e.g., Stark and Parker, 1987). Results to inverse problems of this type are often expressed in terms of a *tradeoff curve* between model roughness and the data misfit (Fig. 5.7). A very smooth model is possible if we don't care about fitting the data. However, minimizing the data misfit to the lowest possible value often results in huge increases in the model roughness. It is pointless to attempt to fit the data perfectly since there is always some error associated with the data. The optimal models are generally considered to lie somewhere in the "corner" of the tradeoff curve, where there is a reasonable fit to the data and the model is fairly smooth (finding the optimal model can be made more quantitative if the statistical uncertainties in the data are reliably known, but they almost never are).

Smooth models often suffer from the same presentation problem that the extremal bound models have – they can be so bland in appearance that it doesn't appear as if one is resolving very much. This, of course, is the point. If a smooth model fits the data, then why put in more complicated structures? However, a word of caution is also in order. One should not begin thinking of the smoothest model as the "best" model. There is usually no a-priori reason to expect the Earth to be smooth; the smoothest model is not the "most probable" model.

5.5　Summary: One-dimensional velocity inversion

We have discussed the one-dimensional velocity inversion problem at some length, partly because of its importance, but also as an illustration of the complexities that are often encountered in seismological and geophysical inverse problems. Despite considerable effort, the problem of determining the best $v(z)$ model, or limits on the best model, from realistic travel time data is far from resolved. There are several reasons for this:

1. Travel times from secondary branches in triplications are rarely available, and first-arrival data are inherently non-unique.
2. Travel time data typically are noisy and contain both random and systematic errors. The systematic errors most often result from lateral velocity variations not included in one-dimensional velocity inversions.

It is unclear how much benefit will result from attempting to push 1-D travel time inversion techniques any further. There is a risk of becoming distracted by statistical arguments and losing sight of the goal of learning about Earth structure using all available constraints. Much more information is generally available from seismograms than travel times. For example, seismic amplitudes are very sensitive to velocity gradients and can be diagnostic of the presence of triplications. The current state of the art in 1-D modeling generally involves synthetic seismogram modeling of the entire waveform. In this case, travel time inversions are used only to produce starting models for more sophisticated techniques. Other phases, such as discontinuity reflected or converted arrivals, often can be used to resolve fine scale structure in much better detail than can be obtained using refracted arrivals.

Attention is also shifting to resolving 3-D structure, and there are clearly limits to the usefulness of 1-D models in interpreting Earth structure. If the main source of scatter in $T(X)$ data is due to lateral heterogeneity, then the problem of finding the "best" 1-D model is not clearly defined, and systematic offsets in the travel times caused by the velocity perturbations will doom any statistical treatments that assume uncorrelated errors in the data.

5.6　Three-dimensional velocity inversion

Observed travel times typically exhibit some scatter compared to the times predicted by even the best reference 1-D Earth model. The travel time *residual* may be computed by subtracting the predicted time from the observed time, $t_{resid} = t_{obs} - t_{pred}$.

Negative residuals result from early arrivals indicative of faster-than-average structure, while positive residuals are late arrivals suggestive of slow structure. Residuals within a selected range window are often plotted as a histogram to show the spread in residuals. If the average residual is non-zero, as in this example, this indicates that the reference 1-D velocity model may require some adjustment.

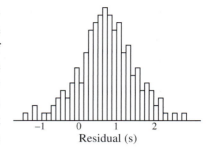

Residual (s)

The spread in the residual histogram can be modeled as the sum of two parts: (1) random scatter in the times due to picking errors and (2) systematic travel time differences due to lateral heterogeneity. The goal of 3-D velocity inversion techniques is to resolve the lateral velocity perturbations. These techniques are now commonly called *seismic tomography* by analogy to medical imaging methods such as CAT scans. However, it is worth noting that 3-D seismic velocity inversion is much more complicated than the medical problem. This is due to several factors: (1) seismic ray paths generally are not straight and are a function of the velocity model itself, (2) the distribution of seismic sources and receivers is sparse and non-uniform, (3) the locations of the seismic sources are not well known and often trade off with the velocity model, and (4) picking and timing errors in the data are common.

Thus the analogy to medical tomography can be misleading when seismologists speak of *imaging* Earth structure, since the term "image" implies a rather direct measurement of the structure, whereas, in practice, seismic velocity inversion usually requires a number of modeling assumptions to deal with the difficulties listed above. It is comparatively easy to produce an image of apparent 3-D velocity perturbations; the more challenging task is to evaluate its statistical significance, robustness, and resolution.

5.6.1 Setting up the tomography problem

Assuming that a reference 1-D model is available, the next step is to parameterize the model of 3-D velocity perturbations. This is commonly done in two different ways: (1) the model is divided into blocks of uniform velocity perturbation or (2) spherical harmonic functions can be used in the case of global models to parameterize lateral velocity perturbations, with either layers or polynomial functions used to describe vertical variations.

As an example, we now illustrate the block parameterization in the case of body waves. Consider a two-dimensional geometry with the model divided into blocks as shown in Figure 5.8. For each travel time residual, there is an associated ray

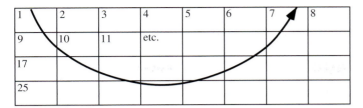

Figure 5.8 An example ray path and cell numbering scheme for a simple 2-D tomography problem.

path that connects the source and receiver. Finding this exact ray path comprises the *two-point ray tracing problem*, and this can be a non-trivial task, particularly in the case of iterative tomography methods in which rays must be traced through 3-D structures. Methods for solving the two-point ray tracing problem include: (1) *ray shooting* in which slightly different takeoff angles at the source are sampled in order to converge on the correct receiver location, (2) *ray bending* in which a nearby ray path is slightly deformed to arrive at the desired receiver location, or (3) finite difference or graph theory techniques that require a grid of points (e.g., Vidale, 1988; Moser, 1991). Fortunately, Fermat's principle suggests that we do not have to get precisely the right ray path to obtain the correct travel time – getting reasonably close should suffice, since, to first order, the travel times are insensitive to perturbations in the ray paths.

Once we have determined the geometry of the ray path, the next step is to find the travel time through each block that the ray crosses (although in principle this is straightforward, programming this on the computer can be a substantial chore!). The total travel time perturbation along the ray path is then given by the sum of the product of each block travel time with the fractional velocity perturbation within the block. In other words, the travel time residual r can be expressed as

$$r = \sum_k b_k v_k, \tag{5.16}$$

where b_k is the ray travel time through the kth block and v_k is the fractional velocity perturbation in the block (note that v_k is unitless, with $v_k = -0.01$ for 1% fast, $v_k = 0.01$ for 1% slow, etc.). The ray paths and the b_k values are assumed to be fixed to the values obtained from ray tracing through the reference model. Note that the velocity perturbations v_k are constant within individual blocks, but the velocity within each block may not be constant if the reference 1-D model contains velocity gradients. Since velocity perturbations will affect the ray paths, Equation (5.16) represents an approximation that is accurate only for small values of v_k.

If we set the ray travel times for the blocks not encountered by the ray to zero, we can express the travel time residual for the ith ray path as:

$$r_i = \sum_{j=1}^{m} b_{ij} v_j, \qquad (5.17)$$

where m is the total number of blocks in the model. Note that most of the values of b_{ij} are zero since each ray will encounter only a small fraction of the blocks in the model. For n travel time measurements, this becomes a matrix equation:

$$
\begin{bmatrix} r_1 \\ r_2 \\ r_3 \\ \cdot \\ \cdot \\ \cdot \\ r_n \end{bmatrix} =
\begin{bmatrix}
0 & 0 & 0 & 0 & 0.8 & \cdots \\
0 & 0.6 & 0 & 1.3 & 0 & \cdots \\
0.1 & 0 & 0 & 0 & 0 & \cdots \\
\cdot & \cdot & \cdot & \cdot & \cdot & \cdots \\
\cdot & \cdot & \cdot & \cdot & \cdot & \cdots \\
\cdot & \cdot & \cdot & \cdot & \cdot & \cdots \\
0 & 0 & 0.7 & 0 & 0 & \cdots
\end{bmatrix}
\begin{bmatrix} v_1 \\ v_2 \\ \cdot \\ \cdot \\ \cdot \\ v_m \end{bmatrix}, \qquad (5.18)
$$

where the numbers are examples of individual ray travel times through particular blocks. This can be written as

$$\mathbf{d} = \mathbf{Gm} \qquad (5.19)$$

using the conventional notation of $\mathbf{d}$ for the data vector, $\mathbf{m}$ for the model vector, and $\mathbf{G}$ for the linear operator that predicts the data from the model. The numbers in $\mathbf{G}$ are the travel times for each ray through each block. $\mathbf{G}$ will generally be extremely sparse with mostly zero elements. In the case shown, the number of travel time observations is greater than the number of model blocks ($n > m$), and, in principle, the problem is overdetermined and suitable for solution using standard techniques. The least squares solution to (5.19) is

$$\mathbf{m} = (\mathbf{G}^T \mathbf{G})^{-1} \mathbf{G}^T \mathbf{d}. \qquad (5.20)$$

In tomography problems this formula can almost never be used since the matrix $\mathbf{G}^T \mathbf{G}$ is invariably singular or extremely ill-conditioned. Some of the ray paths may be nearly identical while some of the blocks may not be sampled by any of the ray paths. These difficulties can be reduced in the case of small matrices with linear algebra techniques such as singular value decomposition (SVD). More commonly, however, m is so large that direct matrix inversion methods cannot be used. In either case, it will typically turn out that there is no unique solution to the problem – there are too many undersampled blocks and/or tradeoffs in the perturbations between different blocks.

A common approach to dealing with ill-posed least squares problems is to impose additional constraints on the problem, a process referred to as regularization. One example of regularization is the *damped least squares* solution in which (5.19) is replaced with

$$\begin{bmatrix} \mathbf{d} \\ 0 \end{bmatrix} = \begin{bmatrix} \mathbf{G} \\ \lambda \mathbf{I} \end{bmatrix} \mathbf{m}, \tag{5.21}$$

where $\mathbf{I}$ is the identity matrix and λ is a weighting parameter that controls the degree of damping. The least squares solution to this problem will minimize the functional

$$\|\mathbf{Gm} - \mathbf{d}\|^2 + \lambda^2 \|\mathbf{m}\|^2,$$

where the first term is the misfit to the data and the second term is the variance of the model. By adjusting the parameter λ we can control the tradeoff between misfit and model variance. These constraints add stability to the inversion – perturbations in blocks that are not sampled by rays will go to zero; anomalies will be distributed equally among blocks that are sampled only with identical ray paths. However, the damped least squares solution will not necessarily lead to a smooth model, since it is the size of the model, not its roughness, that is minimized. Model perturbations in adjacent blocks can be quite different.

A common measure of model roughness for block models is the *Laplacian operator* ∇^2, which can be approximated with a difference operator in both 2-D and 3-D block geometries. To minimize ∇^2 we replace $\mathbf{I}$ with $\mathbf{L}$ in (5.21):

$$\begin{bmatrix} \mathbf{d} \\ 0 \end{bmatrix} = \begin{bmatrix} \mathbf{G} \\ \lambda \mathbf{L} \end{bmatrix} \mathbf{m}, \tag{5.22}$$

where $\mathbf{L}$ is the finite difference approximation to the Laplacian applied over all model blocks. Each row of $\mathbf{L}$ is given by the difference between the target block and the average of the adjacent cells.

For example, in a 2-D model the Laplacian becomes

$$\nabla^2_j \simeq \tfrac{1}{4}(m_{\text{left}} + m_{\text{right}} + m_{\text{up}} + m_{\text{down}}) - m_j,$$

where ∇^2_j is the Laplacian of the jth model point. In this case the least squares inversion will minimize

$$\|\mathbf{Gm} - \mathbf{d}\|^2 + \lambda^2 \|\mathbf{Lm}\|^2,$$

where λ controls the tradeoff between misfit and model roughness. This type of regularization adds stability to the inversion in a different way than damped least squares. The resulting models will be smooth, but not necessarily of minimum variance. Blocks that are not sampled by ray paths will be interpolated between nearby cells, or, more dangerously, extrapolated when they are near the edge of the model.

Both damped least squares and minimum roughness inversions have advantages and disadvantages, and the best regularization method to use will vary from problem to problem. In general, one should distrust damped least squares solutions that contain significant fine-scale structure at scale lengths comparable to the block dimensions, whereas minimum roughness solutions are suspect when they produce large-amplitude anomalies in regions constrained by little data.

We have so far assumed that all of the data are weighted equally. This is not always a good idea in tomography problems since travel time residuals are often non-Gaussian and plagued with outliers. This difficulty has been addressed in different ways. Often the residuals are first windowed to remove the largest outliers. Travel time residuals from similar ray paths are commonly averaged to form *summary ray* residuals before beginning the inversion. In iterative schemes the influence of anomalous data points can be downweighted in subsequent steps, thus simulating a more robust misfit norm than used in least squares.

5.6.2 Solving the tomography problem

For "small" problems (number of blocks in model $m < 500$ or so), conventional linear algebra methods such as Gauss reduction or singular value decomposition can be used to obtain exact solutions to equations (5.21) or (5.22). In these cases, we have a significant advantage in that it is also practical to compute formal resolution and model covariance matrices. However, more commonly m is too large for such calculations to be practical. For example, a 3-D model parameterized by 100 blocks laterally and 20 blocks in depth contains 200 000 model points. Clearly we are not going to be able to invert directly a 200 000 by 200 000 matrix! Indeed we could not even fit such a matrix into the memory of our computer.

Thus, we must turn to iterative methods designed for large sparse systems of equations in order to solve these problems. Fortunately these have proven extremely useful in tomography problems and are found to converge fairly rapidly to close approximations to the true solutions. Examples of iterative methods include names such as ART-backprojection, SIRT, conjugate gradient, and LSQR (see Nolet, 1987, for a detailed discussion of many of these methods). Although it is instructive to see the form of equations such as (5.18) and (5.19), in practice we rarely attempt to construct **G** as a matrix. Rather we treat **G** as a linear operator that acts on the model

Synthetic model   Ray geometry   Inversion result

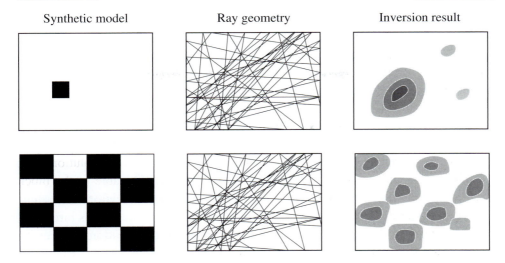

Figure 5.9 The resolution of tomographic models is often evaluated using the impulse response test (top) or the checkerboard test (bottom). In each, a synthetic set of travel times are created for a simple velocity model using the same ray paths present in the real data; then the synthetic times are inverted to see how well the starting model is recovered.

to predict the data. On the computer, this often will take the form of a subroutine. Since the iterative techniques effectively use only one row of **G** at a time, they are sometimes given the name *row action methods*.

A disadvantage of these iterative solutions is that it becomes impossible to compute formal resolution and covariance matrices for the model. As substitutes for these measures, it has become common practice to conduct experiments on synthetic data sets. The synthetic data are generated by assuming a particular model of velocity perturbations and computing travel time anomalies using the same ray paths as the real data. The synthetic data are then inverted to see how well the test model is recovered (Fig. 5.9). One example of this procedure is the *impulse response test*, in which a single localized anomaly is placed in a region of interest to see how well it can be resolved. Another method that is often applied is the *checkerboard test*, in which a model with a regular pattern of alternating fast and slow velocities is examined. In this case, the degree of smearing of the checkerboard pattern will vary with position in the model, giving some indication of the relative resolution in different areas.

It is not always clear that these tests give a reliable indication of the true resolution and uniqueness of the velocity inversions. Impulse response and checkerboard tests can be misleading because they typically assume uniform amplitude anomalies and perfect, noise-free data. In real tomography problems, the data are contaminated by noise to some degree and the velocity models that are obtained contain anomalies of varying amplitude. In these cases it is often only the higher amplitude features that are unambiguously resolved. In principle, some of these problems can be addressed using techniques that randomly resample the data (such as "jackknife" or "bootstrap" methods). However, these require repeating the inversion procedure up to 100 times or more, a significant obstacle in these computationally intensive analyses. Questions regarding the best way to evaluate resolution in tomographic inversions are not fully answered, and this continues to be an active area of research.

5.6.3 Tomography complications

In the preceding discussion it has been assumed that the source locations and origin times were precisely known. However, in the case of earthquakes this is rarely the case, and there is the potential for bias due to errors in the locations. Since the earthquakes are generally located using a reference 1-D velocity model, we would expect the locations to change given a 3-D velocity model, and indeed there is often a tradeoff between velocity anomalies and earthquake locations. This problem can be addressed by *joint hypocenter and velocity inversions* (JHV) that solve for both the earthquake locations and the velocity structure. In practice, for large inversions, this is often an iterative process in which initial earthquake locations are assumed, a velocity model is derived, the earthquakes are relocated using the new model, a new velocity model is derived, etc. Tradeoffs between quake locations and velocity structure will be minimized in this procedure, but only if a wide variety of ray paths are available to locate each quake (we will discuss the earthquake location problem in greater detail in the next section).

Another ambiguity in velocity inversions concerns the shallow structure at each seismic station. Rays generally come up at near-vertical angles beneath individual stations and sample only a very limited lateral area in the uppermost crust. Because of this, and the fact that no information is generally obtained for the shallow structure between stations, times to individual stations in large-scale inversions are usually adjusted using a *station correction*, a time for each station obtained by averaging the residuals from all ray paths to the station. As in the case of earthquake locations, it is important that the station correction be obtained from a wide range of ray paths, to minimize the biasing effect of travel time differences from deeper velocity anomalies.

Seismic tomography works best when a large number of different ray geometries are present and each cell in the model is crossed by rays at a wide range of angles. Unfortunately, this is often not the case, since the sources and receivers are unevenly distributed, and, at least in global tomography problems, largely confined to Earth's surface. Typically, this will result in many blocks being sampled at only a limited range of ray angles. When this occurs, anomalies are smeared along the ray path

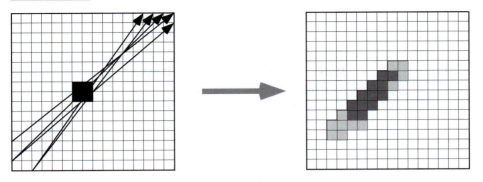

Figure 5.10 When only a limited range of ray angles are available, resolution of velocity anomalies is limited in the direction parallel to the rays.

orientation (Fig. 5.10). This problem cannot be cured by regularization or other numerical methods – only the inclusion of additional ray paths at different angles can improve the resolution.

In some cases, there is the danger that the 3-D velocity perturbations could cause the source–receiver ray paths to deviate significantly from the reference model ray paths. If these ray-path deviations are large enough, then Fermat's principle may not save us and our results could be biased. This concern can be addressed by performing full 3-D ray tracing calculations on the velocity model and iterating until a stable solution is achieved. This requires significantly more work and has not generally been done in global tomography problems where the velocity perturbations are only a few percent. This effect is probably of greater importance in local and regional tomography problems where larger velocity anomalies are found and steep velocity gradients and/or discontinuities are more likely to be present.

There is also a tendency for rays to bend or be diffracted around localized slow anomalies, which may introduce a bias into tomographic inversions by making such features less resolvable than fast velocity anomalies (Nolet and Moser, 1993). More details concerning traditional seismic tomography techniques can be found in the books by Nolet (1987) and Iyer and Hirahara (1993).

5.6.4 Finite frequency tomography

"Classic" seismic tomography assumes the ray theoretical approximation, in which travel-time anomalies are accumulated only along the geometrical ray path. However, at realistic seismic wavelengths there will always be some averaging of structure adjacent to the theoretical ray path. Recently, seismologists have begun computing these finite-frequency effects in the form of kernels (sometimes called *Fréchet derivatives*) that show the sensitivity of the travel time or other observables for

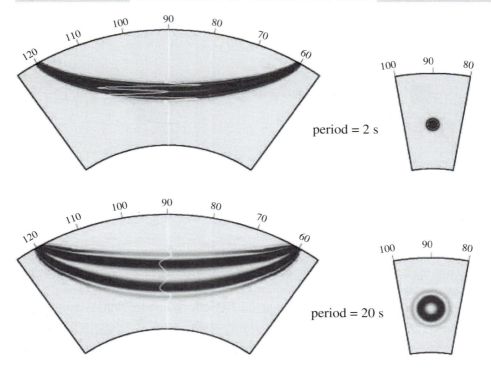

Figure 5.11 Banana-doughnut kernels showing the sensitivity of *P*-wave travel times at 60° epicentral distance to velocity perturbations in the mantle. The right-hand plots show the cross-section perpendicular to the ray direction at its midpoint. Note the much wider kernel at 20 s period compared to 2 s period and the more pronounced "doughnut hole" along the geometrical ray path. Figure from Dahlen *et al.* (2000).

a particular seismic phase and source-receiver geometry to velocity perturbations throughout the Earth (e.g., Dahlen *et al.*, 2000; Hung *et al.*, 2000; Zhao *et al.*, 2000). Examples of these kernels computed for a 1-D reference model for a *P* wave at 60° range are plotted in Figure 5.11. These are sometimes given the name *banana-doughnut kernels*, with "banana" describing the fact they are wider at the middle of the ray path than near its endpoints, and "doughnut" arising from the counterintuitive fact that their sensitivity is zero to velocity perturbations exactly along the geometrical ray path. The width of the kernels shrinks with the frequency of the waves and thus the finite-frequency differences from geometrical ray theory are most important at long periods.

In principle, the use of finite-frequency kernels should improve seismic tomography by properly accounting for the effects of off-ray-path structure. There has been some recent controversy as to how significant these improvements are for the global mantle tomography problem with respect to the imaging of plumes, when compared to differences arising from data coverage and regularization (see Montelli *et al.*, 2004; de Hoop and van der Hilst, 2005a,b; Dahlen and Nolet, 2005). However, it is clear that finite-frequency tomography represents a significant theoretical advance and will eventually become common practice. Researchers are now computing sensitivity kernels based on 3-D Earth models and developing sophisticated algorithms for directly inverting waveforms for Earth structure (e.g., Zhao et al., 2005; Liu and Tromp, 2006). These methods hold the promise of resolving structure using much more of the information in seismograms than simply the travel times of direct phases.

5.7 Earthquake location

The problem of locating earthquakes from travel time data is one of the oldest challenges in seismology and continues to be an important component of seismic research. Earthquakes are defined by their *origin times* and *hypocenters*. The hypocenter is the (x, y, z) location of the event, while the *epicenter* is defined as the (x, y) point on the Earth's surface directly above the hypocenter. Earthquakes are generally treated as point sources in location methods. For large earthquakes that rupture from tens to hundreds of kilometers, the hypocenter is not necessarily the "center" of the earthquake. Rather it is the point at which seismic energy first begins to radiate at the beginning of the event. Since the rupture velocity is less than the P-wave velocity, the hypocenter can be determined from the first arrival times regardless of the eventual size and duration of the event. Earthquake information given in standard catalogs, such as the Preliminary Determination of Epicenters (PDE), is based on travel times of high-frequency body-wave phases. These origin times and hypocenters should not be confused with long-period inversion results, which often give a *centroid* time and location for the event, representing the "average" time and location for the entire event.

Four parameters describe the origin time and hypocenter. Let's call these parameters the model, and define a model vector

$$\mathbf{m} = (m_1, m_2, m_3, m_4) = (T, x, y, z). \qquad (5.23)$$

Now suppose we are given n observations of travel times, t_i, at individual seismic stations. In order to invert these times for the earthquake parameters, $\mathbf{m}$, we first must assume a reference Earth model. For every value of $\mathbf{m}$ we can then calculate ranges to the ith station and compute predicted arrival times,

$$t_i^p = F_i(\mathbf{m}), \qquad (5.24)$$

where $\mathbf{F}$ is the operator that gives the predicted arrival time at each station from $\mathbf{m}$. The difference between the observed and predicted times is

$$r_i = t_i - t_i^p = t_i - F_i(\mathbf{m}), \tag{5.25}$$

where r_i is the residual at the ith station. We wish to find the $\mathbf{m}$ that, in some sense, gives the smallest residuals between the observed and predicted times. Note that $\mathbf{F}$ is a function both of the Earth model and of the individual station locations. Most importantly, $\mathbf{F}$ is a *non-linear* function of the model parameters (with the exception of the origin time T). In practice, for 1-D Earth models, $\mathbf{F}(\mathbf{m})$ is not particularly difficult to calculate, since the arrival times can be interpolated at the appropriate ranges from known travel time tables for the reference velocity model. However, the non-linear dependence of the travel times on the earthquake location parameters greatly complicates the task of inverting for the best earthquake model. This non-linearity is apparent even in the simple example of 2-D location within a plane of uniform velocity. The travel time from a station with coordinates (x_i, y_i) to a point (x, y) is given by

$$t_i = \frac{\sqrt{(x - x_i)^2 + (y - y_i)^2}}{v}, \tag{5.26}$$

where v is the velocity. Clearly t does not scale linearly with either x or y in this equation. The result is that we cannot use standard methods of solving a system of linear equations to obtain a solution. Given a set of travel times to the stations, there is no single-step approach to finding the best event location.

Before discussing practical location strategies, it is instructive to consider what we might do if an infinite amount of computer power were available. In this case, we could perform a grid search over all possible locations and origin times and compute the predicted arrival times at each station. We could then find the particular $\mathbf{m}$ for which the predicted times t_i^p and the observed times t_i were in best agreement. How do we define "best" agreement? A popular choice is least squares, that is, we seek to minimize

$$\epsilon = \sum_{i=1}^{n} \left[t_i - t_i^p \right]^2, \tag{5.27}$$

where n is the number of stations. The average squared residual, ϵ/n, is called the *variance*; thus we are trying to minimize the variance of the residuals. A common term that you may hear in describing models is *variance reduction* ("I got a 50% variance reduction with just two parameters" or "Their model only gives a 5% variance reduction in the raw data"). Here we use the term variance loosely to describe the spread in the residuals, independently of the number of free parameters in the fitting procedure. More formally, in statistics the variance is defined as ϵ/n_{df}, where n_{df} is the number of degrees of freedom (n_{df} is n minus the number of free parameters in the fit). For typical problems the number of fitting parameters is much less than the number of data, and so n and n_{df} are approximately equal.

Least squares is often used as a measure of misfit since it leads to simple analytical forms for the equations in minimization problems. It will tend to give the right answer if the misfit between t and t^p is caused by uncorrelated, random Gaussian noise in t. However, in many instances the errors are non-Gaussian, in which case least squares will give too much weight to the *outliers* in the data (a residual of 2 contributes 4 times more to the misfit than a residual of 1). As an alternative, we could use the sum of the differences

$$\epsilon = \sum_{i=1}^{n} \left| t_i - t_i^p \right|. \tag{5.28}$$

This measure of misfit is called the *L1 norm* and is considered more robust than the *L2 norm* (least squares) when excessive outliers are present in the data. For a distribution of numbers, the minimum *L2* norm yields the mean or average of the numbers, while the minimum *L1* norm gives the median value. The *L1* norm is not often used because the absolute value sign creates great complications in the equations. As an alternative to robust norms such as *L1*, it is possible to weight the residuals in the least squares problem using an iterative procedure that reduces the influence of the outlying points in subsequent steps. Of course in the case of our hypothetical "brute force" grid search it is straightforward to apply any norm that we desire. Once we have defined a measure of misfit, we can find the "best" $\mathbf{m}$ as the one with the smallest misfit, $\epsilon(\mathbf{m})$. The next step is to estimate the probable uncertainties in our location.

Some indication of these uncertainties can be seen in the behavior of the misfit function in the vicinity of its minimum. In our two-dimensional example, suppose that we contour $\epsilon(\mathbf{m})$ as a function of x and y, assuming that the origin time is known (since the t^p are a linear function of the origin time, determination of the best origin time for a given location is trivial). Clearly, if ϵ grows rapidly as we move away from the minimum point, we have resolved the location to better accuracy than when ϵ grows only very slowly away from its minimum.

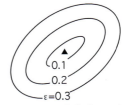

How can we quantify this argument? By far the most common approach is based on least squares and the *L2* norm, since the statistics of Gaussian processes are well

Table 5.1: Percentage points of the χ^2 distribution.			
n_{df}	$\chi^2(95\%)$	$\chi^2(50\%)$	$\chi^2(5\%)$
5	1.15	4.35	11.07
10	3.94	9.34	18.31
20	10.85	19.34	31.41
50	34.76	49.33	67.50
100	77.93	99.33	124.34

understood. In this case we define

$$\chi^2 = \sum_{i=1}^{n} \frac{[t_i - t_i^p]^2}{\sigma_i^2}, \qquad (5.29)$$

where σ_i is the expected standard deviation of the ith residual due to random measurement error. The expected value of χ^2 is approximately the number of degrees of freedom n_{df} (in our case $n_{df} = n - 4$ because $\mathbf{m}$ has 4 components) and 95% confidence limits may be obtained by consulting standard statistical tables (e.g., Table 5.1).

For example, if we locate an earthquake using 14 travel times, then $n_{df} = 10$ and there is a 90% probability that the value of χ^2 computed from the residuals at the best fitting hypocenter will be between 3.94 and 18.31. There is only a 5% chance that the value of χ^2 will exceed 18.31. The value $\chi^2(\mathbf{m})$ will grow as we move away from the best-fitting location, and by contouring values of $\chi^2(\mathbf{m})$ we can obtain an estimate of the 95% error ellipse for the event location.

Note that the σ_i values are critical in this analysis – the statistics are based on the data misfit being caused entirely by random, uncorrelated Gaussian errors in the individual travel time measurements. However, the misfit in earthquake location problems is usually larger than would be expected from timing and picking errors alone. If the σ_i are set significantly smaller than the average residual, then the χ^2 measure may indicate that the solution should be rejected, most likely because unmodeled velocity structure is dominating the misfit. Alternatively, if the σ_i are set significantly larger than the average residual, then the best-fitting hypocenter could be rejected because it fits the data "too well."

To avoid these embarrassments, the estimated data uncertainties σ_i are often estimated from the residuals at the best location,

$$\sigma^2(\mathbf{m}_{best}) = \frac{\sum_{i=1}^{n} [t_i - t_i^p(\mathbf{m}_{best})]^2}{n_{df}}, \qquad (5.30)$$

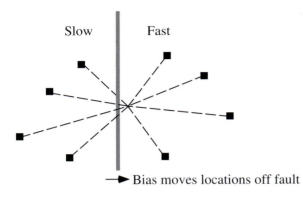

Slow   Fast

Bias moves locations off fault

Figure 5.12 Earthquakes located along a fault will often be mislocated if the seismic velocity changes across the fault.

where $\mathbf{m}_{\text{best}}$ is the best-fitting location, and this constant value of σ^2 is used for all the σ_i^2 in (5.29), that is,

$$\chi^2(\mathbf{m}) = \frac{\sum_{i=1}^{n} \left[t_i - t_i^p(\mathbf{m}) \right]^2}{\sigma^2}. \tag{5.31}$$

Note that $\chi^2(\mathbf{m}_{\text{best}}) = n_{\text{df}}$ so that the χ^2 value at the best-fitting hypocenter is close to the 50% point in the χ^2 distribution. By contouring $\chi^2(\mathbf{m})$, we can then obtain an estimate of the 95% confidence ellipse for the solution; that is, we can approximate the region within which there is a 95% chance that the true location lies.[2]

However, a serious problem with typical confidence limits is that they don't take into account the correlated changes to travel time residuals resulting from unmodeled lateral heterogeneity. For example, consider a model in which a vertical fault separates the crust into two blocks with slightly different velocities (Fig. 5.12). Events occurring on the fault will tend to be mislocated off the fault into the faster velocity block owing to a systematic bias in the travel times. This possibility is not accounted for in the formal error analysis, which, in this case, incorrectly assumes that the travel time uncertainties are *uncorrelated* between different stations. The effects of unmodeled lateral heterogeneity are the dominant source of error for earthquake locations, provided a good station distribution is used in the inversion. Global locations in the ISC and PDE catalogs are typically off by about 25 km in horizontal position and depth (assuming depth phases such as pP are used to constrain the depth; if not, the probable depth errors are much greater). Techniques that can be used to improve earthquake locations include joint hypocenter velocity inversion (see preceding section) and master event methods (see Section 5.7.2).

When a good station distribution is not available, location errors can be quite large. For example, the distance to events occurring outside of a seismic array is

---

[2] The error ellipse is only approximate because the uncertainties in the σ_i estimate are ignored.

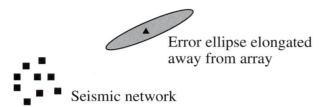

Error ellipse elongated
away from array

Seismic network

Figure 5.13 Earthquake locations for events outside of a network are often not well constrained.

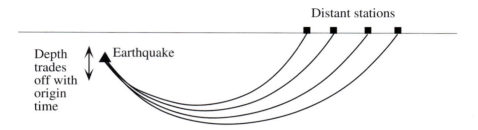

Distant stations

Depth
trades
off with
origin
time

Earthquake

Figure 5.14 Earthquake depth can be hard to determine if only distant stations are available.

not well constrained, since there is a large tradeoff between range and origin time (Fig. 5.13). In this case, the location could be improved dramatically if a travel time was available from a station on the opposite side of the event. Generally it is best to have a good azimuthal distribution of stations surrounding an event to avoid these kinds of location uncertainties. Another problem is the tradeoff between event depth and origin time that occurs when stations are not available at close ranges (Fig. 5.14). Since the takeoff angles of the rays are very similar, changes in the earthquake depth may be compensated for by a shift in the event origin time.

In the preceding examples, we have assumed that only direct P-wave data are available. The addition of other phases recorded at the same stations can substantially improve location accuracy, since the use of *differential times* between phases removes the effect of the earthquake origin time. For example, S arrivals travel at a different speed than P arrivals and can be used to estimate the source–receiver range at each station directly from the $S - P$ time (a convenient rule of thumb for crustal phases is that the distance to the event in kilometers is about 8 times the $S - P$ time in seconds). Even better than S for determining earthquake depths from teleseismic data is the depth phase pP since the differential time $pP - P$ is very sensitive to the earthquake depth.

5.7.1 Iterative location methods

In our discussion so far we have assumed that the minimum ϵ could be found directly by searching over all $\epsilon(\mathbf{m})$. In practice, this often becomes computationally unfeasible and less direct methods must be employed. The standard technique is to *linearize* the problem by considering small perturbations to a target location

$$\mathbf{m} = \mathbf{m}_0 + \Delta\mathbf{m}, \tag{5.32}$$

where $\mathbf{m}_0$ is the current guess as to the best location and $\mathbf{m}$ is a new location a small distance away from $\mathbf{m}_0$. The predicted times at $\mathbf{m}$ may be approximated using the first term in the Taylor series expansion

$$t_i^P(\mathbf{m}) = t_i^P(\mathbf{m}_0) + \frac{\partial t_i^P}{\partial m_j}\Delta m_j. \tag{5.33}$$

The residuals at the new location $\mathbf{m}$ are given by

$$
\begin{aligned}
r_i(\mathbf{m}) &= t_i - t_i^P(\mathbf{m}) \\
&= t_i - t_i^P(\mathbf{m}_0) - \frac{\partial t_i^P}{\partial m_j}\Delta m_j \\
&= r_i(\mathbf{m}_0) - \frac{\partial t_i^P}{\partial m_j}\Delta m_j.
\end{aligned}
\tag{5.34}
$$

In order to minimize these residuals we seek to find $\Delta\mathbf{m}$ such that

$$r_i(\mathbf{m}_0) = \frac{\partial t_i^P}{\partial m_j}\Delta m_j \tag{5.35}$$

or

$$\mathbf{r}(\mathbf{m}_0) = \mathbf{G}\,\Delta\mathbf{m}, \tag{5.36}$$

where $\mathbf{G}$ is the matrix of partial derivatives $G_{ij} = \partial t_i^P/\partial m_j$, $i = 1, 2, ..., n$, $j = 1, ..., 4$. The best fit to (5.36) may be obtained using standard least squares techniques to obtain the location adjustment $\Delta\mathbf{m}$. Next, we set $\mathbf{m}_0$ to $\mathbf{m}_0 + \Delta\mathbf{m}$ and repeat the process until the location converges. This iterative procedure generally converges fairly rapidly provided the initial guess is not too far from the actual location.

5.7.2 Relative event location methods

In the common situation where the location error is dominated by the biasing effects of unmodeled 3-D velocity structure, the relative location among events within a localized region can be determined with much greater accuracy than the absolute location of any of the events. This is because the lateral velocity variations outside the local region, which lead to changes in the measured travel times at distant stations, will have nearly the same effect on all of the events. In other words, the residuals caused by 3-D structure to a given station will be correlated among all of the events. If the ray path to a station is anomalously slow for one event, then it will be slow for the other events as well, provided the local source region is small compared to the heterogeneity. However, the bias in the locations caused by the 3-D structure will vary among the events because they typically do not have picks from exactly the same set of stations.

The simplest way to improve relative location accuracy among nearby earthquakes is to consider differential times relative to a designated *master event*. The arrival times of other events relative to the master event times are

$$t^{\text{rel}} = t - t^{\text{master}}. \tag{5.37}$$

Setting the master event location to $\mathbf{m}_0$ in (5.35), we see that the relative location $\Delta\mathbf{m}$ is given by the best-fitting solution to

$$t_i^{\text{rel}} = t_i^p(\mathbf{m}) - t_i^p(\mathbf{m}_0) = \frac{\partial t_i^p}{\partial m_j} \Delta m_j, \tag{5.38}$$

where the solution will be valid provided $\Delta\mathbf{m}$ is small enough that the linear approximation holds. This approach works because the differential times subtract out any travel-time perturbations specific to a particular station. Note that the absolute location accuracy is limited by the location accuracy of the master event, which is assumed fixed. However, if the absolute location of the master event is known by other means (e.g., a surface explosion), then these relative locations can also be converted to absolute locations.

This approach can be generalized to optimally relocate events within a compact cluster with respect to the cluster centroid by projecting out the part of the travel-time perturbations that are common to particular stations, a method termed *hypocentroidal decomposition* by Jordan and Sverdrup (1981). A simpler technique is to compute station terms by averaging the residuals at each station, recompute the locations after correcting the observed picks for the station terms, and iterate until a stable set of locations and station terms is obtained (e.g., Frohlich, 1979). It can be shown that this iterative approach converges to the same solution as hypocentroidal decomposition (Lin and Shearer, 2005).

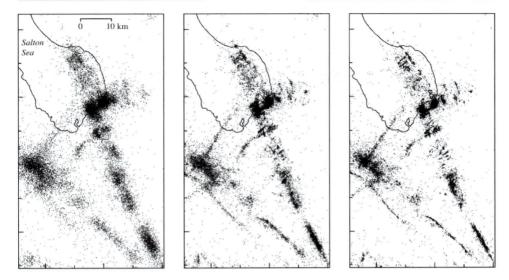

Figure 5.15 Earthquake locations for over 17 000 earthquakes in the Imperial Valley, California (1981–2005), as computed using: (left) single event location, (middle) source-specific station term (SSST) locations, and (right) waveform cross-correlation locations using results from Lin *et al.* (2007).

These ideas can be generalized to distributed seismicity where the effect of 3-D velocity structure on travel times will vary among different source regions. The *double-difference* location algorithm (Waldhauser and Ellsworth, 2000; Waldhauser, 2001) performs simultaneous relocation of distributed events by minimizing the residual differences among nearby events. The *source-specific station term* (SSST) method (Richards-Dinger and Shearer, 2000; Lin and Shearer, 2006) iteratively computes spatially varying time corrections to each station. Further improvements in relative location accuracy can be achieved by using waveform cross-correlation to compute more accurate differential times among nearby events than can be measured using arrival time picks on individual seismograms. Figure 5.15 illustrates the improvement in local earthquake locations that can be achieved using these methods compared to classic single event location. Note the increasingly sharp delineation of seismicity features that is obtained using source-specific station terms and waveform cross-correlation.

5.8 Exercises

1. Project VESE (very expensive seismic experiment) deployed 60 seismometers in a linear array extending 240 km away from a large surface explosion. Despite careful

picking of the resulting seismograms, the first-arrival *P*-wave travel times (plotted in Fig. 5.16 and also given in the supplemental web material) show considerable scatter.

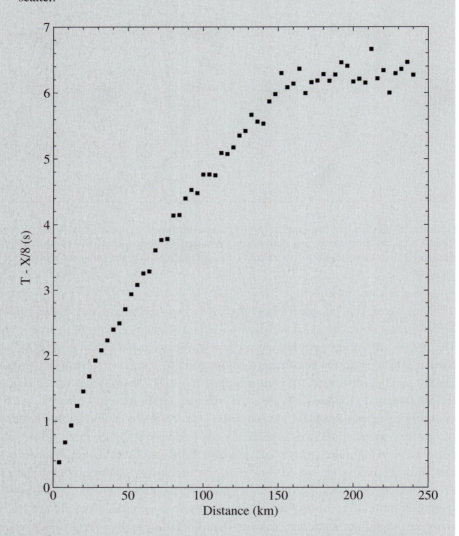

Figure 5.16 *P*-wave travel time data from a very expensive seismic experiment.

Fit these points with a series of straight lines and compute the ray parameter *p* and the delay time τ for each line. The first of these lines should go through the origin (zero time and range). Be sure to take into account the reduction velocity of 8 km/s in computing *p*. Using equation (5.12), invert these results for a layer-cake *P*-velocity model of the crust and uppermost mantle. List your model in a table,

		Quake 1	Quake 2
x (km)	y (km)	t1 (s)	t2 (s)
9.0	24.0	14.189	20.950
24.0	13.2	13.679	21.718
33.0	4.8	13.491	21.467
45.0	10.8	14.406	21.713
39.0	27.0	13.075	20.034
54.0	30.0	15.234	20.153
15.0	39.0	13.270	18.188
36.0	42.0	12.239	16.008
27.0	48.0	12.835	15.197
48.0	48.0	14.574	16.280
15.0	42.0	12.624	16.907
18.0	15.0	13.496	21.312
30.0	36.0	10.578	16.664

Table 5.2: *P*-wave arrival times for two earthquakes.

starting with the surface layer and continuing downward, with each line consisting of a depth (km) and a velocity (km/s). Specify the velocity discontinuities between layers by listing the depth twice, with the first line containing the velocity in the upper layer and the second line the lower layer velocity. Make sure that the first column of your table is absolute depth and not layer thickness. For example, a three-layer model with a 2 km thick top layer of 4 km/s, a 4 km thick middle layer of 6 km/s, and a bottom layer of 8.1 km/s would be written as:

```
0.0  4.0
2.0  4.0
2.0  6.0
6.0  6.0
6.0  8.1
```

What is the *Pn* crossover distance? How thick is the crust in your model? How much uncertainty would you assign to your crustal thickness estimate? Note: Not everyone will get the same answer to this exercise! It's fun to plot the different models to see how well they agree.

2. (COMPUTER) You are given P-wave arrival times for two earthquakes recorded by a 13-station seismic array. The station locations and times are listed in Table 5.2 and also given in the supplemental web material.

    (a) Write a computer program that performs a grid search to find the best location for these events. Try every point in a 100 km by 100 km array ($x = 0$ to 100 km, $y = 0$ to 100 km). At each point, compute the range to each of the 13 stations. Convert these ranges to time by assuming the velocity is 6 km/s (this is a 2-D problem, don't worry about depth). Compute the average sum of the squares of the residuals to each grid point (after finding the best-fitting origin time at the grid point; see below).

    (b) For each quake, list the best-fitting location and origin time.

    (c) From your answers in (b), estimate the uncertainties of the individual station residuals (e.g., σ^2 in 5.30) for each quake.

    (d) For each quake, use (c) to compute χ^2 at each of the grid points. What is χ^2 at the best-fitting point in each case?

    (e) Identify those values of χ^2 that are within the 95% confidence ellipse. For each quake, make a plot showing the station locations, the best quake location, and the points within the 95% confidence region.

    (f) Note: Don't do a grid search for the origin time! Instead assume an origin time of zero to start; the best-fitting origin time at each grid point will be the average of the residuals that you calculate for that point. Then just subtract this time from all of the residuals to obtain the final residuals at each point.

6

Ray theory: Amplitude and phase

Up to this point we have considered only the travel times of rays traveling in Earth, ignoring the amplitude, polarity, and shape of the pulses. Such an analysis is not without its merits in examining real data, since observed travel times are usually more robust and stable than amplitudes. However, amplitude and waveform shape are also important and contain valuable additional information about Earth structure and seismic sources.

To model amplitude variations, ray theory must account for geometrical spreading effects, reflection and transmission coefficients at interfaces, and intrinsic attenuation. We have already seen some aspects of geometrical spreading in the $1/r$ factor in the equations for the spherical wavefront (Section 3.6) and in the eikonal equation (Appendix C). However, because geometrical spreading is most easily understood in terms of the energy density contained in wavefronts, we begin by examining the energy in seismic waves.

6.1 Energy in seismic waves

The energy density $\tilde{E}$ contained in a seismic wave may be expressed as a sum of kinetic energy $\tilde{E}_K$ and potential energy $\tilde{E}_W$:

$$\tilde{E} = \tilde{E}_K + \tilde{E}_W. \tag{6.1}$$

The kinetic energy density is given by

$$\tilde{E}_K = \frac{1}{2}\rho \dot{u}^2, \tag{6.2}$$

where ρ is the density and $\dot{u}$ is the velocity. This is analogous to $E = \frac{1}{2}mv^2$ from elementary physics. The potential energy density $\tilde{E}_W$ is also called *strain energy*

139

and results from the distortion of the material (the strain) working against a restoring force (the stress). From thermodynamic considerations (e.g., see Aki and Richards, 2002, pp. 22–3) it can be shown that

$$\tilde{E}_W = \frac{1}{2}\tau_{ij}e_{ij}, \tag{6.3}$$

where τ_{ij} and e_{ij} are the stress and strain tensors respectively.

Now consider a harmonic plane S wave propagating in the x direction with displacement in the y direction. We have

$$u_y = A\sin(\omega t - kx), \tag{6.4}$$

$$\dot{u}_y = A\omega\cos(\omega t - kx), \tag{6.5}$$

where A is the wave amplitude, ω is the angular frequency, and $k = \omega/\beta$ is the wavenumber where β is the shear velocity. The kinetic energy density is thus

$$\tilde{E}_K = \frac{1}{2}\rho A^2\omega^2\cos^2(\omega t - kx). \tag{6.6}$$

Since the mean value of $\cos^2$ is $1/2$, we may express the average kinetic energy density $\overline{E_K}$ as

$$\overline{E_K} = \frac{1}{4}\rho A^2\omega^2, \tag{6.7}$$

where the average may be taken over either t or x.

Recalling the expression for the strain tensor, $e_{ij} = \frac{1}{2}(\partial_i u_j + \partial_j u_i)$, the only non-zero strains for the S plane wave are

$$e_{12} = e_{21} = \frac{1}{2}\frac{\partial u_y}{\partial x} = -\frac{1}{2}Ak\cos(\omega t - kx). \tag{6.8}$$

Assuming the isotropic stress–strain relationship, the non-zero stresses are given by (2.30) and are

$$\tau_{12} = \tau_{21} = 2\mu e_{12} = -Ak\mu\cos(\omega t - kx). \tag{6.9}$$

Substituting into (6.3), we obtain the strain energy density

$$\tilde{E}_W = \frac{1}{2}A^2k^2\mu\cos^2(\omega t - kx) \tag{6.10}$$

and its average

$$\overline{E_W} = \frac{1}{4}A^2k^2\mu. \tag{6.11}$$

Using $k = \omega/\beta$ and $\beta^2 = \mu/\rho$ this becomes

$$\overline{E_W} = \frac{1}{4}\rho A^2 \omega^2, \tag{6.12}$$

which is seen to be identical to the equation for the kinetic energy (6.7). Similar expressions may be derived for a harmonic plane P wave; thus, in general

$$\overline{E_K} = \overline{E_W} = \frac{1}{4}\rho A^2 \omega^2 \tag{6.13}$$

and from (6.1) we get

$$\overline{E} = \frac{1}{2}\rho A^2 \omega^2. \tag{6.14}$$

The mean energy density is proportional to the square of both the amplitude and the frequency; for the same wave amplitude, higher frequency waves carry more energy. The energy flux density in the propagation direction per unit time per unit area (perpendicular to the wavefront) is given by

$$\tilde{E}^{\text{flux}} = \frac{1}{2}c\rho A^2 \omega^2, \tag{6.15}$$

where c is the wave velocity ($= \alpha$ for P waves, $= \beta$ for S waves).

To account for geometrical spreading in ray theory, consider a small patch of surface area dS_1 on a wavefront at time t_1. The total energy flux through this patch is given by

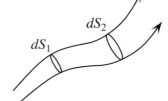

$$E^{\text{flux}}(t_1) = \frac{1}{2}c\rho A_1^2 \omega^2 \, dS_1. \tag{6.16}$$

As the wavefront propagates, the edges of this patch are bounded by the rays that define the local propagation direction.

Assuming that energy only travels along the rays (a high-frequency approximation), then the energy flux within this ray tube must remain constant, although the area of the patch may change. Thus at time t_2

$$E^{\text{flux}}(t_1) = E^{\text{flux}}(t_2) = \frac{1}{2}c\rho A_2^2 \omega^2 \, dS_2, \tag{6.17}$$

and from (6.16) and (6.17) we have, for constant c and ρ,

$$\frac{A_2}{A_1} = \sqrt{\frac{dS_1}{dS_2}}. \tag{6.18}$$

The amplitude varies inversely as the square root of the surface area of the patch bounded by the ray tube. Amplitudes decrease when the wavefront spreads out in area and increase if the wavefront is focused to a smaller area. For a spherical wave front, in which the surface area grows as r^2, the amplitude scales as $1/r$, consistent with our previous results for spherical waves (Section 3.6) and amplitudes predicted by the eikonal equation (Appendix C).

If the density ρ and wave speed c vary with position, then the amplitude will also change. In the absence of any geometrical spreading (i.e., $dS_1 = dS_2$), we then have

$$\frac{A_2}{A_1} = \sqrt{\frac{\rho_1 c_1}{\rho_2 c_2}}. \tag{6.19}$$

The product ρc is termed the *impedance* of the material; the amplitude varies inversely as the square root of impedance. This means that seismic amplitudes will increase as waves move into slower, less dense solids. This is an important factor in strong motion seismology where it is commonly observed that shaking from large earthquakes is more intense at sites on top of sediment compared with nearby sites on bedrock. Another important site amplification effect, not included in (6.19), involves resonance and reverberations within near-surface layers.

6.2 Geometrical spreading in 1-D velocity models

Now let us examine the amplitude of the ray ar-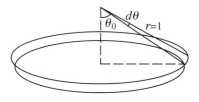
rivals in a one-dimensional Earth model. Suppose
the source is isotropic (uniform in all directions)
and that the total seismic energy radiated is E_s.
Let the surface slowness be u_0 and consider the
rays leaving the source between the angles θ_0 and
$\theta_0 + d\theta$. The energy between these angles leaves the source in a band around the unit sphere. The circumference of the band is $2\pi \sin \theta_0$ and its area is $2\pi \sin \theta_0 \, d\theta$. Since the unit sphere has a total area of 4π, the energy in the band, $E_{d\theta}$, may be expressed as

$$E_{d\theta} = \frac{1}{2} \sin \theta_0 \, d\theta \, E_s. \tag{6.20}$$

Next, consider where these rays strike the surface. The rays leaving the source between θ_0 and $\theta_0 + d\theta$ intersect the surface in a ring with area $2\pi X \, dX$. The corresponding area on the wavefront is $2\pi X \cos \theta_0 \, dX$ since the rays arrive at the surface at the angle θ_0 and the wavefronts are perpendicular to the rays. The energy

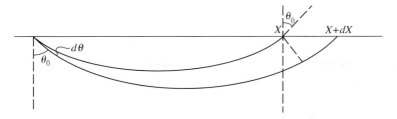

in the wavefront is given by the product of the wavefront area and the energy density $\tilde{E}$ on the wavefront:

$$E_{dX} = 2\pi X \cos\theta_0 \, dX \tilde{E}(X). \qquad (6.21)$$

From conservation of energy along the rays, we have $E_{d\theta} = E_{dX}$ and thus

$$\tilde{E}(X) = \frac{\sin\theta_0}{4\pi X \cos\theta_0} \left| \frac{d\theta}{dX} \right| E_{\text{s}}.$$

Substituting for $\sin\theta_0$ and $d\theta$ using $p = u_0 \sin\theta_0$ and $dp = u_0 \cos\theta_0 \, d\theta_0$, we have

$$\tilde{E}(X) = \frac{p}{4\pi u_0^2 X \cos^2\theta_0} \left| \frac{dp}{dX} \right| E_{\text{s}}. \qquad (6.22)$$

The amplitude at X is proportional to the square root of this expression. This equation assumes an isotropic source and a flat Earth; in the case of a spherical Earth, the corresponding equation for a source at radius r_1 and a receiver at radius r_2 is

$$\tilde{E}(\Delta) = \frac{p_{\text{sph}}}{4\pi u_1^2 r_1^2 r_2^2 \sin\Delta \cos\theta_1 \cos\theta_2} \left| \frac{dp_{\text{sph}}}{d\Delta} \right| E_{\text{s}}, \qquad (6.23)$$

where p_{sph} is the spherical Earth ray parameter, u_1 is the slowness at the source, Δ is the source–receiver range (in radians), and θ_1 and θ_2 are the incidence angles at source and receiver, respectively. It should be noted that amplitudes computed directly from (6.22) and (6.23) are based upon (6.18) and consider only geometrical spreading effects; impedance differences along the ray path will also affect the amplitudes and can be accounted for by applying (6.19).

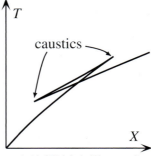

Equation (6.22) could also be written in terms of the factor $1/|dX/dp|$, illustrating that large amplitudes are predicted whenever dX/dp is small, so that a large number of rays (with different p values) all land close together. When $dX/dp = 0$ an energy catastrophe is predicted and the amplitudes become infinite at this singular point. These are called caustics in ray theory and occur at the cusps in the triplications in travel time curves. This result is only true in the limit of infinitely high frequency; at finite frequencies of interest the wavelengths will be long enough that they will tend to average over points in space where $dX/dp = 0$. Thus, the amplitudes will become large, but not infinite near caustics. In practice, the infinite amplitudes predicted at caustics with geometrical ray theory do not usually cause numerical problems since the caustics are single points along the travel time curves and the integrated energy remains finite.

6.3 Reflection and transmission coefficients

So far we have limited our analysis to models in which the velocity changes smoothly with depth – we have not considered what happens at velocity discontinuities. We now extend our discussion to include the complications due to reflected and transmitted waves at interfaces in the model. Specifically, we will compute the reflection and transmission coefficients that result from a plane wave incident on a horizontal interface. Recalling (3.35), we can write the general expression for plane wave displacement as

$$\mathbf{u}(\mathbf{x}, t) = \mathbf{f}(t - \mathbf{s} \cdot \mathbf{x}), \qquad (6.24)$$

where $\mathbf{s}$ is the slowness vector ($= u\hat{\mathbf{s}}$ where u is the slowness and $\hat{\mathbf{s}}$ is a unit vector in the direction of propagation) and $\mathbf{x} = (x, y, z)$ is the position. Now let us assume that velocity varies only in the z (vertical) direction and define our coordinate system such that z grows in the downward direction and $\mathbf{s}$ is in the xz plane. Recalling that the horizontal component of slowness is the ray parameter p and that the vertical slowness is $\eta = (u^2 - p^2)^{1/2}$, we can write

$$\mathbf{u} \;=\; \mathbf{f}(t - px - \eta z) \qquad \text{downgoing wave}, \qquad (6.25)$$

$$\;=\; \mathbf{f}(t - px + \eta z) \qquad \text{upgoing wave}, \qquad (6.26)$$

where we have restricted ourselves to waves that travel in the positive x direction. We often will work with harmonic waves, in which case we can express the downgoing wave as

$$\mathbf{u}(t) \;=\; \mathbf{A}(\omega) e^{-i\omega(t - px - \eta z)} \qquad (6.27)$$

$$\;=\; \mathbf{A}(\omega) e^{-i(\omega t - k_x x - k_z z)}, \qquad (6.28)$$

where $k_x = \omega p$ is termed the *horizontal wavenumber* and $k_z = \omega \eta$ is the *vertical wavenumber.*

For a vertically stratified medium, it can be shown (e.g., Aki and Richards, 2002, p. 220) that the wave equation separates into two types of solutions: a *P–SV* system in which the *S* waves are polarized in the vertical plane containing the slowness vectors (normally the vertical source–receiver plane) and an *SH* system in which the *S* waves are polarized in the horizontal direction. Within homogenous layers, the *P* and *SV* solutions are separate, but they are coupled at velocity discontinuities and within velocity gradients. The coupling within velocity gradients is stronger at low frequencies; in the ray theoretical approximation the frequencies are assumed to be sufficiently high that no coupling occurs. The *SH* polarized waves never couple with either *P* or *SV* for any laterally homogeneous medium.

6.3.1 *SH*-wave reflection and transmission coefficients

Let us now consider *SH*-wave propagation in a vertically stratified medium. The particle motion is perpendicular to the *xz* plane; thus $\mathbf{u} = (0, u_y, 0)$ and the displacement for downgoing waves within a homogeneous layer is

$$u_y = \grave{A} f(t - px - \eta z), \tag{6.29}$$

where $\grave{A}$ is the amplitude of the downgoing wave, p is the horizontal slowness, and η is the vertical slowness.

Now consider a discontinuous change in properties at a horizontal interface between two homogeneous layers, with a downgoing *SH* wave in the top layer. There will be both a reflected upgoing wave in the upper layer and a transmitted downgoing wave in the bottom layer. We will use the terms *incident* for the waves moving toward the interface and *scattered* for the waves moving away from the interface. We know from ray theory that the horizontal phase velocity p must be preserved at the interface; thus $\grave{\theta}_1 = \acute{\theta}_1$ and $u_1 \sin \grave{\theta}_1 = u_2 \sin \grave{\theta}_2$, where u_1 and u_2 are the layer slownesses. Thus, we know the ray angles of the scattered waves as a function of the angle of the incident wave. However, this does not give us any amplitude information. To obtain the amplitudes of the scattered waves, we need to consider the boundary conditions at the interface.

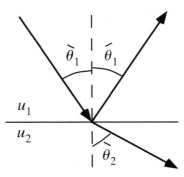

For a solid–solid interface, we must have continuity of displacement and traction across the interface. The displacement in layer 1 immediately above the interface is

$$u_y^+ = \grave{A}_1 f(t - px) + \acute{A}_1 f(t - px), \tag{6.30}$$

where $\acute{A}_1$ is the amplitude of the upgoing wave and we have assumed that the interface is at $z = 0$. The displacement in layer 2 immediately below the interface is

$$u_y^- = \grave{A}_2 f(t - px). \tag{6.31}$$

Since $u_y^+ = u_y^-$ from the continuity of displacement, we have

$$\grave{A}_1 + \acute{A}_1 = \grave{A}_2. \tag{6.32}$$

Assuming that the amplitude of the incident wave, $\grave{A}_1$, is known we now have one equation and two unknowns. We are still not able to express $\acute{A}_1$ or $\grave{A}_2$ entirely in terms of $\grave{A}_1$ and the material properties. To obtain another equation, we must consider the second boundary condition, the continuity of traction across the interface.[1]

Recalling the definition of the stress tensor (equation 2.1), we can write the traction across a horizontal plane (the plane normal to $\hat{\mathbf{z}}$ in our coordinate system) as

$$\mathbf{t}(\hat{\mathbf{z}}) = \begin{bmatrix} \tau_{xz} \\ \tau_{yz} \\ \tau_{zz} \end{bmatrix}. \tag{6.33}$$

Our equations for the *SH* plane wave have been entirely in terms of displacement **u**. Now recall (3.13), which relates stress to displacement for isotropic media:

$$\tau_{ij} = \lambda \delta_{ij} \partial_k u_k + \mu (\partial_i u_j + \partial_j u_i). \tag{6.34}$$

In our two-dimensional *SH* problem, $u_x = u_z = 0$ and $\partial/\partial y = 0$ since displacement is constant in the y direction. Thus, the first term drops out entirely, and the only non-zero components of τ_{ij} are:

$$\tau_{yx} = \tau_{xy} = \mu \frac{\partial u_y}{\partial x}, \tag{6.35}$$

$$\tau_{yz} = \tau_{zy} = \mu \frac{\partial u_y}{\partial z}. \tag{6.36}$$

Note that the components of $\boldsymbol{\tau}$ are related to the spatial derivatives of the displacement by the shear modulus μ. The continuity of $\mathbf{t}(\hat{\mathbf{z}})$ requires that τ_{yz} be continuous

---

[1] It is often misstated that *stress* must be continuous across the interface. This is incorrect. In general there will be components of the stress tensor τ_{ij} that are discontinuous across the interface. However, the force acting on the plane from above must be equal to the force acting on the plane from below – this is the continuity of traction boundary condition.

but not τ_{yx}. From (6.29) for the displacement of the downgoing *SH* wave, we can write

$$\grave{\tau}_{yz} = -\mu\eta\grave{A}f'(t - px - \eta z), \tag{6.37}$$

and for the upgoing wave, we obtain

$$\acute{\tau}_{yz} = \mu\eta\acute{A}f'(t - px + \eta z). \tag{6.38}$$

Immediately above the interface at $z = 0$, we have

$$\tau_{yz}^{+} = -(\grave{A}_1 - \acute{A}_1)\mu_1\eta_1 f'(t - px), \tag{6.39}$$

and immediately below the interface, we have

$$\tau_{yz}^{-} = -\grave{A}_2\mu_2\eta_2 f'(t - px). \tag{6.40}$$

Continuity of traction requires that $\tau_{yz}^{+} = \tau_{yz}^{-}$ and thus

$$(\grave{A}_1 - \acute{A}_1)\mu_1\eta_1 = \grave{A}_2\mu_2\eta_2. \tag{6.41}$$

Together with (6.29) we now have two equations that relate $\grave{A}_1$, $\acute{A}_1$, and $\grave{A}_2$. If we set the incident wave amplitude to unity ($\grave{A}_1 = 1$), we have two equations and two unknowns:

$$1 + \acute{A}_1 = \grave{A}_2, \tag{6.42}$$

$$(1 - \acute{A}_1)\mu_1\eta_1 = \grave{A}_2\mu_2\eta_2. \tag{6.43}$$

Substituting for $\grave{A}_2$ in the second equation, we have

$$(1 - \acute{A}_1)\mu_1\eta_1 = (1 + \acute{A}_1)\mu_2\eta_2,$$

$$\acute{A}_1(\mu_1\eta_1 + \mu_2\eta_2) = \mu_1\eta_1 - \mu_2\eta_2,$$

$$\acute{A}_1 = \frac{\mu_1\eta_1 - \mu_2\eta_2}{\mu_1\eta_1 + \mu_2\eta_2}, \tag{6.44}$$

and substituting this into (6.42), we have

$$\grave{A}_2 = 1 + \frac{\mu_1\eta_1 - \mu_2\eta_2}{\mu_1\eta_1 + \mu_2\eta_2}$$

$$= \frac{2\mu_1\eta_1}{\mu_1\eta_1 + \mu_2\eta_2}. \tag{6.45}$$

In seismology we will generally find it more convenient to work in terms of velocity, density, and incidence angle. From $\beta^2 = \mu/\rho$ and $\eta = u\cos\theta = \cos\theta/\beta$, where θ

is the ray angle from vertical, we may replace $\mu\eta$ in (6.44) and (6.45) with $\rho\beta\cos\theta$ and we have

$$\grave{S}\acute{S} \equiv \acute{A}_1 = \frac{\rho_1\beta_1\cos\theta_1 - \rho_2\beta_2\cos\theta_2}{\rho_1\beta_1\cos\theta_1 + \rho_2\beta_2\cos\theta_2}, \tag{6.46}$$

$$\grave{S}\grave{S} \equiv \grave{A}_2 = \frac{2\rho_1\beta_1\cos\theta_1}{\rho_1\beta_1\cos\theta_1 + \rho_2\beta_2\cos\theta_2}. \tag{6.47}$$

These are the standard expressions for the *SH*-wave *reflection* and *transmission coefficients*, where $\grave{S}\acute{S}$ is the reflection coefficient and $\grave{S}\grave{S}$ is the transmission coefficient (adopting the notation used in Aki and Richards, note that $\grave{S}\acute{S}$ is the reflection coefficient for downgoing *S* to upgoing *S*; it is not the product between $\grave{S}$ and $\acute{S}$). Before continuing, let us explore what these formulae mean. First, notice that they are a function not just of velocity but also of the density. This is something new. All of the travel times that we considered in geometric ray theory (and ray-theoretical amplitudes due to geometrical spreading) depended only upon the seismic velocities. Indeed, many types of seismic data are not sensitive to density. Reflection and transmission coefficients, if they can be measured inside Earth, provide a way to place constraints on density.

The *SH* formulae in (6.46) and (6.47) are fairly simple because there exist only two scattered waves, the reflected and transmitted *SH* pulses. The *P/SV* system is far more complicated, since it involves four types of waves – the downgoing and upgoing *P* and *SV* waves. For example, a *P*-wave incident on an interface will generate both *P* and *SV* reflections and transmitted waves (Fig. 6.1). The *P/SV*

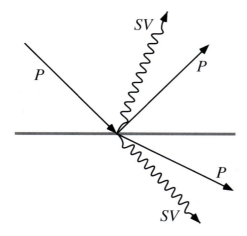

Figure 6.1 A non-vertical *P* wave striking a horizontal interface will generate four different scattered waves: the reflected and transmitted *P* and *SV* waves

coefficients can be derived in a similar way as the *SH* coefficients, but the algebra is fairly involved. A computer subroutine for computing the *P/SV* coefficients is listed in Appendix D.

6.3.2 Example: Computing *SH* coefficients

A downgoing *SH* wave with $p = 0.2$ s/km strikes a horizontal interface with $\beta_1 = 3.2$ km/s, $\rho_1 = 2.6$ Mg/m³ in the top layer and $\beta_2 = 3.9$ km/s, $\rho_2 = 2.9$ Mg/m³ in the bottom layer. What are the amplitudes of the reflected and transmitted waves? From the definition of the ray parameter ($p = u \sin \theta$) and $u_1 = 1/\beta_1 = 0.3125$ s/km and $u_2 = 1/\beta_2 = 0.2564$ s/km, we have $\theta_1 = 39.8°$ and $\theta_2 = 51.3°$. From (6.46) and (6.47), we have

$$\grave{S}\acute{S} = \frac{2.6 \cdot 3.2 \cos 39.8° - 2.9 \cdot 3.9 \cos 51.3°}{2.6 \cdot 3.2 \cos 39.8° + 2.9 \cdot 3.9 \cos 51.3°} = -0.05$$

$$\grave{S}\grave{S} = \frac{2 \cdot 2.6 \cdot 3.2 \cos 39.8°}{2.6 \cdot 3.2 \cos 39.8° + 2.9 \cdot 3.9 \cos 51.3°} = 0.95$$

Note that we don't need to worry about the units (as long as they are consistent in the different terms) because they cancel out in these equations. Assuming that the incident *SH* wave has unit amplitude, the upgoing reflected wave has amplitude -0.05 and the downgoing transmitted wave has amplitude 0.95. The signs of the coefficients indicate that the incident wave polarity is flipped for the reflected wave and preserved for the transmitted wave.

6.3.3 Vertical incidence coefficients

For the case of vertical incidence ($\theta = 0$), the *SH* coefficients are

$$\grave{S}\acute{S}_{\text{vert}} = \frac{\rho_1\beta_1 - \rho_2\beta_2}{\rho_1\beta_1 + \rho_2\beta_2}, \tag{6.48}$$

$$\grave{S}\grave{S}_{\text{vert}} = \frac{2\rho_1\beta_1}{\rho_1\beta_1 + \rho_2\beta_2}. \tag{6.49}$$

There is no distinction between *SH* and *SV* at vertical incidence, and therefore these expressions are also valid for *SV*. For vertical *P* waves there is no conversion from *P* to *SV* and the *P*-wave coefficients also have a simple form:

$$\grave{P}\acute{P}_{\text{vert}} = -\frac{\rho_1\alpha_1 - \rho_2\alpha_2}{\rho_1\alpha_1 + \rho_2\alpha_2}, \tag{6.50}$$

$$\grave{P}\grave{P}_{\text{vert}} = \frac{2\rho_1\alpha_1}{\rho_1\alpha_1 + \rho_2\alpha_2}. \tag{6.51}$$

The product ρc, where c is the velocity, is the impedance. Note that the reflection coefficient is half of the impedance contrast; for example, a 10% change in $\rho\beta$ (relative to the mean $\rho\beta$) results in a reflection coefficient of 5%. The difference in the sign of the reflection coefficient between P and SH results from the fact that the polarity of SH is defined independently of the ray direction, whereas P polarities are relative to the ray direction, which changes sign upon the reflection.

At near-vertical incidence, the reflection coefficient is negative for SH and positive for P when the impedance increase is positive. An example of this would be a reflection off the top of the Moho. The reflected pulse is phase reversed for SH relative to the incident pulse (Fig. 6.2). In contrast, for underside reflections off the Moho, it is the P reflection that experiences the polarity reversal. The Earth's surface represents a different boundary condition; in this case all the components of the traction vector are zero since virtually no force is exerted by the air on the ground. For upgoing waves incident on a free surface, there is no transmitted pulse and the reflection coefficient is 1 for SH and -1 for P. The transmission coefficient is greater than one for a near-vertical SH wave that propagates into a layer with smaller impedance. For example if $\rho_1\beta_1 = (1.1)\rho_2\beta_2$, then $\grave{A}_2 = 2.2/2.1 \simeq 1.05$. This example highlights the difference at an interface between amplitude (which need not be conserved) and energy (which always must be conserved).

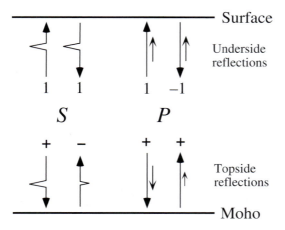

Figure 6.2 Examples of the polarity changes for near-vertical P and S waves at horizontal interfaces. When the impedance decreases, as for underside reflections off the free surface, the P wave flips polarity whereas the S-wave polarity is unchanged. In contrast, when the impedance increases, as for topside reflections off the Moho, the reflected S pulse is of opposite polarity, while the P pulse is the same polarity as the incident wave. This difference between S and P results from the fact that P-wave particle motions are measured relative to the ray direction; the small arrows indicate the first motion of the P waves.

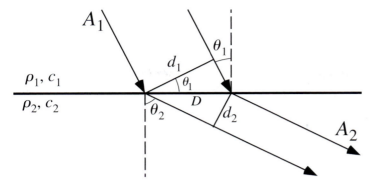

Figure 6.3 The ray and wavefront geometry for a plane wave crossing an interface. In this case, the velocity is higher in the lower layer.

6.3.4 Energy-normalized coefficients

It is sometimes more convenient to work with energy-normalized reflection and transmission coefficients. Consider an incident wave of amplitude A_1 in layer 1 and a scattered wave of amplitude A_2 in layer 2 (Fig. 6.3). The energy flux per unit wavefront area $\tilde{E}_1^{\text{flux}}$ in the incident wave is given by (6.15):

$$\tilde{E}_1^{\text{flux}} = \frac{1}{2}c_1\rho_1 A_1^2 \omega^2, \tag{6.52}$$

where c_1 is the velocity (either P or S) and ρ_1 is the density in layer 1. The incident energy flux density on the interface, $\tilde{E}_1^{\text{flux}}$, is reduced by the angle of the wave,

$$\tilde{E}_1^{\text{flux}} = \frac{1}{2}c_1\rho_1 A_1^2 \omega^2 \cos\theta_1, \tag{6.53}$$

where θ_1 is the incident angle of the incoming wave. Similarly, the energy flux density on the interface from the scattered wave A_2 is given by

$$\tilde{E}_2^{\text{flux}} = \frac{1}{2}c_2\rho_2 A_2^2 \omega^2 \cos\theta_2. \tag{6.54}$$

The sum of the energy flux density on the interface from all of the scattered waves must equal the energy flux density from the incident waves. It is therefore convenient to define

$$A_{\text{norm}} = \sqrt{\frac{\tilde{E}_2^{\text{flux}}}{\tilde{E}_1^{\text{flux}}}} = \frac{A_2}{A_1}\sqrt{\frac{\rho_2 c_2 \cos\theta_2}{\rho_1 c_1 \cos\theta_1}} = A_{\text{raw}}\sqrt{\frac{\rho_2 c_2 \cos\theta_2}{\rho_1 c_1 \cos\theta_1}}, \tag{6.55}$$

where A_{norm} is the *energy-normalized* reflection/transmission coefficient and A_{raw} is the ratio of the scattered amplitude to the incident amplitude (e.g., for S waves, $\grave{S}\acute{S}$ in 6.47). With this definition, the sum of the squares of A_{norm} for all the scattered waves will be one.

For the *SH*-wave reflection and transmission coefficients (6.46, 6.47), the energy-normalized equations are:

$$\grave{S}\acute{S}_{norm} = \frac{\rho_1\beta_1\cos\theta_1 - \rho_2\beta_2\cos\theta_2}{\rho_1\beta_1\cos\theta_1 + \rho_2\beta_2\cos\theta_2} = \grave{S}\acute{S}_{raw}, \qquad (6.56)$$

$$\grave{S}\grave{S}_{norm} = \frac{(\rho_2\beta_2\cos\theta_2)^{1/2}}{(\rho_1\beta_1\cos\theta_1)^{1/2}} \frac{2\rho_1\beta_1\cos\theta_1}{\rho_1\beta_1\cos\theta_1 + \rho_2\beta_2\cos\theta_2}. \qquad (6.57)$$

Notice that the energy-normalized *SH* reflection coefficient is unchanged since $\rho c \cos\theta$ is equal for the incident and reflected wave. This is also true in the P/SV system for reflected waves of the same type as the incident wave (e.g., $\grave{P}\acute{P}$, $\grave{S}\acute{S}$), but it is not true when the wave type changes (e.g., $\grave{P}\acute{S}$, $\grave{S}\acute{P}$).

6.3.5 Dependence on ray angle

Let us now explore how the reflection and transmission coefficients change as a function of ray angle. To motivate the discussion, we will consider a specific example – the response of the Moho to downgoing *SH* waves. We take as our starting point the PREM velocity and density values at the Moho: $\rho_1 = 2.9$, $\rho_2 = 3.38$, $\beta_1 = 3.9$, $\beta_2 = 4.49$ with density in g/cm^3 and velocity in km/s. Note that $\rho_1\beta_1 = 11.31$, $\rho_2\beta_2 = 15.18$, and that the increase in $\rho\beta$ across the Moho is about 30% (relative to the mean $\rho\beta$).

Using equations (6.46) and (6.47) we can compute the reflection and transmission coefficients as a function of ray angle for the incident wave. These results are plotted in Figure 6.4. First, consider the results for near-vertical incidence ($\theta = 0°$). Here the reflected wave has an amplitude of -0.146 (as predicted by the 30% impedance change), while the transmitted wave amplitude is 0.854. Note that energy flux is conserved since $1^2(11.31) \simeq (-0.146)^2(11.31) + 0.854^2(15.18)$.

We also plot the phase of the reflected and transmitted waves; at vertical incidence the reflected pulse has experienced a π phase shift (180°) while the transmitted pulse is unaltered. The reflection and transmission coefficients for $\theta < 30°$ change very little and are well approximated by the vertical incidence values. As θ increases to larger values and the incident ray becomes more horizontal, the amplitude of the transmitted wave increases and the reflected amplitude approaches zero. At

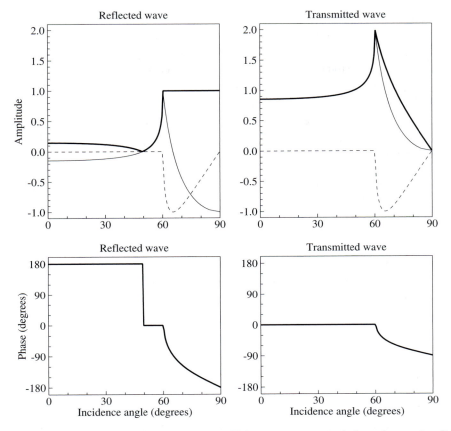

Figure 6.4 Reflection and transmission coefficients versus ray angle for a downgoing *SH*-wave incident on the Moho. In the top plots, the real part of the reflection coefficient is shown with a thin solid line, the imaginary part with a dashed line, and the magnitude with a heavy line. The lower plots show the change in the phase angle for a harmonic wave. The sign of the imaginary part of the reflection coefficients plotted here assumes that a phase shift of $-90°$ represents a $\pi/2$ phase advance (see text).

an incidence angle near 49°, the reflected amplitude is zero and the transmitted amplitude is one. Despite crossing a significant change in density and velocity, there is no reflected wave for an incident ray at this angle.

Beyond 49°, the amplitude of the transmitted wave continues to increase. This amplitude increase in the transmitted wave results from the increasingly horizontal orientation of the transmitted ray. Near-horizontal rays contribute much less to the vertical energy flux than near-vertical rays of the same amplitude. Thus, to balance the energy flux, the amplitude of the near-horizontal rays must increase.

The reflected wave is now positive in amplitude (no phase shift) and grows as θ increases in order to balance the displacements on both sides of the Moho.

This continues as θ approaches the *critical angle* (near $60°$ in our example), where the transmitted ray is horizontal. At the critical angle, the transmitted *SH* wave has an amplitude of two and the reflected wave amplitude is one. All reflections at angles less than the critical angle are termed *precritical reflections*. To go beyond the critical angle, we face a problem in our calculations. This arises from the $\cos\theta_2$ term in the denominator of (6.46). Beyond the critical angle at which $\theta_2 = 90°$ there is no transmitted ray in the lower layer. How do we handle this situation? Recall that the vertical slowness is given by

$$\eta = u\cos\theta = \sqrt{u^2 - p^2}. \tag{6.58}$$

When $u = p$, then $\theta = 90°$ and the ray is horizontal. When $p > u$, the vertical slowness becomes imaginary. It turns out that waves with imaginary vertical slownesses do exist and satisfy the wave equation (see Aki and Richards, 2002, p. 149). They are termed *inhomogeneous* or *evanescent* waves and their amplitude grows or decays with depth. They have no vertical energy flux, and their energy-normalized transmission coefficient is zero. In contrast, the *homogeneous waves* with real vertical slownesses that we have studied so far have oscillatory behavior with depth and non-zero values of vertical energy flux.

Our formulae for reflection and transmission coefficients remain valid for imaginary vertical slowness, provided we use

$$\cos\theta = \sqrt{1 - p^2c^2}, \tag{6.59}$$

where c is the velocity for the appropriate wave type. Past the critical angle, both the reflected and transmitted amplitude coefficients become complex. No energy is transmitted downward in the lower layer; all of the energy is reflected upward ($|\grave{S}\acute{S}| = 1$). This is termed *total internal reflection*. Unlike precritical reflections, the shape of the reflected pulse at postcritical angles is distorted. This arises from multiplication with a complex reflection coefficient. In the frequency domain this is equivalent to multiplying by $e^{i\theta}$ where θ is the phase of $\grave{S}\acute{S}$, introducing a frequency-independent phase shift. These phase shifts are plotted as a function of ray angle in Figure 6.4. A phase shift of π or $-\pi$ is simply a polarity reversal. However, intermediate phase shifts between 0 and π and between 0 and $-\pi$ cause pulse distortion.

A reflection coefficient of $-i$ corresponds to a phase advance of $\pi/2$ and produces what is called the *Hilbert transform* (see Appendix E). The Hilbert transform of a delta function looks like the diagram at the top of the next page. It can be shown that any phase shift can be expressed as a linear combination of a function and its Hilbert transform (e.g., see Aki and Richards, 2002, Section 5.3).

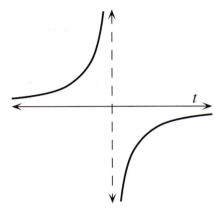

A complication that arises when reflection and transmission coefficients are complex is that the sign of the imaginary part will vary depending upon the sign convention used for the Fourier transform. Here we assume that the time series is obtained through an integral over $e^{-i\omega t}$ (see Appendix E), and this produces the $-i$ coefficient for the $\pi/2$ phase advance of the Hilbert transform. As noted by Aki and Richards (2002, p. 151), the term *phase advance* is more fundamental and less ambiguous than the more commonly used "phase shift" because of the sign differences in phase shift definitions. It's easy to check which sign convention is appropriate for a particular set of subroutines by computing the inverse Fourier transform of $-i$. If the result resembles the Hilbert transform, then the $e^{-i\omega t}$ convention applies; if the result looks polarity reversed, then the $e^{i\omega t}$ convention applies and a coefficient of $+i$ is required to yield the Hilbert transform.

Now consider the travel time curve that results from the *SH* Moho reflection for a simple model of the crust and upper mantle (Fig. 6.5). The reflected arrivals form a retrograde branch. The critical angle occurs at the caustic where the *Sn* branch emerges. At closer ranges the reflections are precritical and the amplitudes are much

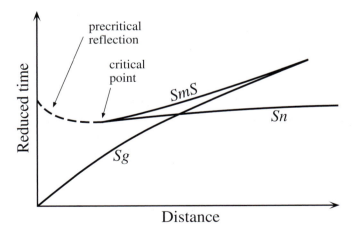

Figure 6.5 The different arrivals for *S* waves in the crust and upper mantle.

smaller. At vertical incidence the pulse is phase reversed. At some point along the precritical branch the reflection coefficient goes to zero and then increases rapidly to 1 at the critical distance. Along this part of the reflected branch the pulse is not phase reversed. The pulse shape is distorted along the postcritical branch, changing gradually from a phase shift of 0 at the critical point to $-\pi$ (phase reversed) at the far caustic.

We have gone into this example in some detail in order to illustrate the range of behavior that occurs with reflection and transmission coefficients. Other boundary conditions will lead to different behavior but many of the general concepts remain the same. The reflection and transmission coefficients are all real provided the ray angles for the waves are real. However, as soon as at least one of the scattered waves becomes inhomogeneous, then all of the coefficients become complex.

6.4 Turning points and Hilbert transforms

We now make a brief digression to consider some of the subtleties involved in plane waves and ray theoretical versus homogeneous layer modeling. In our previous discussions of plane waves, we have considered only homogeneous models in which wavefronts are planes extending to infinity or models with an interface between two uniform half spaces, in which case the wavefronts have a kink at the interface but are planar on either side. We now generalize the idea of a plane wave to include models with vertical velocity gradients. In this case the plane wave is defined by the wavefronts orthogonal to the family of rays with constant horizontal slowness (the ray parameter p). In regions of velocity gradients the rays will curve and the wavefronts will no longer be planar (Fig. 6.6); however, we will continue to use the term "plane wave" since the wavefronts are locally planar in any homogeneous part of the model. Monochromatic plane waves are thus waves at constant frequency and horizontal wavenumber ($k_x = \omega p$). In order to solve the wave equation for laterally homogeneous models, it is often convenient to first transform the equations into the (ω, k_x) domain from the (X, t) domain. The solution at the surface is then expressed as a sum of plane waves at different values of ω and k_x.

Figure 6.6 Waves of constant ray parameter turning within a vertical velocity gradient.

Now consider a vertical velocity gradient that is approximated by a staircase model with very fine layer spacing. We wish to consider what happens at the turning point for an *SH* plane wave in such a model. A downgoing *SH* plane wave with ray parameter p will be totally reflected at the top of the layer in which the velocity exceeds $\beta = 1/p$.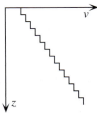

Let the velocity jump at this interface be from $\beta^- = \beta - d\beta$ to $\beta^+ = \beta + d\beta$ and the density jump to be from $\rho^- = \rho - d\rho$ to $\rho^+ = \rho + d\rho$. The *SH* reflection coefficient (6.46) may then be expressed as

$$\grave{S}\acute{S} = \frac{\rho^-\beta^-\sqrt{1 - \frac{1}{\beta^2}(\beta - d\beta)^2} - \rho^+\beta^+\sqrt{1 - \frac{1}{\beta^2}(\beta + d\beta)^2}}{\rho^-\beta^-\sqrt{1 - \frac{1}{\beta^2}(\beta - d\beta)^2} + \rho^+\beta^+\sqrt{1 - \frac{1}{\beta^2}(\beta + d\beta)^2}}, \qquad (6.60)$$

where we have used $\cos\theta = \sqrt{1 - p^2\beta^2}$ from (6.59).

Evaluating this expression as the layer spacing becomes infinitely fine $(d\rho, d\beta \to 0)$, we may ignore terms of $(d\beta)^2$, and thus

$$\sqrt{1 - \frac{1}{\beta^2}(\beta - d\beta)^2} = \sqrt{1 - \frac{1}{\beta^2}(\beta^2 - 2\beta d\beta)}$$

$$= \sqrt{\frac{2d\beta}{\beta}}$$

and

$$\sqrt{1 - \frac{1}{\beta^2}(\beta + d\beta)^2} = \sqrt{\frac{-2d\beta}{\beta}}.$$

Ignoring higher-order terms in $d\rho$ and $d\beta$, we have

$$\grave{S}\acute{S} = \frac{\rho\beta\sqrt{\frac{2d\beta}{\beta}} - \rho\beta\sqrt{\frac{-2d\beta}{\beta}}}{\rho\beta\sqrt{\frac{2d\beta}{\beta}} + \rho\beta\sqrt{\frac{-2d\beta}{\beta}}}$$

$$= \frac{1 - i}{1 + i}$$

$$= -i. \qquad (6.61)$$

This result for a staircase model suggests that plane waves that turn in a vertical velocity gradient should experience a $\pi/2$ phase advance at the turning point. In fact, this is the case and can be shown more formally using *WKBJ* theory (Aki and

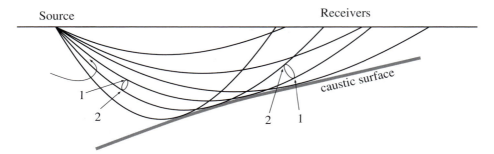

Figure 6.7 Along a retrograde branch of a travel time curve, the rays turn back sharply toward the source. The crossing of the ray paths forms an internal caustic surface that produces a $\pi/2$ phase advance in the waveforms (diagram after Choy and Richards, 1975).

Richards, 2002, pp. 434–6). This is a high frequency approximation that is not valid near the turning point itself, but can be used to connect the downgoing and upgoing plane-wave solutions some distance above the turning point depth.

Because of the $\pi/2$ phase advance, the upgoing plane wave is the Hilbert transform of the downgoing wave and its pulse shape is altered. Why does this happen? Some insight may be obtained by considering what happens to the wavefront defined by adjacent rays in a small ray tube near the turning point (Choy and Richards, 1975). The ray paths cross and the wavefront folds over itself at the turning point. This is termed an *internal caustic surface* in ray theory and also occurs along retrograde branches for rays from a point source (Fig. 6.7). The folding of the wavefront along the caustic surface is what produces the $\pi/2$ phase advance.

Generally the rays from a point source do not cross at the turning point and no phase shift occurs. How can we reconcile this result with the $\pi/2$ phase advance at the turning point for the plane wave? The answer involves the fact that adjacent rays in the case of the point source have different ray parameters, unlike the constant ray parameter that prevails for the plane wave. To obtain the point source solution from a sum of plane waves of different ray parameters, we must perform an inverse transform from the (ω, k_x) domain to the (t, X) domain. The inverse transformation involves a $\pi/2$ phase shift that cancels the $-\pi/2$ phase shift produced at the plane wave turning points.

Although in most cases a point source does not produce an internal caustic surface, some ray geometries will produce such a result. Examples of this include retrograde branches from a steep velocity gradient and surface reflected phases such as PP (Fig. 6.8). In each case, adjacent ray paths cross themselves and a $\pi/2$ phase shift occurs. For example, PP is observed to be the Hilbert transform of P (plus another phase shift of π that occurs at the surface reflection point).

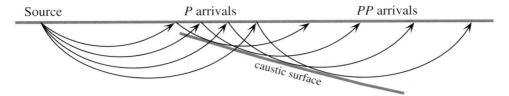

Figure 6.8 Ray paths for the surface reflected phase *PP* also form an internal caustic; this causes *PP* to be Hilbert transformed (as well as of opposite polarity owing to the surface reflection) relative to direct *P*.

In the case of a steep velocity gradient producing a retrograde branch, ray theory predicts that the arrivals along this branch will be Hilbert transformed compared to the prograde branches. However, this result is only valid provided the velocity gradient is not too steep. In the limit, an infinitely steep velocity gradient becomes a step discontinuity. As we have seen (see Fig. 6.4), the post critical reflection coefficients for such an interface involve a continuous change in phase with ray angle, rather than a constant $\pi/2$ phase shift.

6.5 Matrix methods for modeling plane waves[†]

Our emphasis for much of this book has been motivated by understanding wave propagation in terms of simple ray theory. However, many problems are more easily addressed when the wave field is decomposed into plane waves (an example would be computing the effect of resonance within a near-surface layer), and powerful techniques have been developed for calculating the plane wave response of horizontally layered models. Here we provide an introduction to a method for modeling plane wave propagation in which we set up a system of equations for displacement and stress for the different wave types. This method is quite general and is widely used in seismology to model both body- and surface-wave propagation.

Once again we will consider the *SH* system, but the method is easily generalized to the *P/SV* system. The technique is almost always set up in the frequency domain. The displacement due to *SH* waves at a given frequency is

$$u_y = \grave{S}e^{-i\omega(t-px-\eta z)} = \grave{S}e^{-i\omega(t-px)}e^{i\omega\eta z} \quad \text{(downgoing)}, \quad (6.62)$$

$$u_y = \acute{S}e^{-i\omega(t-px+\eta z)} = \acute{S}e^{-i\omega(t-px)}e^{-i\omega\eta z} \quad \text{(upgoing)}, \quad (6.63)$$

where z points downward and we are considering only waves propagating in the $+x$ direction. We will assume that ω and p are fixed; our solution is for steady state, monochromatic plane waves. Note that the depth dependence can be separated from the $e^{-i\omega(t-px)}$ term in each case. Next we define a vector, the *stress-displacement vector* or *Haskell vector*, that contains the displacements and stress components that must be continuous with depth:

$$\mathbf{H}(z) \equiv \begin{bmatrix} u_y \\ \tau_{yz} \end{bmatrix} = \mathbf{f}(z)e^{-i\omega(t-px)}, \qquad (6.64)$$

where the $\mathbf{f}$ vector contains the depth-dependent part of $\mathbf{H}$. Recalling our expression for τ_{yz} (6.36), we have

$$\tau_{yz} = \mu\frac{\partial u_y}{\partial z} = +i\omega\mu\eta u_y \quad \text{(downgoing)}, \qquad (6.65)$$

$$= -i\omega\mu\eta u_y \quad \text{(upgoing)}, \qquad (6.66)$$

and thus we can express $\mathbf{f}$ as

$$\mathbf{f}(z) = \begin{bmatrix} \grave{S}e^{i\omega\eta z} + \acute{S}e^{-i\omega\eta z} \\ \grave{S}i\omega\mu\eta e^{i\omega\eta z} - \acute{S}i\omega\mu\eta e^{-i\omega\eta z} \end{bmatrix}$$

$$= \begin{bmatrix} e^{i\omega\eta z} & e^{-i\omega\eta z} \\ i\omega\mu\eta e^{i\omega\eta z} & -i\omega\mu\eta e^{-i\omega\eta z} \end{bmatrix} \begin{bmatrix} \grave{S} \\ \acute{S} \end{bmatrix} \qquad (6.67)$$

$$\equiv \mathbf{Fw}. \qquad (6.68)$$

where $\mathbf{F}$ is termed the *solution matrix* or *layer matrix*, and the $\mathbf{w}$ vector contains the amplitudes of the downgoing and upgoing waves. The matrix $\mathbf{F}$ may be factored as

$$\mathbf{F}(z) = \begin{bmatrix} 1 & 1 \\ i\omega\mu\eta & -i\omega\mu\eta \end{bmatrix} \begin{bmatrix} e^{i\omega\eta z} & 0 \\ 0 & e^{-i\omega\eta z} \end{bmatrix} \qquad (6.69)$$

$$\equiv \mathbf{E}\mathbf{\Lambda}. \qquad (6.70)$$

Now let us explore some of the power of our new notation. Applying boundary conditions at a solid–solid interface is simple. First we match displacement and traction at the interface between layer 1 and layer 2:

$$\mathbf{H}_1 = \mathbf{H}_2. \qquad (6.71)$$

Next we use (6.64) to obtain

$$\mathbf{f}_1 = \mathbf{f}_2 \qquad (6.72)$$

after dividing through by the common factor of $e^{-i\omega(t-px)}$. From (6.67)–(6.70) we may write

$$\begin{bmatrix} 1 & 1 \\ \mu_1\eta_1 & -\mu_1\eta_1 \end{bmatrix} \begin{bmatrix} \grave{S}_1 \\ \acute{S}_1 \end{bmatrix} = \begin{bmatrix} 1 & 1 \\ \mu_2\eta_2 & -\mu_2\eta_2 \end{bmatrix} \begin{bmatrix} \grave{S}_2 \\ \acute{S}_2 \end{bmatrix} \qquad (6.73)$$

after setting $z = 0$ so that $\boldsymbol{\Lambda} = \mathbf{I}$. To compute reflection and transmission coefficients we need to rearrange these equations to separate the incident and scattered waves. The result is

$$\begin{bmatrix} 1 & -1 \\ -\mu_1\eta_1 & -\mu_2\eta_2 \end{bmatrix} \begin{bmatrix} \acute{S}_1 \\ \grave{S}_2 \end{bmatrix} = \begin{bmatrix} -1 & 1 \\ -\mu_1\eta_1 & -\mu_2\eta_2 \end{bmatrix} \begin{bmatrix} \grave{S}_1 \\ \acute{S}_2 \end{bmatrix}. \qquad (6.74)$$

We then have

$$\begin{bmatrix} \acute{S}_1 \\ \grave{S}_2 \end{bmatrix} = \begin{bmatrix} 1 & -1 \\ -\mu_1\eta_1 & -\mu_2\eta_2 \end{bmatrix}^{-1} \begin{bmatrix} -1 & 1 \\ -\mu_1\eta_1 & -\mu_2\eta_2 \end{bmatrix} \begin{bmatrix} \grave{S}_1 \\ \acute{S}_2 \end{bmatrix}$$

$$= \mathbf{R} \begin{bmatrix} \grave{S}_1 \\ \acute{S}_2 \end{bmatrix} \qquad (6.75)$$

$$= \begin{bmatrix} \grave{S}\acute{S} & \acute{S}\acute{S} \\ \grave{S}\grave{S} & \acute{S}\grave{S} \end{bmatrix} \begin{bmatrix} \grave{S}_1 \\ \acute{S}_2 \end{bmatrix}, \qquad (6.76)$$

where $\mathbf{R}$ is the matrix of reflection and transmission coefficients. It is easily verified that the coefficients obtained are identical to our previous expressions (6.46) and (6.47).

Next, let us see how we can propagate a solution to a different depth. Within a homogeneous layer, the amplitude factors $\mathbf{w}$ are constant. At depth z_1 we have

$$\mathbf{f}(z_1) = \mathbf{F}_1\mathbf{w} \qquad \longrightarrow \qquad \mathbf{w} = \mathbf{F}_1^{-1}\mathbf{f}(z_1). \qquad (6.77)$$

At depth z_2 we have

$$\begin{aligned} \mathbf{f}(z_2) &= \mathbf{F}_2\mathbf{w} = \mathbf{F}_2\mathbf{F}_1^{-1}\mathbf{f}(z_1) \\ &= \mathbf{E}\boldsymbol{\Lambda}_2(\mathbf{E}\boldsymbol{\Lambda}_1)^{-1}\mathbf{f}(z_1) = \mathbf{E}\boldsymbol{\Lambda}_2\boldsymbol{\Lambda}_1^{-1}\mathbf{E}^{-1}\mathbf{f}(z_1) \\ &= \mathbf{E}\begin{bmatrix} e^{i\omega\eta(z_2-z_1)} & 0 \\ 0 & e^{-i\omega\eta(z_2-z_1)} \end{bmatrix}\mathbf{E}^{-1}\mathbf{f}(z_1) \\ &\equiv \mathbf{P}(z_1, z_2)\mathbf{f}(z_1). \end{aligned} \qquad (6.78)$$

The matrix $\mathbf{P}$ is termed the *propagator matrix* because it propagates the solution from z_1 to z_2. In this way the displacements and stresses at the top of a homogeneous layer may be propagated to the bottom of the layer. Because the displacements and

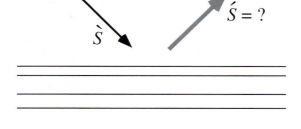

$\grave{S}$

$\acute{S} = ?$

Figure 6.9 A downgoing *SH*-wave incident on a stack of homogeneous layers will generate upgoing scattered *SH* waves that are a product of all of the internal reflections and reverberations within the stack.

stresses must be continuous at an interface between layers, the solution may be propagated downward (or upward) through a series of layers by using a different propagator for each layer. Note that this does not require the use of reflection and transmission coefficients at each interface – it is the displacements and stresses that are being propagated, not the wave amplitudes.

This sounds too easy; doesn't it? Why can't we use this method to compute synthetic seismograms in layer-cake models without any fuss? The answer lies in the fact that a complete knowledge of **H** at any depth in the model implies that we know both the upgoing and downgoing wave amplitudes:

$$\mathbf{H} = \begin{bmatrix} u_y \\ \tau_{yz} \end{bmatrix} = \mathbf{F}\mathbf{w}e^{-i\omega(t-px)} = \mathbf{F}\begin{bmatrix} \grave{S} \\ \acute{S} \end{bmatrix} e^{-i\omega(t-px)}. \tag{6.79}$$

Since the layer matrix **F** is a fixed function of the layer properties, **w** can be computed from **H** and vice versa. But we don't normally know both the upgoing and downgoing wave amplitudes. For example, we might want to compute the response of a stack of layers to a downgoing plane wave (Fig. 6.9). We know the amplitude, $\grave{S}$, of the downgoing *SH* wave at $z = z_{\text{ref}}$. The problem is determining the amplitude of the upgoing wave, $\acute{S}$, at the same depth, since the stack will produce a complicated series of reflections and reverberations. Once $\acute{S}(z_{\text{ref}})$ is known, the problem of determining the displacements and stresses, **H**, or the wave amplitudes, **w**, at any other depth in the stack is essentially solved.

Fortunately, there are methods for determining the complete response of a series of layers. A technique using the propagator matrix was developed by Thomson and Haskell in the 1950s and has been used extensively to model surface waves. Another method computes generalized reflection and transmission coefficients for the entire stack of layers and forms the basis for *reflectivity* synthetic seismogram algorithms (e.g., Fuchs and Müller, 1971; Kennett, 1974, 1983; Mallick and Frazer, 1987). Note that the equations in this section apply to harmonic plane waves with a specific frequency and ray parameter. By solving for the plane-wave response over a variety of values of ω and p, one obtains a complex frequency-wavenumber spectrum, which can then be integrated over frequency and ray parameter to compute synthetic seismograms (which are a function of time at specific distances). The frequency integral is simply an inverse Fourier transform ($\omega \rightarrow t$), but the integral over ray parameter can be quite tricky to perform and much of the challenge in writing reflectivity codes comes from computing this integral.

Our treatment in this section has been for SH waves. However, all of the methods and notation are readily generalized to P/SV, in which case

$$\mathbf{w} = \begin{bmatrix} \grave{P} \\ \grave{S} \\ \acute{P} \\ \acute{S} \end{bmatrix}, \quad \mathbf{H} = \begin{bmatrix} u_x \\ u_z \\ \tau_{xz} \\ \tau_{zz} \end{bmatrix}. \tag{6.80}$$

Note that for P/SV the displacement is confined to the xz plane ($\mathbf{u} = (u_x, 0, u_z)$) and hence $\tau_{yz} = \mu \partial u_y / \partial z = 0$. Chapter 5 of Aki and Richards (2002) contains a complete treatment of the P/SV equations, including expressions for the reflection/transmission coefficients and the 4×4 matrix $\mathbf{E}$.

As a final note, the results for the matrix methods can also be derived from formulations that begin by casting the problem in the form

$$\frac{\partial \mathbf{f}}{\partial z} = \mathbf{A}\mathbf{f}, \tag{6.81}$$

where $\mathbf{f}$ is defined as above and the $\mathbf{A}$ matrix depends only upon the elastic properties of the medium, p and ω. The elements of $\mathbf{A}$ can be obtained directly from the wave equation and the stress–strain relationship (Aki and Richards contains complete expressions for $\mathbf{A}$). Solutions to coupled first-order differential equations such as (6.81) are obtained by computing the eigenvalues and eigenvectors of $\mathbf{A}$. The eigenvalues turn out to be related to the vertical slownesses η for the different wave types; the eigenvectors are related to the columns in the layer matrix $\mathbf{F}$.

6.6 Attenuation

So far we have considered the changes in seismic wave amplitude that result from geometrical spreading of wavefronts and the reflection and transmission coefficients that occur at discontinuities. A third factor that affects amplitudes is energy loss due to anelastic processes or internal friction during wave propagation. This *intrinsic attenuation* may be distinguished from *scattering attenuation*, in which amplitudes in the main seismic arrivals are reduced by scattering off small-scale heterogeneities, but the integrated energy in the total wavefield remains constant. The word *attenuation* is also sometimes used simply to describe the general decrease in amplitude of seismic waves with distance, which is primarily a result of geometrical spreading, but in this section we will use the term only to refer to intrinsic attenuation. The strength of intrinsic attenuation is given by the dimensionless quantity Q in terms of the fractional energy loss per cycle

$$\frac{1}{Q(\omega)} = -\frac{\Delta E}{2\pi E},$$ (6.82)

where E is the peak strain energy and $-\Delta E$ is the energy loss per cycle. Q is sometimes called the *quality factor*, a term borrowed from the engineering literature. Note that Q is inversely related to the strength of the attenuation; low-Q regions are more attenuating than high-Q regions. For seismic waves in the Earth, the energy loss per cycle is very small, in which case one may derive an approximation (valid for $Q \gg 1$) that is better suited than (6.82) for seismic applications:

$$A(x) = A_0 e^{-\omega x/2cQ},$$ (6.83)

where x is measured along the propagation direction and c is the velocity (e.g., $c = \alpha$ for P waves with attenuation Q_α and $c = \beta$ for S waves with attenuation Q_β). The amplitude of harmonic waves may then be written as a product of a real exponential describing the amplitude decay due to attenuation and an imaginary exponential describing the oscillations

$$A(x, t) = A_0 e^{-\omega x/2cQ} e^{-i\omega(t-x/c)}.$$ (6.84)

Sometimes the exponentials in this equation are combined, and the effect of Q is included directly in $e^{-i\omega(t-x/c)}$ by adding a small imaginary part to the velocity c (the imaginary part is small because $Q \gg 1$). This provides an easy way to incorporate the effects of attenuation into homogeneous layer techniques (e.g., reflectivity) for computing synthetic seismograms, since these methods typically operate in the frequency domain. However, as discussed below, velocity dispersion must also be included to obtain accurate pulse shapes using this approach.

6.6.1 Example: Computing intrinsic attenuation

A wave propagates for 50 km through a material with velocity 6 km/s and $Q = 100$. What is the amplitude reduction for a wave at 1 Hz and at 10 Hz? From (6.83) and recalling that $\omega = 2\pi f$, we have

$$A(1\,\text{Hz}) = e^{-2\pi 50/2\cdot 6\cdot 100} = 0.77,$$

$$A(10\,\text{Hz}) = e^{-20\pi 50/2\cdot 6\cdot 100} = 0.0073.$$

The wave retains 77% of its original amplitude at 1 Hz, but only 0.73% of its amplitude at 10 Hz.

6.6.2 t^* and velocity dispersion

In ray theoretical methods, attenuation is often modeled through the use of t^*, defined as the integrated value of $1/Q$ along the ray path

$$t^* = \int_{\text{path}} \frac{dt}{Q(\mathbf{r})}, \tag{6.85}$$

where $\mathbf{r}$ is the position vector. Equation (6.83) then becomes

$$A(\omega) = A_0(\omega)e^{-\omega t^*/2}. \tag{6.86}$$

The amplitude reduction at each frequency is obtained by multiplying by $e^{-\omega t^*/2}$. The factor of ω means that the high frequencies are attenuated more than the low frequencies; thus a pulse that travels through an attenuating region will gradually lose its high frequencies. This property can be used to measure t^* from observed spectra (assuming the starting spectrum generated at the source is flat, i.e., a delta-function source) by plotting $\log |A(\omega)|$ versus f.

$$\log_{10} |A(\omega)| = \log_{10} A_0 - \frac{1}{2}\omega t^* \log_{10}(e) \tag{6.87}$$

$$= \log_{10} A_0 - 0.2171\omega t^* \tag{6.88}$$

or

$$\log_{10} |A(f)| = \log_{10} A_0 - 1.364 f t^*, \tag{6.89}$$

where $f = \omega/2\pi$. Assuming Q is constant with frequency, $\log |A(f)|$ will plot as a straight line with a slope that is directly proportional to t^*. If Q is frequency dependent, as will be discussed in the next section, then the slope of $\log |A(f)|$ is proportional to what is often termed apparent t^* or $\overline{t^*}$, defined as

$$\overline{t^*} = -\frac{d(\ln A)}{\pi df} = t^* + f\frac{dt^*}{df}. \tag{6.90}$$

In the time domain, attenuation can be modeled by convolving the original pulse by a t^* *operator*, defined as the inverse Fourier transform of the $A(\omega)$ function in the frequency domain (Fourier transforms and convolution are discussed in Appendix E). This convolution will cause the pulse broadening that is observed in the time domain.

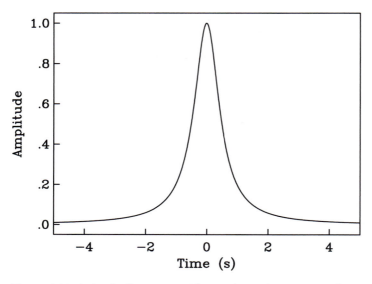

Figure 6.10 A simple t^* operator with no velocity dispersion predicts a physically unrealistic symmetric pulse.

However, there is a complication that we have so far neglected that involves the phase part of the spectrum. The spectrum, $A(\omega)$, of a time series is complex and has real and imaginary parts, which define both an amplitude and a phase at each frequency point. Attenuation will reduce the amplitude part of the spectrum according to (6.86). If we assume that the phase part of the spectrum is unchanged (i.e., velocity is constant as a function of frequency) then the t^* operator will take the form of a symmetric pulse, as shown in Figure 6.10. For a δ-function source, this represents the expected pulse shape resulting from propagation through the attenuating medium. The pulse is broadened because of the removal of the higher frequencies, but the symmetric shape of the pulse violates causality because the leading edge of the pulse will arrive ahead of the ray-theoretical arrival time. It turns out that the existence of attenuation requires that velocity vary with frequency, even if Q itself is not frequency dependent over the same frequency band. This follows both from the necessity to maintain causality in the attenuated pulse and from considerations of physical mechanisms for attenuation. A full discussion of these topics may be found in Aki and Richards (2002, pp. 163–75); one of the most important results is that over a frequency interval in which $Q(\omega)$ is constant, the velocity c may be expressed as

$$c(\omega) = c(\omega_0) \left(1 + \frac{1}{\pi Q} \ln \frac{\omega}{\omega_0} \right), \tag{6.91}$$

where ω_0 is a reference frequency. The predicted velocity dispersion in the Earth is fairly small at typically observed values of Q. For example, at $Q = 150$, the velocities at periods of 1 and 10 s differ by only about 0.5%.

Let us now consider the effects of this velocity dispersion on the spectrum for the case where $Q(\omega)$ is constant. From (6.84) we have

$$A(x, \omega, t) = A_0 e^{-\omega x/2c(\omega)Q} e^{-i\omega(t - x/c(\omega))}. \tag{6.92}$$

From (6.91) and using the approximation $1/(1 + \epsilon) \approx 1 - \epsilon$ for $\epsilon \ll 1$ (valid since $Q \gg 1$) we can express the $1/c(\omega)$ term as

$$\frac{1}{c(\omega)} = \frac{1}{c_0}\left[1 - \frac{1}{\pi Q}\ln\frac{\omega}{\omega_0}\right] \tag{6.93}$$

$$= \frac{1}{c_0} - \frac{\ln(\omega/\omega_0)}{\pi c_0 Q}, \tag{6.94}$$

where $c_0 = c(\omega_0)$. Substituting into (6.92) and ignoring terms of $1/Q^2$, we have

$$A(x, \omega, t) = A_0 e^{-\omega x/2c_0 Q} e^{-i\omega(t - x/c_0)} e^{-i\omega x \ln(\omega/\omega_0)/\pi c_0 Q}. \tag{6.95}$$

Defining $t_0^* = x/c_0 Q$ we have

$$A(x, \omega, t) = A_0 e^{-\omega t_0^*/2} e^{-i\omega(t - x/c_0)} e^{-i\omega t_0^* \ln(\omega/\omega_0)/\pi}. \tag{6.96}$$

Here $e^{-\omega t_0^*/2}$ is the amplitude reduction term, $e^{-i\omega(t - x/c_0)}$ is the phase shift due to propagation for a distance x at the reference velocity c_0, and the final term gives the additional phase advance or delay for frequencies that deviate from the reference frequency (note that this phase shift is zero when $\omega = \omega_0$). The shapes of the resulting t^* operators are shown in Figure 6.11, where the times are given relative to the predicted arrival time at 10 Hz (i.e., the phase shift given by the final term in (6.96) when $\omega_0 = 20\pi$). These operators represent the pulses that would be observed from a delta-function source after traveling through attenuating material. Larger values of t^* describe more attenuation, reduced high-frequencies, and increased pulse broadening.

These t^* operators have more impulsive onsets than the symmetric pulse plotted in Figure 6.10 and preserve causality because no energy arrives before the time predicted by the largest value of $c(\omega)$. Note that energy at frequencies greater than 10 Hz (which would plot at negative times in front of the pulses shown) is not visible because it is severely attenuated by the $e^{-\omega t_0^*/2}$ term.

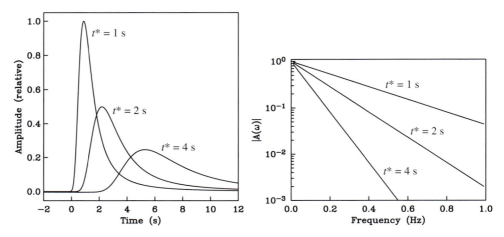

Figure 6.11 (left) Pulse shapes (t^* operators) for t^* values of 1, 2 and 4 s, computed using equation (6.96) for a reference frequency of 10 Hz. (right) The corresponding amplitude spectra plot as straight lines on a linear-log plot, according to equation (6.89).

6.6.3 The absorption band model[†]

Although Q is often approximated as being constant over a frequency band of interest, it is clear from (6.91) that this approximation must fail at very high or low frequencies. If Q were completely constant, then $c(\omega)$ becomes negative for small ω and $c(\omega) \to \infty$ for $\omega \to \infty$. This is not observed in the Earth so it must be the case that Q decreases at both low and high frequencies. This behavior is commonly modeled in terms of an *absorption band operator* (e.g., Liu *et al.*, 1976; Lundquist and Cormier, 1980; Doornbos, 1983), in which case

$$Q^{-1}(\omega) = 2Q_m^{-1}D_Q(\omega),\tag{6.97}$$

where Q_m^{-1} is the peak value of $Q^{-1}(\omega)$, and

$$D_Q(\omega) = \frac{1}{\pi}\tan^{-1}\left[\frac{\omega(\tau_2 - \tau_1)}{1 + \omega^2\tau_1\tau_2}\right],\tag{6.98}$$

where τ_1 and τ_2 are the lower and upper relaxation times that define the edges of the absorption band (see Fig. 6.12). These times correspond to lower and upper frequency limits, $f_1 = 1/2\pi\tau_2$ and $f_2 = 1/2\pi\tau_1$ (yes, the ones and twos should be switched!), between which the attenuation is approximately constant. Because $\tan^{-1}\theta \approx \theta$ for $\theta \ll 1$, $Q^{-1}(\omega)$ will vary as ω for frequencies well below f_1 and will vary as ω^{-1} for frequencies well above f_2.

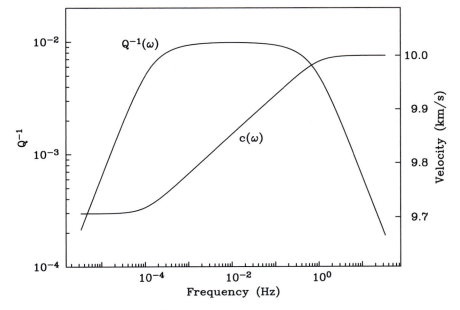

Figure 6.12 Attenuation, Q^{-1}, and velocity, c, as a function of frequency for an absorption band attenuation model in which $f_1 = 0.0001$ Hz and $f_2 = 1$ Hz, peak attenuation $Q_m = 100$, and $c_\infty = 10$ km/s.

For this model, the velocity dispersion is given by

$$c(\omega) = c_\infty \left[1 - \frac{D_c(\omega)}{Q_m} \right]. \qquad (6.99)$$

where c_∞ is the velocity at infinite frequency and

$$D_c(\omega) = \frac{1}{2\pi} \ln \left(\frac{1 + 1/\omega^2 \tau_1^2}{1 + 1/\omega^2 \tau_2^2} \right). \qquad (6.100)$$

As shown in Figure 6.12, the velocity increases to c_∞ over the absorption band from the low-frequency limit of $c = c_\infty [1 - \ln(\tau_2/\tau_1)/\pi Q_m]$. Now rewrite (6.84) for frequency-dependent $Q(\omega)$ and $c(\omega)$

$$A(x, \omega, t) = A_0 e^{-\omega x/2c(\omega)Q(\omega)} e^{-i\omega(t - x/c(\omega))} \qquad (6.101)$$

For $Q_m \gg 1$, we can express the $1/c(\omega)$ term as

$$\frac{1}{c(\omega)} = \frac{1}{c_\infty} + \frac{D_c}{c_\infty Q_m}. \qquad (6.102)$$

Substituting into (6.101) and ignoring the $1/Q_m^2$ terms, we have

$$A(x, \omega, t) = A_0 e^{-\omega t^* D_Q(\omega)} e^{i\omega t^* D_c(\omega)} e^{-i\omega(t - x/c_\infty)}, \qquad (6.103)$$

where we define t^* as

$$t^* = \frac{x}{c_\infty Q_m}. \qquad (6.104)$$

Behavior in the middle of the band

Now consider the case when we are near the middle of the absorption band so that $Q(\omega) \approx Q_m$ is locally constant and $\omega\tau_1 \ll 1$ and $\omega\tau_2 \gg 1$. We then have

$$D_c(\omega) \approx \frac{1}{2\pi} \ln\left(\frac{1}{\omega^2 \tau_1^2}\right) \qquad (6.105)$$

and (6.102) becomes

$$\frac{1}{c(\omega)} = \frac{1}{c_\infty} + \frac{1}{2\pi c_\infty Q_m} \ln\left(\frac{1}{\omega^2 \tau_1^2}\right) \qquad (6.106)$$

$$= \frac{1}{c_\infty} + \frac{-2\ln(\omega\tau_1)}{2\pi c_\infty Q_m} \qquad (6.107)$$

$$= \frac{1}{c_\infty} - \frac{\ln(\tau_1)}{\pi c_\infty Q_m} - \frac{\ln(\omega)}{\pi c_\infty Q_m}. \qquad (6.108)$$

Again substituting into (6.101), we obtain

$$A(x, \omega, t) = A_0 e^{-\omega t^*/2} e^{-i\omega(t - x/c_\infty)} e^{-i\omega t^* \ln(\tau_1)/\pi} e^{-i\omega t^* \ln(\omega)/\pi}. \qquad (6.109)$$

Notice that in this case there is a phase shift, $e^{-i\omega t^* \ln(\tau_1)/\pi}$, associated purely with the value of the upper-frequency cutoff ($f_2 = 1/2\pi\tau_1$) of the absorption band. Increasing f_2 will not affect the shape of the pulse but will increase the delay because we are assuming c_∞ is constant so $c(\omega)$ will decrease for any fixed ω if f_2 increases. Because f_2 is not constrained by the pulse shape (and in general may be poorly known), it is convenient to combine terms to obtain

$$A(x, \omega, t) = A_0 e^{-\omega t^*/2} e^{-i\omega(t - t_0)} e^{-i\omega t^* \ln(\omega)/\pi}, \qquad (6.110)$$

where the total time delay is given by

$$t_0 = \frac{x}{c_\infty} - \frac{t^* \ln(\tau_1)}{\pi}.$$ (6.111)

Note the similarity of this equation to the constant Q result (6.96).

6.6.4 The standard linear solid[†]

Most of the physical mechanisms that have been proposed to explain intrinsic attenuation (grain boundary processes, crystal defect sliding, fluid-filled cracks, etc.) can be parameterized in terms of a *standard linear solid*, a simple model that exhibits viscoelastic behavior (see Fig. 6.13). The displacement of the mass is not simply proportional to the applied force (as was the case for the purely elastic stress–strain relations we described in Chapter 3) because the resistance of the dashpot scales with the velocity rather than the displacement. The result is that the response is different depending upon the time scale of the applied force. In the high-frequency limit, the dashpot does not move and the response is purely elastic and defined by the second spring. In the low-frequency limit, the dashpot moves so slowly that there is no significant resistance and the response is again purely elastic and given by the sum of the two springs. But at intermediate frequencies, the dashpot will move and dissipate energy, producing a viscoelastic response and attenuation.

The stress–strain relationship for a standard linear solid is often expressed as

$$\sigma + \tau_\sigma \dot{\sigma} = M_R(\epsilon + \tau_\epsilon \dot{\epsilon}),$$ (6.112)

where σ is stress (we don't use τ for the stress tensor as we did in Chapter 3 because here we are using τ for the relaxation times), ϵ is strain, the dots represent time derivatives, τ_ϵ is the relaxation time for the strain under an applied step in stress, τ_σ is the relaxation time for stress given a step change in strain, and M_R is the *relaxed*

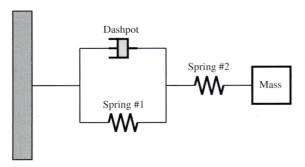

Figure 6.13 A mechanical model of the behavior of a standard linear solid, consisting of a spring and dashpot in parallel, connected in series to a second spring.

modulus, which gives the ratio of stress to strain at infinite time. It can be shown (e.g., Aki and Richards, 2002, pp. 172–3) that Q for such a model is given by

$$Q^{-1}(\omega) = \frac{\omega(\tau_\epsilon - \tau_\sigma)}{1 + \omega^2 \tau_\sigma \tau_\epsilon} \tag{6.113}$$

and the velocity is given by

$$c(\omega) = \sqrt{\frac{M_U}{\rho}}\left[1 + \left(\frac{M_U}{M_R} - 1\right)\frac{1}{1 + \omega^2 \tau_\epsilon^2}\right]^{-1/2}, \tag{6.114}$$

where ρ is density and M_U is the *unrelaxed modulus*

$$M_U = M_R \frac{\tau_\epsilon}{\tau_\sigma} \tag{6.115}$$

Note that the minimum velocity occurs at low frequencies and is a function of the relaxed modulus ($c_{\min} = \sqrt{M_R/\rho}$) and the maximum velocity occurs at high frequencies and is given by the unrelaxed modulus ($c_{\max} = \sqrt{M_U/\rho}$), and that for $Q \gg 1$ the ratio M_R/M_U is close to unity. For this model, the attenuation peaks near $\omega = (\tau_\sigma \tau_\epsilon)^{-1/2}$ (termed the *Debye peak*) and the velocity $c(\omega)$ increases from $(M_R/\rho)^{1/2}$ at low frequencies to $(M_U/\rho)^{1/2}$ at high frequencies. This is shown in

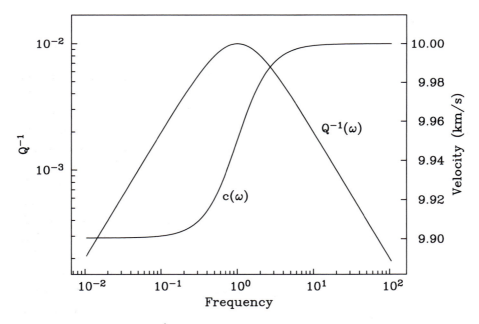

Figure 6.14 Attenuation, Q^{-1}, and velocity, c, as a function of frequency for a standard linear solid in which $\tau_\epsilon = 0.1607$ s, $\tau_\sigma = 0.1576$ s, and $c_\infty = 10$ km/s.

the example plotted in Figure 6.14 in which the relaxation times are chosen to give $Q = 100$ for an attenuation peak at 1 Hz. The Debye peak represents the behavior of a solid with a single relaxation mechanism – attenuation is sharply peaked at a resonant frequency. In contrast, Earth's attenuation is nearly constant between 0.001 and 0.1 Hz, implying that a range of relaxation mechanisms operating at different scales is occurring. As shown by Liu *et al.* (1976), a superposition of these relaxation peaks can produce an absorption band over which Q^{-1} is constant.

Of course, there is no reason to expect that Q^{-1} in the Earth should be perfectly flat over a wide band. There is some evidence from mineral physics experiments that Q may have a power-law dependence on frequency, i.e., that $Q^{-1} \sim \omega^\gamma$ where $\gamma \approx -0.2$ to -0.3 (see Baig and Dahlen, 2004, who reference Berckhemer *et al.*, 1982; Jackson *et al.*, 1992, 2002). Seismic observations are not clear enough yet to resolve the details of the decrease in attenuation at high frequencies. However, the shape of the Debye peak shows that the falloff of Q^{-1} should never exceed ω^{-1}.

6.6.5 Earth's attenuation

The equations in the previous sections apply equally to either P-wave attenuation Q_α or S-wave attenuation Q_β, which are related to bulk attenuation Q_κ and shear attenuation Q_μ by the relationships

$$Q_\alpha^{-1} = L Q_\mu^{-1} + (1 - L) Q_\kappa^{-1}, \tag{6.116}$$

$$Q_\beta = Q_\mu, \tag{6.117}$$

where $L = (4/3)(\beta/\alpha)^2$. Shear attenuation is observed to be much stronger than bulk attenuation in the deep Earth where Q_κ is generally assumed to be infinite except in the inner core. Note that for $Q_\kappa = \infty$ and $\alpha/\beta = \sqrt{3}$ (Poisson solid), $Q_\alpha^{-1} = (4/9) Q_\beta^{-1}$ and $t_\alpha^* = (4/9\sqrt{3}) t_\beta^* \approx (1/4) t_\beta^*$, which are commonly used approximations for the relative strength of P and S attenuation in the mantle. Long-period P waves at epicentral distances between $30°$ and $90°$ are observed to have t^* values of roughly 1 s; corresponding S waves have t^* values of about 4 s.

Earth's attenuation is observed to be approximately constant for frequencies between about 0.001 and 0.1 Hz. This part of the seismic spectrum is observed with normal modes and long-period surface and body waves (10 to 1000 s period). The PREM model (see Fig. 6.15) provides a representative constant Q model for this band. However, attenuation falls off sharply at higher frequencies and observed body-wave amplitudes at 1 Hz are much higher than predicted by PREM. This behavior can be approximately fitted with an absorption band model in which the upper frequency cutoff, f_2, is between 0.5 and 2 Hz (see, e.g., Sipkin and Jordan, 1979; Warren and Shearer, 2000). Figure 6.15 compares Q_α^{-1} values between

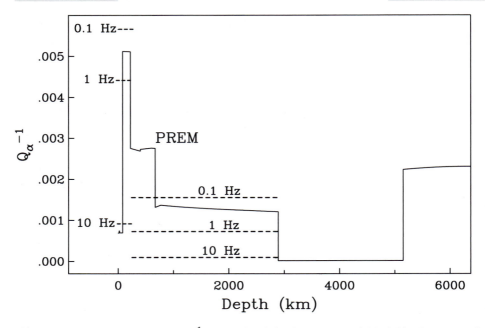

Figure 6.15 *P*-wave attenuation, Q_α^{-1}, versus depth for the PREM model (solid line), compared to the two-layer, frequency-dependent mantle Q_α^{-1} of Warren and Shearer (2000) plotted at frequencies of 0.1, 1, and 10 Hz (dashed lines). The jumps in the PREM *Q* model occur at 80, 220, and 670 km. The change in the Warren and Shearer model occurs at 220 km.

PREM (derived from normal modes) and the frequency-dependent, two-layer mantle model WS2000 (Warren and Shearer, 2000, derived from short-period *P* and *PP* observations). Note the approximate agreement between the models at low frequency (0.1 Hz) but the much lower attenuation for WS2000 at frequencies of 1 Hz and greater.

Attenuation is strongest in the upper mantle and inner core; typically assumed long-period values are provided by the PREM model, for which Q_μ is 600 above 80 km depth, drops to 80 between 80 and 220 km, increases to 143 from 220 to 670 km, and is 312 for the lower mantle (below 670 km). Shear waves do not propagate in the fluid outer core; the PREM model assigns $Q_\mu = 85$ and $Q_\kappa = 1328$ for the inner core.

In the crust, attenuation is concentrated near the surface where the effects of cracks and/or fluids can be significant. Unlike mantle attenuation, Q_κ^{-1} is significant in the crust and *P*-wave attenuation Q_α^{-1} can equal or exceed *S*-wave attenuation Q_β^{-1}. Attenuation decreases rapidly with depth in the crust and typical mid-crustal values of Q_α are 500 to 1000. Attenuation increases again in the upper mantle, particularly in the aesthenosphere where a seismic low-velocity zone is often observed and partial melt may be present. Attenuation is lower in the mid to lower mantle but still generally exceeds that observed for most of the crust. Significant lateral variations in crust and upper-mantle attenuation are observed, with higher attenuation seen in tectonically active areas such as the western United States compared with more stable continental shield and platform regions. The result is that earthquakes in the eastern United States (rare, but they do happen) are felt and cause damage over much wider areas than for the same magnitude earthquakes in California.

Attenuation causes a small amount of velocity dispersion between long and short-period seismic waves. Velocities in any global seismic model are always with respect to a fixed frequency. Velocities in PREM are tabulated for a reference period of 1 s, but can be converted to other periods using equation (6.91). Although the velocity difference between short and long-period body waves is generally less than 1%, this can cause travel-time differences in global phases of several seconds, which are easily observable. In the 1960s, this velocity dispersion due to Q was not widely understood (despite attempts by Harold Jeffreys to call attention to this effect!) and many seismologists puzzled for some time over a discrepancy between normal-mode and body-wave velocity models. The normal-mode models predicted body-wave travel times longer than those actually observed by an amount that was termed the *baseline correction*. It was recognized by several authors (e.g., Liu *et al.*, 1976) in the mid 1970s that the bulk of the baseline correction could be explained as physical dispersion. There remains some debate, however, whether dispersion can account for all of the baseline correlation (also sometimes called the *S discrepancy*), or if some other effect might be involved, such as 3-D velocity heterogeneity (e.g., Nolet and Moser, 1993; Baig and Dahlen, 2004).

6.6.6 Observing Q

At long periods, Q can be observed from analysis of Earth's normal modes (see Chapter 8), which can be modeled as decaying harmonic functions. Attenuation causes a broadening of the line spectrum that would be obtained from pure sine waves; thus the Q of a mode can be measured from the width of its spectral peak. By taking into account the different depth and lateral sensitivities of the various normal modes, these measurements can be inverted for 1-D and/or 3-D attenuation models. Surface-wave attenuation studies have mostly used amplitude measurements. Here the tricky part is separating the amplitude variations caused by focusing and defocusing of energy related to lateral velocity changes from the amplitude reductions caused by attenuation. Body-wave Q studies often measure t^* for specific seismic phases by computing the spectrum and measuring the falloff rate (i.e., using equation (6.89)). However, earthquake source spectra also fall off at high frequencies (see Chapter 9) so corrections must be applied for the source spectral content. This can be done by assuming a particular source model or empirically for each source if enough data are available. Alternatively, differential spectra between phase pairs (such as S and SS) can be studied; these spectral ratios are insensitive to the source, provided the ray takeoff angles from the event are sufficiently similar that radiation pattern differences can be ignored. As an alternate to spectral methods, some body-wave Q studies examine the *rise time* or the width of the displacement pulse, but, just as in the spectral methods, these measurements must be corrected for source effects.

At short periods the effects of scattering become increasingly important and it can be difficult to separate the effects of intrinsic attenuation (the focus of our discussion in this section) from scattering attenuation. Both have the effect of reducing the high frequencies in transmitted body waves. Measurements of S coda decay rates from local earthquakes are often used to estimate a parameter called *coda Q*, the meaning of which has been debated, but which seems to represent primarily an average of intrinsic attenuation within the scattering volume that is contributing to the coda.

In principle, attenuation measurements can be inverted for 2-D or 3-D Q structure using the tomographic methods discussed in Chapter 5, but Q studies have generally lagged behind velocity inversions because of the much greater scatter exhibited by observed amplitudes and t^* and other measurements. This scatter is likely caused by source directivity effects, interference from other seismic phases, and multipathing and focusing/defocusing from general 3-D structure. However, attenuation studies are worth pursuing despite their difficulty because they provide important independent constraints on Earth properties since their sensitivity to temperature, fluids, and compositional differences is distinct from that provided by P- and S-wave velocities.

6.6.7 Non-linear attenuation

Linear stress–strain theory is generally valid at the low strains typical of most seismic waves. However, there are some circumstances in which non-linear deformation is important, including the following:

1. There is a non-linear regime close to underground nuclear explosions in which permanent fracturing and other rock damage occurs. These effects must be considered in developing a complete theory of how to recover nuclear source properties from far-field seismic records.
2. Strong ground accelerations from large earthquakes can produce a non-linear response in shallow soils. This can be studied by comparing surface and borehole seismic records for earthquakes of different sizes. When a non-linear site response is present, then the

shaking from large earthquakes cannot be predicted by simple scaling of records from small earthquakes. This is an active area of research in strong motion and engineering seismology.

6.6.8 Seismic attenuation and global politics

Beginning in 1976, a nuclear threshold test ban treaty was observed by both the United States and the Soviet Union, which limited underground nuclear tests to 150 kilotons or less. The yield of underground tests can be estimated from seismic observations using an empirical magnitude versus yield curve for nuclear tests of known size. The United States calibrated its "official" magnitude-yield relationship mostly using data from nuclear tests at the Nevada Test Site (NTS). The Soviets detonated a number of large explosions at during the 1980s that appeared to exceed 150 kilotons when their yields were estimated using this formula. This led the Reagan administration to charge that the Soviets were "likely" in violation of the treaty. However, more careful analysis showed that seismic results from NTS were biased by the relatively strong upper-mantle attenuation beneath the tectonically active western United States compared to the stable continental regions of the Soviet tests. This reduced the amplitudes of seismic waves from NTS explosions and their computed magnitudes. Thus, the Soviet tests generated larger seismic magnitudes than would be produced by a 150 kiloton NTS shot. But when the difference in Q structure was taken into account, the evidence for Soviet cheating disappeared. For more about this controversy, see http://www.ciaonet.org/conf/nya02/nya02ah.html.

6.7 Exercises

1. Derive equation (6.23), the geometrical spreading formula for a spherical Earth. For $r_1 = r_2 = 1$ and small Δ, show that this equation is identical to the flat-earth expression (6.22).

2. Using the P-wave travel times listed in Table 6.1 (values from the IASP91 tables for a source at zero depth, Kennett, 1991), apply the geometrical spreading equation (6.23) to predict the relative amplitude of P waves recorded at the surface from an isotropic surface source at ranges of $30°$, $60°$, and $90°$. Assume that the surface P velocity is 5.8 km/s. If the amplitude at $30°$ is unity, what is the amplitude at $60°$ and $90°$? Hints: Be sure to convert the p_{sph} values in Table 6.1 to s/radian in order to obtain correct units. Check the values you compute for the ray takeoff angle θ_1 to be sure that they look reasonable.

Δ (deg.)	T (s)	p_{sph} (s/deg.)
Table 6.1: *P* travel times from iasp91.		
29	361.41	8.89
30	370.27	8.85
31	379.10	8.81
59	601.38	6.95
60	608.29	6.88
61	615.13	6.80
89	776.67	4.69
90	781.35	4.66
91	786.00	4.64

3. Solve for **R** in (6.75) and show that your result agrees with equations (6.44) and (6.45).

4. (COMPUTER) Use subroutine RTCOEF (Appendix D) to tabulate the amplitude and phase of reflected and transmitted waves for downgoing *P* waves incident on the core–mantle boundary. Use PREM values (Appendix A) for the velocities and densities across the interface. Plot your results as a function of ray incidence angle. Note: RTCOEF is for a solid–solid boundary condition and will blow up for a fluid–solid boundary. To get around this, use a very small, but non-zero, value for the outer core shear velocity. Ignore the result that you obtain for the transmitted shear wave; it should contain very little energy.

5. What is the phase shift $(0, \pi/2, \pi, \text{ or } 3\pi/2)$ for the following phases compared to *P*: *PP*, *PPP*, *PPPP*, *pP*, *pPP*, *PcP* (near normal incidence), *PcP* (near intersection with *P*)? What is the phase shift for the following *SH* phases compared to *S*: *SS*, *SSS*, *SSSS*, *sS*, *sSS*, *ScS* (near normal incidence), *ScS* (near intersection with *S*)?

6. Use the values for Q_α plotted in Figure 6.15 for the PREM model to:

   (a) Estimate the attenuation for waves of 30 s period (0.033 Hz) for *PcP* ray paths at vertical incidence. Consider only the intrinsic attenuation along the ray paths; do not include the reflection coefficient at the core–mantle boundary or geometrical spreading. Be sure to count both the surface-to-CMB and CMB-to-surface legs. You can get the PREM *P* velocities from Figure 1.1 or the values tabulated in Appendix A. You may estimate the travel times through the different *Q* layers in PREM by assuming a fixed velocity within each layer (e.g., use the average velocity between 3 and 80 km, the average between 80 and 220 km, etc.). Note that in this case the "surface" is assumed to be the bottom of the ocean at 3 km depth. Compute t^* for PcP. Finally give the attenuated amplitude of the PcP ray assuming the ray had an initial amplitude of one.

(b) Using the approximation $t_\beta^* = 4t_\alpha^*$, compute the attenuated amplitude of ScS at 30 s period.

(c) Repeat (a) and (b) for 1 s period (1 Hz) PcP and ScS waves.

(d) Repeat your calculations for the PcP and ScS attenuated amplitudes at 1 Hz, but this time use the Warren and Shearer (2000) Q_α values from Figure 6.15. How do the predicted amplitudes compare to the PREM model predictions? What do these results predict for the observability of teleseismic S arrivals at 1 Hz?

7. Assume that $Q_\mu = 100$ for the inner core and is constant with frequency, that all of the attenuation is in shear ($Q_\kappa = \infty$), that $\alpha = 11$ km/s and $\beta = 3.5$ km/s (at 1 s period) throughout the inner core, and that the inner-core radius is 1221.5 km.

(a) What are Q_α and Q_β for the the inner core?

(b) For a P wave that travels straight through the Earth (i.e., travels a distance twice the radius), how much will amplitudes be attenuated in the inner core for waves of 100 s, 10 s, and 1 s period? By how many seconds will the total travel time through the inner core vary as a function of these periods?

(c) Repeat (b) for the case of S waves.

7

Reflection seismology

One of the most important applications of seismology involves the probing of Earth's internal structure by examining energy reflected at steep incidence angles from subsurface layers. This technique may loosely be termed *reflection seismology* and has been used extensively by the mining and petroleum industries to study the shallow crust, generally using portable instruments and artificial sources. However, similar methods can be applied to the deeper Earth using recordings of earthquakes or large explosions. Because reflected seismic waves are sensitive to sharp changes in velocity or density, reflection seismology can often provide much greater lateral and vertical resolution than can be obtained from study of direct seismic phases such as P and S (analyses of these arrivals may be termed *refraction seismology*). However, mapping of reflected phases into reflector depths requires knowledge of the average background seismic velocity structure, to which typical reflection seismic data are only weakly sensitive. Thus refraction experiments are a useful complement to reflection experiments when independent constraints on the velocity structure (e.g., from borehole logs) are unavailable.

Reflection seismic experiments are typically characterized by large numbers of sources and receivers at closely spaced and regular intervals. Because the data volume generally makes formal inversions too costly for routine processing, more practical approximate methods have been widely developed to analyze the results. Simple time versus distance plots of the data can produce crude images of the subsurface reflectors; these images become increasingly accurate as additional processing steps are applied to the data.

Our discussion in this chapter will be limited to P-wave reflections, as the sources and receivers in most reflection seismic experiments are designed to produce and record P waves. Our focus will also mainly be concerned with the travel time rather than the amplitude of seismic reflections. Although amplitudes are sometimes studied, historically amplitude information has assumed secondary importance in reflection processing. Indeed often amplitudes are self-scaled prior to plotting using *automatic gain control* (AGC) techniques. Finally, we will consider a two-dimensional geometry, for which the sources, receivers, and reflectors are assumed to lie within a vertical plane. Recently, an increasing number of reflection surveys involve a grid of sources and receivers on the surface that are capable of resolving three-dimensional Earth structure. Most of the concepts described in this chapter, such as common midpoint stacking and migration, are readily generalized to three dimensions, although the data volume and computational requirements are much greater in this case.

Reflection seismology is a big topic, and only a brief outline can be presented here. For additional details, the reader is referred to texts such as Yilmaz (1987), Sheriff and Geldart (1995), and Claerbout (1976, 1985).

7.1 Zero-offset sections

Consider a collocated source and receiver at the surface above a horizontally layered velocity structure (Fig. 7.1). The downward propagating P waves from the source are reflected upward by each of the interfaces. The receiver will record a series of pulses at times determined by the two-way P travel time between the surface and the interfaces. If the velocity structure is known, these times can easily be converted to depths.

Now imagine repeating this as the source and receiver are moved to a series of closely spaced points along the surface. At each location, the receiver records the reflected waves from the underlying structure. By plotting the results as a function of time and distance, an image can be produced of the subsurface structure (Fig. 7.2). For convenience in interpreting the results, these record sections are plotted with downward increasing time (i.e., upside down compared with the plots

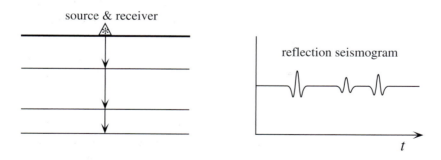

Figure 7.1 Seismic waves from a surface source are reflected by subsurface layers, producing a seismogram with discrete pulses for each layer. In this example, the velocity contrasts at the interfaces are assumed to be small enough that multiple reflections can be ignored.

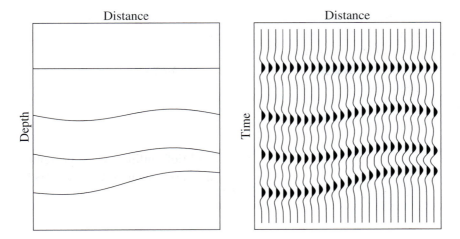

Figure 7.2 The structural cross-section on the left is imaged by an idealized zero-offset seismic reflection profile on the right. Here we have assumed that velocity is approximately constant throughout the model (except for thin reflecting layers) so that time and depth scale linearly.

in Chapter 4). Another convention, commonly used in reflection seismology, is to darken the positive areas along the seismograms, increasing the visibility of the reflected pulses.

If the large-scale P velocity is constant throughout the region of interest, then time in this image scales linearly with depth and we can readily convert the vertical axis to depth. If velocity increases with depth, a somewhat more complicated transformation is necessary. Because the velocity structure is often not known very accurately, most reflection seismic results are plotted as time sections, rather than depth sections.

This example is termed a *zero-offset section* because there is assumed to be no separation between the sources and receivers. More generally, reflection data are recorded at a variety of source–receiver offsets, but the data are processed in order to produce an equivalent zero-offset section that is easier to interpret than the original data. In the idealized example shown in Figure 7.2, the zero-offset section provides a clear, unbiased image of the reflectors. However, in practice there are several factors that can hinder construction and interpretation of such a section:

1. Single records are often noisy and zero-offset data may be contaminated by near-source reverberations. Improved results may be obtained by including different source–receiver offsets to increase the number of data and then averaging or *stacking* the records to increase the signal-to-noise ratio of the reflected pulses.
2. The layer spacing may be short compared to the source duration, producing overlapping arrivals that make it difficult to distinguish the individual reflectors. This may

be addressed through a process termed *deconvolution*, which involves removing the properties of the source from the records, providing a general sharpening of the image.

3. Lateral variations in structure or dipping layers may result in energy being scattered away from purely vertical ray paths. These arrivals can bias estimates of reflector locations and depths. By summing along possible sources of scattered energy, it is possible to correct the data for these effects; these techniques are termed *migration* and can result in a large improvement in image quality.

4. Uncertainties in the overall velocity structure may prevent reliable conversion between time and depth and hinder application of stacking and migration techniques. Thus it is critical to obtain the most accurate velocity information possible; in cases where outside knowledge of the velocities are unavailable, the velocities must be estimated directly from the reflection data.

We now discuss each of these topics in more detail.

7.2 Common midpoint stacking

Consider a source recorded by a series of receivers at increasing distance. A pulse reflected from a horizontal layer will arrive earliest for the zero-offset receiver, while the arrivals at longer ranges will be delayed (Fig. 7.3). If the layer has thickness h and a uniform P velocity of v, then the minimum travel time, t_0, defined by the two-way vertical ray path, is

$$t_0 = \frac{2h}{v}. \tag{7.1}$$

More generally, the travel time as a function of range, x, may be expressed as

$$t(x) = \frac{2d}{v}, \tag{7.2}$$

where d is the length of each leg of the ray path within the layer. From the geometry, we have

$$
\begin{aligned}
d^2 &= h^2 + (x/2)^2, \\
4d^2 &= 4h^2 + x^2.
\end{aligned}
\tag{7.3}
$$

Squaring (7.2) and substituting for $4d^2$, we may write

$$v^2 t^2 = x^2 + 4h^2 \tag{7.4}$$

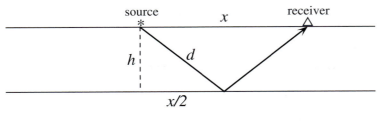

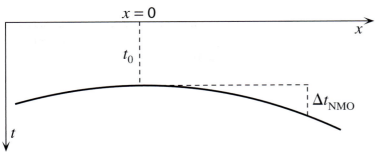

Figure 7.3 A reflected ray path (top) and the corresponding travel time curve as a function of source—receiver separation (bottom). For a constant velocity model, the travel times form a hyperbola.

or

$$\frac{v^2 t^2}{4h^2} - \frac{x^2}{4h^2} = 1,$$ (7.5)

and we see that the travel time curve has the form of a hyperbola with the apex at $x = 0$. This is often expressed in terms of t_0 rather than h by substituting $4h^2 = v^2 t_0^2$ from (7.1) to obtain

$$\frac{t^2}{t_0^2} - \frac{x^2}{v^2 t_0^2} = 1.$$ (7.6)

Solving for t we have

$$t(x) = \sqrt{t_0^2 + \frac{x^2}{v^2}}$$

$$= t_0 \sqrt{1 + \left(\frac{x}{v t_0}\right)^2}.$$ (7.7)

For small offsets ($x \ll vt_0$) we may approximate the square root as

$$t(x) \approx t_0 \left[1 + \tfrac{1}{2}(x/vt_0)^2 \right].\tag{7.8}$$

The difference in time between the arrival at two different distances is termed the *moveout* and may be expressed as

$$\Delta t = t(x_2) - t(x_1) = t_0\sqrt{1 + (x_2/vt_0)^2} - t_0\sqrt{1 + (x_1/vt_0)^2}\tag{7.9}$$

$$\approx t_0 \left[1 + \tfrac{1}{2}(x_2/vt_0)^2 \right] - t_0 \left[1 + \tfrac{1}{2}(x_1/vt_0)^2 \right]$$

$$\approx \frac{x_2^2 - x_1^2}{2v^2 t_0},\tag{7.10}$$

where the approximate form is valid at small offsets. The *normal moveout* (NMO) is defined as the moveout from $x = 0$ and is given by

$$\Delta t_{\text{NMO}} = t_0\sqrt{1 + (x/vt_0)^2} - t_0\tag{7.11}$$

$$\approx \frac{x^2}{2v^2 t_0}.\tag{7.12}$$

These equations are applicable for a single homogeneous layer. More complicated expressions can be developed for a series of layers overlying the target reflector or for dipping layers (e.g., see Sheriff and Geldart, 1995). Alternatively, the ray tracing theory developed in Chapter 4 can be applied to solve for the surface-to-reflector travel time for any arbitrary velocity versus depth function $v(z)$. Thus, a general form for the NMO equation is

$$\Delta t_{\text{NMO}}(x) = 2[t(z, x/2) - t(z, 0)],\tag{7.13}$$

where $t(z, x)$ is the travel time from the surface to a point at depth z and horizontal offset x.

A typical seismic reflection experiment deploys a large number of seismometers to record each source (these instruments are often called *geophones* in these applications, or *hydrophones* in the case of pressure sensors for marine experiments). This is repeated for many different source locations. The total data set thus consists of nm records, where n is the number of instruments and m is the number of sources. The arrival time of reflectors on each seismogram depends on the source–receiver offset as well as the reflector depth. To display these results on a single plot, it is desirable to combine the data in a way that removes the offset dependence in the travel times so that any lateral variability in reflector depths can be seen clearly.

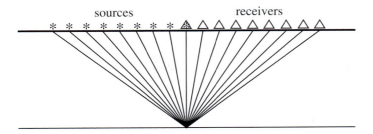

Figure 7.4 The source and receiver locations for a common midpoint (CMP) gather.

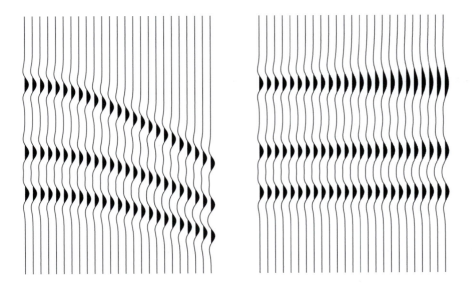

Figure 7.5 The left plot shows reflection seismograms at increasing source–receiver distance. The right plot shows the same profile after applying a NMO correction to each time series. Note that this removes the range dependence in the arrival times. The NMO corrected records can then be stacked to produce a single composite zero-offset record.

This is done by summing subsets of the data along the predicted NMO times to produce a composite zero-offset profile. Data are generally grouped by predicted reflector location as illustrated in Figure 7.4.

Seismograms with common source–receiver midpoints are selected into what is termed a *gather*. A NMO correction is then applied to the records that shifts the times to their zero-offset equivalent (as illustrated in Fig. 7.5). Notice that this correction is not constant for each record but varies with time within the trace. This results in pulse broadening for the waveforms at longer offsets, but for short pulse lengths and small offsets this effect is not large enough to cause problems. Finally the NMO corrected data are summed and averaged to produce a single composite record that represents the zero-offset profile at the midpoint location. This is called *common midpoint (CMP) stacking*, or sometimes common depth point (CDP) stacking. The number of records, n, that are stacked is called the *fold*. For data with random noise, stacking can improve the signal-to-noise ratio of the records by a factor of $\sqrt{n}$. CMP stacking can also minimize the influence of contaminating arrivals, such as direct body waves or surface waves (Rayleigh waves, termed *ground roll* by reflection seismologists, are often the strongest arrival in reflection records), that do not travel along the predicted NMO travel time curves and thus do not stack coherently.

CMP stacking has proven to be very successful in practice and is widely used to produce reflection profiles at a minimum of computational expense. However, it requires knowledge of the velocity-depth function to compute the NMO times and it does not explicitly account for the possibility of energy reflected or scattered from non-horizontal interfaces. We will discuss ways to address some of these limitations later in this chapter, but first we examine source effects.

7.3 Sources and deconvolution

The ideal source for reflection seismology would produce a delta function or a very short impulsive wavelet that would permit closely spaced reflectors to be clearly resolved. In practice, however, more extended sources must be used and the finite source durations can cause complications in interpreting the data. For example, an airgun is often used for marine seismic reflection profiling. This device is towed behind a ship and fires bursts of compressed air at regular intervals. This creates a bubble that oscillates for several cycles before dissipating, producing a complicated "ringy" source-time function (e.g., Fig. 7.6). The reflection seismograms produced by such a source will reproduce this source-time function for each reflector. This is not too confusing in the case where there are only a few, widely separated reflectors. However, if several closely spaced reflectors are present then it becomes difficult to separate the real structure from the source.

The combination of the Earth response with the source-time function is termed *convolution* (see Appendix E) and may be written as

$$u(t) = s(t) * G(t) \equiv \int_0^{t_s} s(\tau)G(t - \tau)\, d\tau, \tag{7.14}$$

where $u(t)$ is the recorded seismogram, $s(t)$ is the effective source-time function (i.e., what is actually recorded by the receiver; we assume that $s(t)$ includes the receiver response and any near-source attenuation), $G(t)$ is the Earth response,

(a)

(b)

(c)

Figure 7.6 (a) A schematic example of a typical source-time function $s(t)$ produced by an airgun in a marine experiment. A series of bubble pulses are produced by pressure reverberations within the water. (b) An idealized example of the Earth response function $G(t)$ showing a number of reflected pulses. (c) The result of convolving (a) and (b). While single isolated reflectors can still be identified, closely spaced reflectors produce a complex time series that cannot easily be unraveled.

and t_s is the duration of the source. Recovering $G(t)$ from $u(t)$ in this case is termed *deconvolution* and is often an important part of reflection seismic processing. However, it is not always clear how best to perform deconvolution and this has been the subject of considerable research. The problem appears simpler in the frequency domain (see Appendix E) where convolution is expressed as a product, that is,

$$u(\omega) = s(\omega)G(\omega), \tag{7.15}$$

where $u(\omega)$, $s(\omega)$, and $G(\omega)$ are the Fourier transforms of $u(t)$, $s(t)$, and $G(t)$. Thus, in principle, frequency-domain deconvolution is straightforward:

$$G(\omega) = \frac{u(\omega)}{s(\omega)}. \tag{7.16}$$

The desired time series $G(t)$ can then be obtained from the inverse Fourier transform of $G(\omega)$. The difficulty with this approach is that (7.16) is exact and stable only for noiseless data and when $s(\omega)$ does not go to zero. In practice, some noise is present and the effective source-time function is usually band-limited so that $s(\omega)$ becomes very small at the low- and high-frequency limits. These complications can cause (7.16) to become unstable or produce artifacts in the deconvolved waveform. To address these difficulties, various methods for stabilizing deconvolution have been developed. Often a time-domain approach is more efficient for data processing, in which case a filter is designed to perform the deconvolution directly on the data.

Although deconvolution is an important part of reflection data processing, no deconvolution method is perfect, and some information is invariably lost in the process of convolution with the source-time function that cannot be recovered. For this reason, it is desirable at the outset to obtain as impulsive a source-time function as possible. Modern marine profiling experiments use airgun arrays that are designed to minimize the amplitudes of the later bubble pulses, resulting in much cleaner and less ringy pulses than the example plotted in Figure 7.6a.

Another important source-time function is produced by a machine that vibrates over a range of frequencies. This is the most common type of source for shallow crustal profiling on land and is termed *vibroseis* after the first commercial application of the method. The machine produces ground motion of the form of a modulated sinusoid, termed a *sweep*,

$$v(t) = A(t) \sin[2\pi(f_0 + bt)t]. \tag{7.17}$$

The amplitude $A(t)$ is normally constant except for a taper to zero at the start and end of the sweep. The sweep lasts from about 5 to 40 s with frequencies ranging from about 10 to 60 Hz. The sweep duration is long enough compared with the interval between seismic reflections that raw vibroseis records are difficult to interpret. To obtain clearer records, the seismograms, $u(t)$, are cross-correlated with the vibroseis sweep function.

The cross-correlation $f(t)$ between two real functions $a(t)$ and $b(t)$ is defined as

$$f(t) = a(t) \star b(t) = \int_{-\infty}^{\infty} a(\tau - t)b(\tau) \, d\tau, \tag{7.18}$$

where, following Bracewell (1978), we use the five-pointed star symbol $\star$ to denote cross-correlation; this should not be confused with the asterisk $*$ that indicates convolution. The cross-correlation integral is very similar to the convolution integral but without the time reversal of (7.14). Note that

$$a(t) * b(t) = a(-t) \star b(t) \tag{7.19}$$

and that, unlike convolution, cross-correlation is not commutative:

$$a(t) \star b(t) \neq b(t) \star a(t). \tag{7.20}$$

Cross-correlation of the vibroseis sweep function $v(t)$ with the original seismogram $u(t)$ yields the processed time series $u'(t)$:

$$u'(t) = v(t) \star u(t) = \int_0^{t_s} v(\tau - t)u(\tau)\,d\tau, \tag{7.21}$$

where t_s is the sweep duration. From (7.14) and replacing $s(t)$ with $v(t)$, we obtain

$$\begin{aligned} u'(t) &= v(t) \star [v(t) * G(t)] &(7.22)\\ &= v(-t) * [v(t) * G(t)] &(7.23)\\ &= [v(-t) * v(t)] * G(t) &(7.24)\\ &= [v(t) \star v(t)] * G(t) &(7.25)\\ &= v'(t) * G(t), &(7.26) \end{aligned}$$

where we have used (7.19) and the associative rule for convolution. The cross-correlation of $v(t)$ with itself, $v'(t) = v(t) \star v(t)$, is termed the *autocorrelation* of $v(t)$. This is a symmetric function, centered at $t = 0$, that is much more sharply peaked than $v(t)$. Thus, by cross-correlating the recorded seismogram with the vibroseis sweep function $v(t)$, one obtains a time series that represents the Earth response convolved with an effective source that is relatively compact. These relationships are illustrated in Figure 7.7. Cross-correlation with the source function is a simple form of deconvolution that is sometimes termed *spiking deconvolution*. Notice that the resulting time series is only an approximation to the desired Earth response function $G(t)$. More sophisticated methods of deconvolution can achieve better results, but $G(t)$ can never be recovered perfectly since $v(t)$ is band-limited and the highest and lowest frequency components of $G(t)$ are lost in the convolution.

7.4 Migration

Up to this point, we have modeled reflection seismograms as resulting from reflections off horizontal interfaces. However, in many cases lateral variations in structure are present; indeed, resolving these features is often a primary goal of reflection profiling. Dipping, planar reflectors can be accommodated by modifying the NMO equations to adjust for differences between the updip and downdip directions. However, more complicated structures will produce scattered and diffracted arrivals that cannot be modeled by simple plane-wave reflections, and accurate interpretation

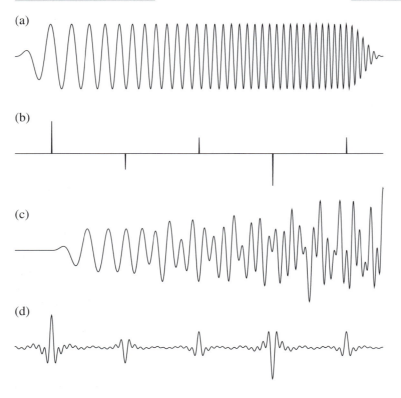

Figure 7.7 (a) An example of a vibroseis sweep function $v(t)$. (b) A hypothetical Earth response function $G(t)$. (c) The result of convolving (a) and (b). (d) The result of cross-correlating (a) with (c).

of data from such features requires a theory that takes these arrivals into account. Most of the analysis techniques developed for this purpose are based on the idea that velocity perturbations in the medium can be thought of as generating secondary seismic sources in response to the incident wavefield, and the reflected wavefield can be modeled as a sum of these secondary wavelets.

7.4.1 Huygens' principle

Huygens' principle, first described by Christiaan Huygens (*c.* 1678), is most commonly mentioned in the context of light waves and optical ray theory, but it is applicable to any wave propagation problem. If we consider a plane wavefront traveling in a homogeneous medium, we can see how the wavefront can be thought to propagate through the constructive interference of secondary wavelets (Fig. 7.8). This simple idea provides, at least in a qualitative sense, an explanation for the behavior of waves when they pass through a narrow aperture.

(a)

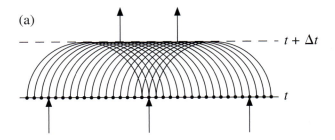

(b)

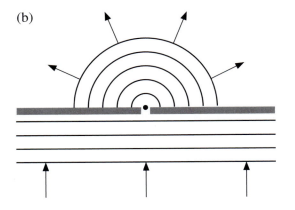

Figure 7.8 Illustrations of Huygens' principle. (a) A plane wave at time $t + \Delta t$ can be modeled as the coherent sum of the spherical wavefronts emitted by point sources on the wavefront at time t. (b) A small opening in a barrier to incident waves will produce a diffracted wavefront if the opening is small compared to the wavelength.

The bending of the ray paths at the edges of the gap is termed *diffraction*. The degree to which the waves diffract into the "shadow" of the obstacle depends upon the wavelength of the waves in relation to the size of the opening. At relatively long wavelengths (e.g., ocean waves striking a hole in a jetty), the transmitted waves will spread out almost uniformly over 180°. However, at short wavelengths the diffraction from the edges of the slot will produce a much smaller spreading in the wavefield. For light waves, very narrow slits are required to produce noticeable diffraction. These properties can be modeled using Huygens' principle by computing the effects of constructive and destructive interference at different wavelengths.

7.4.2 Diffraction hyperbolas

We can apply Huygens' principle to reflection seismology by imagining that each point on a reflector generates a secondary source in response to the incident wavefield. This is sometimes called the "exploding reflector" model. Consider a single point scatterer in a zero-offset section (Fig. 7.9). The minimum travel time is given by

$$t_0 = \frac{2h}{v}, \tag{7.27}$$

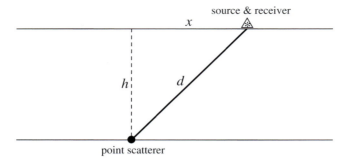

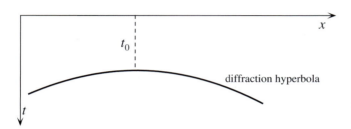

Figure 7.9 A point scatterer will produce a curved "reflector" in a zero-offset section.

where h is the depth of the scatterer and v is the velocity (assumed constant in this case). More generally, the travel time as a function of horizontal distance, x, is given by

$$t(x) = \frac{2\sqrt{x^2 + h^2}}{v}. \tag{7.28}$$

Squaring and rearranging, this can be expressed as

$$\frac{v^2 t^2}{4h^2} - \frac{x^2}{h^2} = 1 \tag{7.29}$$

or

$$\frac{t^2}{t_0^2} - \frac{4x^2}{v^2 t_0^2} = 1 \tag{7.30}$$

after substituting $4h^2 = v^2 t_0^2$ from (7.27). The travel time curve for the scattered arrival has the form of a hyperbola with the apex directly above the scattering point. Note that this hyperbola is steeper and results from a different ray geometry than the NMO hyperbola discussed in Section 7.2 (equation (7.5)). The NMO hyperbola describes travel time for a reflection off a horizontal layer as a function of source–receiver distance; in contrast (7.30) describes travel time as a function of distance

Model   ## Zero-offset section

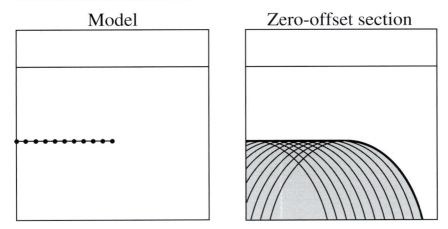

Figure 7.10 The endpoint of a horizontal reflector will produce a diffracted arrival in a zero-offset section. The reflector itself can be modeled as the coherent sum of the diffraction hyperbola from individual point scatterers. The diffracted phase, shown as the curved heavy line, occurs at the boundary of the region of scattered arrivals.

away from a point scatterer at depth for zero-offset data (the source and receiver are coincident).

7.4.3 Migration methods

Consider a horizontal reflector that is made up of a series of point scatterers, each of which generates a diffraction hyperbola in a zero-offset profile (Fig. 7.10). Following Huygens' principle, these hyperbolas sum coherently only at the time of the main reflection; the later contributions cancel out. However, if the reflector vanishes at some point, then there will be a diffracted arrival from the endpoint that will show up in the zero-offset data. This creates an artifact in the section that might be falsely interpreted as a dipping, curved reflector.

Techniques for removing these artifacts from reflection data are termed *migration* and a number of different approaches have been developed. The simplest of these methods is termed *diffraction summation migration* and involves assuming that each point in a zero-offset section is the apex of a hypothetical diffraction hyperbola. The value of the time series at that point is replaced by the average of the data from adjacent traces taken at points along the hyperbola. In this way, diffraction artifacts are "collapsed" into their true locations in the migrated section. In many cases migration can produce a dramatic improvement in image quality (e.g., Fig. 7.11).

A proper implementation of diffraction summation migration requires wave propagation theory that goes beyond the simple ideas of Huygens' principle. In particular, the scattered amplitudes vary as a function of range and ray angle, and the

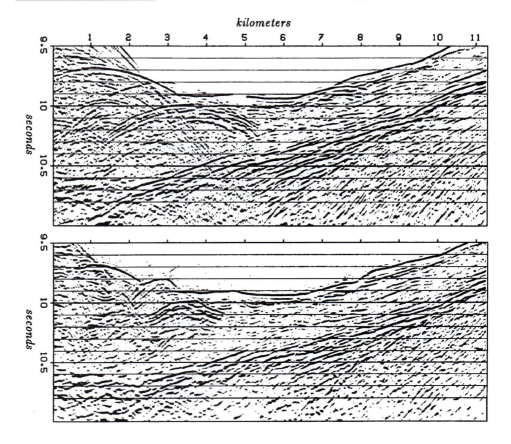

Figure 7.11 Original (top) and migrated (bottom) reflection data from a survey line across the Japan trench (figure modified from Claerbout, 1985; data from the Tokyo University Oceanographic Research Institute).

Huygens secondary sources are given, for a three-dimensional geometry, by the time derivative of the source-time function (in the frequency domain this is described by the factor $-i\omega$, a $\pi/2$ (90-degree) phase shift with amplitude proportional to frequency). In the case of a two-dimensional geometry, the secondary sources are the "half-derivative" of the source function (a 45-degree phase shift with amplitude scaled by the square root of frequency). These details are provided by Kirchhoff theory, which is discussed later in this chapter. The diffraction hyperbola equation assumes a uniform velocity structure, but migration concepts can be generalized to more complicated velocity models. However, it is important to have an accurate velocity model, as use of the wrong model can result in "undermigrated" or "overmigrated" sections.

In common practice, data from seismic reflection experiments are first processed into zero-offset sections through CMP stacking. The zero-offset section is then migrated to produce the final result. This is termed *poststack migration*. Because CMP stacking assumes horizontal layering and may blur some of the details of the original data, better results can be obtained if the migration is performed prior to stacking. This is called *prestack migration*. Although prestack migration is known to produce superior results, it is not implemented routinely owing to its much greater computational cost.

7.5 Velocity analysis

Knowledge of the large-scale background seismic velocity structure is essential for reflection seismic processing (for both stacking and migration) and for translating observed events from time to depth. Often this information is best obtained from results derived independently of the reflection experiment, such as from borehole logs or from a collocated refraction experiment. However, if such constraints are not available a velocity profile must be estimated from the reflection data themselves. This can be done in several different ways.

One approach is to examine the travel time behavior of reflectors in CMP gathers. From (7.7), we have for a reflector overlain by material of uniform velocity v:

$$t^2(x) = t_0^2 + \frac{x^2}{v^2} \tag{7.31}$$

$$= t_0^2 + u^2 x^2, \tag{7.32}$$

where $u = 1/v$ is the slowness of the layer. From observations of the NMO offsets in a CMP gather, one can plot values of t^2 versus x^2. Fitting a straight line to these points then gives the intercept t_0^2 and the slope $u^2 = 1/v^2$. Velocity often is not constant with depth, but this equation will still yield a velocity, which can be shown to be approximately the root-mean-square (rms) velocity of the overlying medium, that is, for n layers

$$v^2 \approx \frac{\sum_{i=1}^{n} v_i^2 \Delta t_i}{\sum_{i=1}^{n} \Delta t_i}, \tag{7.33}$$

where Δt_i is the travel time through the ith layer.

Another method is to plot NMO corrected data as a function of offset for different velocity models to see which model best removes the range dependence in the data or produces the most coherent image following CMP stacking. As in the case of the

$t^2(x^2)$ plotting method, this will only resolve the velocities accurately if a reasonable spread in source–receiver offsets are available. Zero-offset data have no direct velocity resolution; the constraints on velocity come from the NMO offsets in the travel times with range. Thus, wider source–receiver profiling generally produces better velocity resolution, with the best results obtained in the case where receiver ranges are extended far enough to capture the direct refracted phases. However, even zero-offset data can yield velocity information if diffraction hyperbolas are present in the zero-offset profiles, as the curvature of these diffracted phases can be used to constrain the velocities. One approach is to migrate the section with different migration velocities in order to identify the model that best removes the artifacts in the profile.

7.5.1 Statics corrections

Often strong near-surface velocity heterogeneity produces time shifts in the records that can vary unpredictably between sources and stations. This could be caused by topography/bathymetry or a sediment layer of variable thickness. The resulting "jitter" in the observed reflected pulses (Fig. 7.12) can hinder application of stacking and migration techniques and complicate interpretation of the results. Thus, it is desirable to remove these time shifts prior to most processing of the results. This is done by applying timing corrections, termed *statics corrections*, to the data. In the case of the receivers, these are analogous to the station terms (the

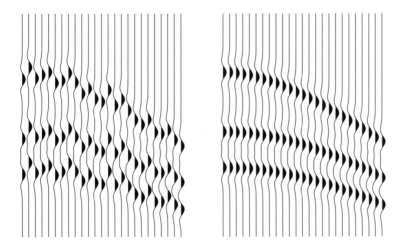

Figure 7.12 Small time shifts on individual records produce offsets in reflectors in CMP gathers (left plot) that prevent coherent stacking of these phases in data processing. These shifts can be removed by applying static corrections (right plot).

average travel time residual at a particular station for many different events) used in travel time inversions for Earth structure. Statics may be computed by tracking the arrival time of a reference phase, such as a refracted arrival. Often automatic methods are applied to find the time shifts that best smooth the observed reflectors. The goal is to shift the timing of the individual records such that reflectors will stack coherently. This problem is tractable since the time shifts are generally fairly small, and solutions for the time shifts are overdetermined in typical reflection experiments (multiple receivers for each source, multiple sources for each receiver).

7.6 Receiver functions

A wide-used technique in global seismology that has many parallels to reflection seismology is the seismic *receiver function* (Langston, 1977). The method exploits the fact that upcoming *P* waves beneath seismic stations will generate *P-to-S* converted phases at any sharp interfaces below the receiver. This *SV*-polarized wave is termed the *Ps* phase (see Figure 7.13) and will follow the direct *P* phase by a time that increases with the depth of the interface.

Thus in principle we could estimate the depth of the discontinuity from the *Ps* − *P* time, but in practice *Ps* is only rarely observed clearly on individual seismograms because it is usually obscured by the coda of the *P* wave. However, *Ps* should have the same shape as the direct *P* pulse (essentially the source-time function of the

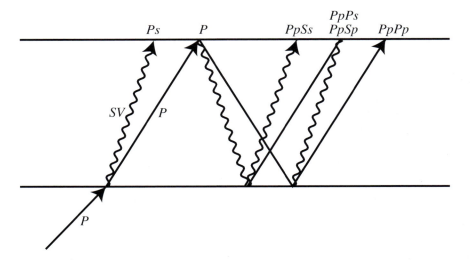

Figure 7.13 An upcoming *P* wave incident on a near-surface velocity discontinuity will generate a number of first-order converted and reflected phases.

earthquake as modified by attenuation along the ray path), and thus, just as in the vibroseis example discussed above, Ps can often be revealed by deconvolving the P pulse from the rest of the seismogram. The deconvolved waveform is termed the receiver function. For the steeply incident ray paths of distant teleseisms, P appears most strongly on the vertical component and Ps on the radial component. Thus, the simplest approach extracts the direct P pulse from the vertical channel and performs the deconvolution on the radial component. However, somewhat cleaner results can be obtained by applying a transformation to estimate the upcoming P and SV wave field from the observed vertical and radial components. This transformation must include the effect of the free-surface reflected phases (i.e., the downgoing P and SV pulses) on the observed displacement at the surface.

Following Kennett (1991) and Bostock (1998), we may express the upcoming P and SV components at the surface as

$$
\begin{bmatrix} P \\ SV \end{bmatrix} = \begin{bmatrix} p\beta^2/\alpha & (1 - 2\beta^2 p^2)/2\alpha\eta_\alpha \\ (1 - 2\beta^2 p^2)/2\beta\eta_\beta & -p\beta \end{bmatrix} \begin{bmatrix} U_R \\ U_Z \end{bmatrix}, \qquad (7.34)
$$

where U_R and U_Z are the radial and vertical components, p is the ray parameter, α and β are the P and S velocities at the surface, and the P and S vertical slownesses are given by $\eta_\alpha = \sqrt{\alpha^{-2} - p^2}$ and $\eta_\beta = \sqrt{\beta^{-2} - p^2}$. Note that here we adopt the sign convention, opposite to that in Bostock (1998), that the incident P wave has the same polarity on the vertical and radial seismometer components.

This transformation has been applied to the vertical and radial channels in Figure 7.14 and shows how the resulting P and SV components isolate the different phases. In particular, note that the direct P arrival appears only on the P component, while Ps appears only on the SV component. The reverberated phases are seen separately on the P and S components according to whether the upcoming final leg beneath the receiver is P or SV. The resulting receiver function is plotted at the bottom of Figure 7.14, and in this case has pulses at times given by the differential times of Ps, $PpPs$ and $PpSs$ with respect to the direct P arrival. In general, the receiver-function pulse shapes are more impulsive and symmetric than that of the P waveform, but their exact shape will depend on the deconvolution method and the bandwidth of the data.

Analysis of receiver functions is similar to reflection seismology in many respects. Both methods study seismic phases resulting from velocity jumps at interfaces beneath receivers and require knowledge of the background seismic velocities to translate the timing of these phases into depth. Both often use deconvolution and stacking to improve the signal-to-noise ratio of the results. When closely spaced seismic receivers are available (for example, a seismometer array), migration methods can be used to image lateral variations in structure and correct for scattering

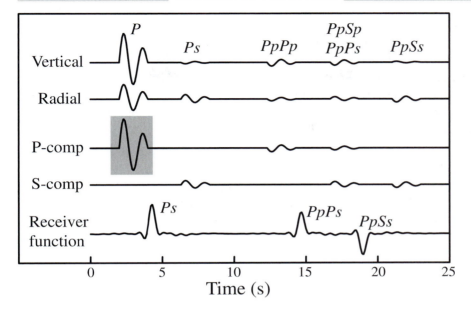

Figure 7.14 Seismograms showing the response of a simple model of a 35 km thick crust to an upcoming mantle P wave. The bottom trace shows the receiver function computed by deconvolving the windowed P pulse on the P-component trace from the S-component trace. The time scale for the top four traces is the same, but the receiver function timing is relative to the P arrival, which is why the receiver function pulses are shifted compared to the pulses in the other traces.

artifacts. However, receiver function analysis is complicated by the multiple arrivals generated by single interfaces (see Figure 7.14), and when several discontinuities are present at different depths, it can be difficult to separate out the effects of Ps phases from possible reverberations from shallower discontinuities.

If results are available for sources at different epicentral distances from the receiver (i.e., which will arrive at different ray parameters), it is sometimes possible to distinguish the reverberated phases from Ps by noting their different moveout with distance. In particular, the differential time $T_{Ps} - T_P$ shrinks with epicentral distance (i.e., decreasing ray parameter) while $T_{PpPs} - T_P$ and $T_{PpSs} - T_P$ increase with epicentral distance. For shallow discontinuities (less than about 150 km), the timing of these arrivals can safely be computed using a plane-wave approximation for the incident P wave, which assumes that all of the arrivals have the same ray parameter. However, for deeper features, such as the transition-zone discontinuities near 410 and 660 km depth, the curvature of the wavefront should be taken into account (Lawrence and Shearer, 2006).

The P-to-SV converted Ps phase is sensitive almost entirely to the S velocity jump at interfaces, whereas the reverberated phases are also sensitive to the P

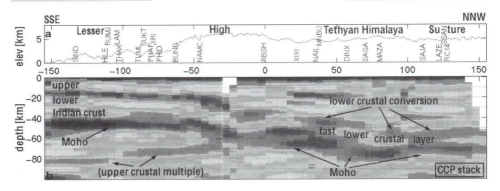

Figure 7.15 A cross-section of the Himalayan crust, produced from a common conversion point (CCP) stack of receiver functions computed for a line of seismic stations crossing Nepal. A clear Moho conversion is seen, indicating that the crustal thickness increases from ~45 km under India to ~75 km beneath Tibet. Figure adapted from Schulte-Pelkum *et al.* (2005).

velocity and density jumps. Thus, in principle integrated analysis of the complete *P* coda wavefield can provide much more information than simple receiver function studies of *Ps* alone. Bostock *et al.* (2001) describe a theory for how this can be done to image crust and upper-mantle structure by processing three-component data from seismic arrays.

Receiver functions have become one of the most popular methods in the global seismologist's toolbox because they are relatively simple to compute and can provide valuable results even from only a single station. They have been used to resolve crustal thickness and depths to the 410 and 660 km discontinuities beneath seismic stations all over the world, and, in subduction zones, to resolve the top of the subducting slab. Where seismic arrays are present, often from temporary seismic experiments, they can produce dramatic images of crust and lithospheric structure as illustrated in Figure 7.15 for a cross-section under the Himalayas.

7.7 Kirchhoff theory[†]

A more rigorous treatment of Huygens' principle was given by Kirchhoff and forms the basis for a number of important techniques for computing synthetic seismograms. Descriptions of applications of Kirchhoff methods to seismology may be found in Scott and Helmberger (1983) and Kampmann and Müller (1989). Kirchhoff theory was first developed in optics and our derivation until equation (7.56) largely follows that of Longhurst (1967). Consider the scalar wave equation (e.g., equation (3.31) where ϕ is the P-wave potential)

$$\nabla^2\phi = \frac{1}{c^2}\frac{\partial^2\phi}{\partial t^2}, \tag{7.35}$$

where c is the wave velocity. Now assume a harmonic form for ϕ, that is, at a particular frequency ω we have the monochromatic function

$$\phi = \psi(\mathbf{r})e^{-i\omega t} = \psi(\mathbf{r})e^{-ikct}, \tag{7.36}$$

where $\mathbf{r}$ is the position and $k = \omega/c$ is the wavenumber. Note that we have separated the spatial and temporal parts of ϕ. We then have

$$\nabla^2\phi = e^{-ikct}\nabla^2\psi \tag{7.37}$$

and

$$\frac{\partial^2\phi}{\partial t^2} = -k^2c^2\psi e^{-ikct} \tag{7.38}$$

and (7.35) becomes

$$\nabla^2\psi = -k^2\psi. \tag{7.39}$$

This is a time-independent wave equation for the space-dependent part of ϕ. It is also sometimes termed the Helmholtz equation.

Next, recall Green's theorem from vector calculus. If ψ_1 and ψ_2 are two continuous single-valued functions with continuous derivatives, then for a closed surface S

$$\int_V (\psi_2\nabla^2\psi_1 - \psi_1\nabla^2\psi_2)\,dv = \int_S \left(\psi_2\frac{\partial\psi_1}{\partial n} - \psi_1\frac{\partial\psi_2}{\partial n}\right)dS, \tag{7.40}$$

where the volume integral is over the volume enclosed by S, and $\partial/\partial n$ is the derivative with respect to the outward normal vector to the surface. Now assume that both ψ_1 and ψ_2 satisfy (7.39), that is,

$$\nabla^2\psi_1 = -k^2\psi_1, \tag{7.41}$$

$$\nabla^2\psi_2 = -k^2\psi_2. \tag{7.42}$$

In this case, the left part of (7.40) vanishes and the surface integral must be zero:

$$\int_S \left(\psi_2\frac{\partial\psi_1}{\partial n} - \psi_1\frac{\partial\psi_2}{\partial n}\right)dS = 0. \tag{7.43}$$

Now suppose that we are interested in evaluating the disturbance at the point P, which is enclosed by the surface S (Fig. 7.16). We set $\psi_1 = \psi$, the amplitude of

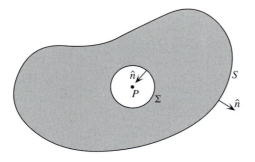

Figure 7.16 A point P surrounded by a surface S of arbitrary shape. Kirchhoff's formula is derived by applying Green's theorem to the volume between S and an infinitesimal sphere Σ surrounding P.

the harmonic disturbance. We are free to choose any function for ψ_2, provided it also satisfies (7.39). It will prove useful to define ψ_2 as

$$\psi_2 = \frac{e^{ikr}}{r}, \tag{7.44}$$

where r is the distance from P. This function has a singularity at $r = 0$ and so the point P must be excluded from the volume integral for Green's theorem to be valid. We can do this by placing a small sphere Σ around P. Green's theorem may now be applied to the volume between Σ and S; these surfaces, together, make up the integration surface. On the surface of the small sphere the outward normal to this volume is opposite to the direction of r and thus $\partial/\partial n$ can be replaced with $-\partial/\partial r$ and the surface integral over Σ may be expressed as

$$\int_{\Sigma} \left(\psi_2 \frac{\partial \psi}{\partial n} - \psi \frac{\partial \psi_2}{\partial n} \right) dS = \int_{\Sigma} \left[\frac{-e^{ikr}}{r} \frac{\partial \psi}{\partial r} + \psi \frac{\partial}{\partial r} \left(\frac{e^{ikr}}{r} \right) \right] dS$$

$$= \int_{\Sigma} \left[\frac{-e^{ikr}}{r} \frac{\partial \psi}{\partial r} + \psi \left(\frac{-e^{ikr}}{r^2} + \frac{ike^{ikr}}{r} \right) \right] dS. \tag{7.45}$$

Now let us change this to an integral over solid angle Ω from the point P, in which dS on Σ subtends $d\Omega$ and $dS = r^2 \, d\Omega$. Then

$$\int_{\Sigma} = \int_{\text{around } P} \left(-re^{ikr} \frac{\partial \psi}{\partial r} - \psi e^{ikr} + rik\psi e^{ikr} \right) d\Omega. \tag{7.46}$$

Now consider the limit as r goes to zero. Assuming that ψ does not vanish, then only the second term in this expression survives. Thus as $r \to 0$

$$\int_{\Sigma} \to \int -\psi e^{ikr} \, d\Omega. \tag{7.47}$$

As the surface Σ collapses around P, the value of ψ on the surface may be assumed to be constant and equal to ψ_P, its value at P. Thus

$$\int_{\Sigma} \rightarrow \int -\psi_P e^{ikr} d\Omega \qquad (7.48)$$

$$= -\psi_P \int e^{ikr} d\Omega \qquad (7.49)$$

$$= -\psi_P \int d\Omega \text{ since } e^{ikr} \rightarrow 1 \text{ as } r \rightarrow 0 \qquad (7.50)$$

$$= -4\pi\psi_P. \qquad (7.51)$$

From (7.43) we know that $\int_{S+\Sigma} = 0$, so we must have $\int_S = +4\pi\psi_P$, or

$$4\pi\psi_P = \int_S \left[\frac{e^{ikr}}{r} \frac{\partial \psi}{\partial n} - \psi \frac{\partial}{\partial n} \left(\frac{e^{ikr}}{r} \right) \right] dS \qquad (7.52)$$

$$= \int_S \left[\frac{e^{ikr}}{r} \frac{\partial \psi}{\partial n} - \psi e^{ikr} \frac{\partial}{\partial n} \left(\frac{1}{r} \right) - \frac{ik\psi e^{ikr}}{r} \frac{\partial r}{\partial n} \right] dS, \qquad (7.53)$$

since $\frac{\partial}{\partial n} = \frac{\partial}{\partial r} \frac{\partial r}{\partial n}$. This is often called Helmholtz's equation. Since $\phi = \psi e^{-ikct}$ (7.35), we have

$$\psi = \phi e^{ikct} \qquad (7.54)$$

and (7.53) becomes

$$\phi_P = \frac{1}{4\pi} e^{-ikct} \int_S \left[\frac{e^{ikr}}{r} \frac{\partial \psi}{\partial n} - \psi e^{ikr} \frac{\partial}{\partial n} \left(\frac{1}{r} \right) - \frac{ik\psi e^{ikr}}{r} \frac{\partial r}{\partial n} \right] dS$$

$$= \frac{1}{4\pi} \int_S \left[\frac{e^{-ik(ct-r)}}{r} \frac{\partial \psi}{\partial n} - \psi e^{-ik(ct-r)} \frac{\partial}{\partial n} \left(\frac{1}{r} \right) - \frac{ik\psi e^{-ik(ct-r)}}{r} \frac{\partial r}{\partial n} \right] dS. \qquad (7.55)$$

This expression gives $\phi(t)$ at the point P. Note that the term $\psi e^{-ik(ct-r)} = \psi e^{-ikc(t-r/c)}$ is the value of ϕ at the element dS at the time $t - r/c$. This time is referred to as the *retarded value* of ϕ and is written $[\phi]_{t-r/c}$. In this way, we can express (7.55) as

$$\phi_P = \frac{1}{4\pi} \int_S \left(\frac{1}{r} \left[\frac{\partial \phi}{\partial n} \right]_{t-r/c} - \frac{\partial}{\partial n} \left(\frac{1}{r} \right) [\phi]_{t-r/c} + \frac{1}{cr} \frac{\partial r}{\partial n} \left[\frac{\partial \phi}{\partial t} \right]_{t-r/c} \right) dS, \qquad (7.56)$$

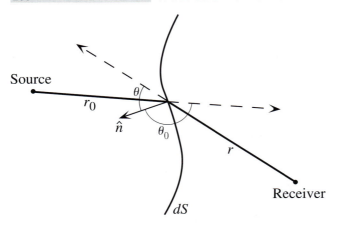

Figure 7.17 The ray geometry for a single point on a surface dS separating a source and receiver.

where we have used $\partial \phi / \partial t = -ikc\psi e^{-ikct}$. Equation (7.56) is a standard form for what is often termed Kirchhoff's formula; it is found in many optics textbooks and is also given in Scott and Helmberger (1983). We see that the disturbance at P can be computed from the conditions of ϕ over a closed surface surrounding P where r/c represents the time taken for the disturbance to travel the distance r from dS to P. We need to know both the value of ϕ and its normal derivative on dS to compute this integral.

This is not an especially convenient form to use directly in most seismic applications. Suppose the value of ϕ at each point on the surface could be obtained from a source time function $f(t)$ a distance r_0 from dS. Then on dS we have

$$\phi = \frac{1}{r_0} f(t - r_0/c), \tag{7.57}$$

$$\frac{\partial \phi}{\partial t} = \frac{1}{r_0} f'(t - r_0/c), \tag{7.58}$$

where the $1/r_0$ term comes from the geometrical spreading of the wavefront.

If θ_0 and θ are the angles of the incoming and outgoing ray paths from the surface normal (Fig. 7.17), then

$$\frac{\partial r_0}{\partial n} = \cos \theta_0 \quad \text{and} \quad \frac{\partial r}{\partial n} = \cos \theta, \tag{7.59}$$

$$\frac{\partial \phi}{\partial n} = \frac{\partial \phi}{\partial r_0} \frac{\partial r_0}{\partial n} \tag{7.60}$$

$$= \frac{\partial \phi}{\partial r_0} \cos \theta_0, \tag{7.61}$$

and

$$\frac{\partial}{\partial n}\left(\frac{1}{r}\right) = \frac{\partial r}{\partial n}\frac{\partial}{\partial r}\left(\frac{1}{r}\right) \qquad (7.62)$$

$$= -\frac{1}{r^2}\cos\theta. \qquad (7.63)$$

We can evaluate $\partial\phi/\partial r_0$ using the chain rule:

$$\frac{\partial}{\partial r_0}\left(\frac{1}{r_0}f(t - r_0/c)\right) = -\frac{1}{r_0^2}f(t - r_0/c) - \frac{1}{cr_0}f'(t - r_0/c) \qquad (7.64)$$

since $\frac{\partial}{\partial r_0} = \frac{\partial t}{\partial r_0}\frac{\partial}{\partial t} = -\frac{1}{c}\frac{\partial}{\partial t}$. Putting (7.57)–(7.64) into (7.56), we have

$$\phi_P(t) = \frac{1}{4\pi}\int_S\left(\frac{-1}{rr_0^2}\cos\theta_0 + \frac{1}{r^2 r_0}\cos\theta\right)f(t - r/c - r_0/c)\,dS$$

$$+ \frac{1}{4\pi}\int_S\left(\frac{-1}{crr_0}\cos\theta_0 + \frac{1}{crr_0}\cos\theta\right)f'(t - r/c - r_0/c)\,dS. \qquad (7.65)$$

The negative signs arise from our definition of $\hat{n}$ in the direction opposing r_0; these terms are positive since $\cos\theta_0$ is negative. Equation (7.65) may also be expressed in terms of convolutions with $f(t)$ and $f'(t)$:

$$\phi_P(t) = \frac{1}{4\pi}\int_S\delta\left(t - \frac{r + r_0}{c}\right)\left(\frac{-1}{rr_0^2}\cos\theta_0 + \frac{1}{r^2 r_0}\cos\theta\right)dS * f(t)$$

$$+ \frac{1}{4\pi}\int_S\delta\left(t - \frac{r + r_0}{c}\right)\left(\frac{-1}{crr_0}\cos\theta_0 + \frac{1}{crr_0}\cos\theta\right)dS * f'(t).$$

$$(7.66)$$

Notice that the $f(t)$ terms contain an extra factor of $1/r$ or $1/r_0$. For this reason they are most important close to the surface of integration and can be thought of as near-field terms. In practice, the source and receiver are usually sufficiently distant from the surface (i.e., $\lambda \ll r, r_0$) that ϕ_P is well approximated by using only the far-field $f'(t)$ terms. In this case we have

$$\phi_P(t) = \frac{1}{4\pi c}\int_S\delta\left(t - \frac{r + r_0}{c}\right)\frac{1}{rr_0}(-\cos\theta_0 + \cos\theta)\,dS * f'(t). \qquad (7.67)$$

This formula is the basis for many computer programs that compute Kirchhoff synthetic seismograms.

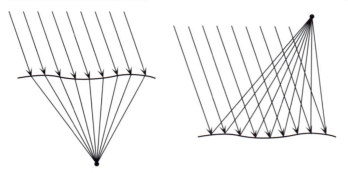

Figure 7.18 Kirchhoff theory can be used to compute the effect of an irregular boundary on both transmitted and reflected waves.

7.7.1 Kirchhoff applications

Probably the most common seismic application of Kirchhoff theory involves the case of an irregular interface between simpler structure on either side. Kirchhoff theory can be used to provide an approximate solution for either the transmitted or reflected wavefield due to this interface (Fig. 7.18). For example, we might want to model the effect of irregularities on the core–mantle boundary, the Moho, the sea floor, or a sediment–bedrock interface. In each case, there is a significant velocity contrast across the boundary.

Let us consider the reflected wave generated by a source above an undulating interface. Assume that the incident wavefield is known and can be described with geometrical ray theory. Then we make the approximation that the reflected wavefield just above the interface is given by the plane-wave reflection coefficient for the ray incident on the surface. This approximation is sometimes called the Kirchhoff, physical optics, or tangent plane hypothesis (Scott and Helmberger, 1983). Each point on the surface reflects the incident pulse as if there were an infinite plane tangent to the surface at that point. Considering only the far-field terms, we then have

$$\phi_P(t) = \frac{1}{4\pi c} \int_S \delta\left(t - \frac{r + r_0}{c}\right) \frac{R(\theta_0)}{r r_0} (\cos\theta_0 + \cos\theta)\, dS * f'(t), \qquad (7.68)$$

where $R(\theta_0)$ is the reflection coefficient, θ_0 is the angle between the incident ray and the surface normal, and θ is the angle between the scattered ray and the surface normal (see Fig. 7.19).

If the overlying layer is not homogeneous, then the $1/r$ and $1/r_0$ terms must be replaced with the appropriate source-to-interface and interface-to-receiver geometrical spreading coefficients. In some cases, particularly for obliquely arriving rays,

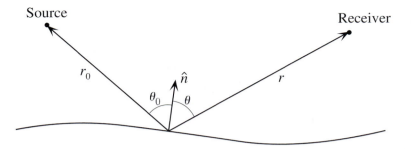

Figure 7.19 Ray angles relative to the surface normal for a reflected wave geometry.

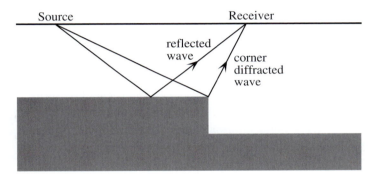

Figure 7.20 This structure will produce both a direct reflected pulse and a diffracted pulse for the source–receiver geometry shown.

the reflection coefficient R may become complex. If this occurs, then this equation will have both a real and an imaginary part. The final time series is obtained by adding the real part to the Hilbert transform of the imaginary part.

The Kirchhoff solution will correctly model much of the frequency dependence and diffracted arrivals in the reflected wavefield. These effects are not obtained through geometrical ray theory alone, even if 3-D ray tracing is used. For example, consider a source and receiver above a horizontal interface containing a vertical fault (Fig. 7.20). Geometrical ray theory will produce only the main reflected pulse from the interface, while Kirchhoff theory will provide both the main pulse and the secondary pulse diffracted from the corner. However, Kirchhoff theory also has its limitations, in that it does not include any multiple scattering or diffractions along the interface; these might be important in more complicated examples.

7.7.2 How to write a Kirchhoff program

As an illustration, let us list the steps involved in writing a hypothetical Kirchhoff computer program to compute the reflected wavefield from a horizontal interface with some irregularities.

1. Specify the source and receiver locations.
2. Specify the source-time function $f(t)$.
3. Compute $f'(t)$, the derivative of the source-time function.
4. Initialize to zero a time series $J(t)$, with sample interval dt, that will contain the output of the Kirchhoff integral.
5. Specify the interface with a grid of evenly spaced points in x and y. At each grid point, we must know the height of the boundary z and the normal vector to the surface $\hat{n}$. We also require the surface area, dA, corresponding to the grid point. This is approximately $dx\, dy$ where dx and dy are the grid spacings in the x and y directions, respectively (if a significant slope is present at the grid point, the actual surface area is greater and this correction must be taken into account). The grid spacing should be finer than the scale length of the irregularities.
6. At each grid point, trace rays to the source and receiver. Determine the travel times to the source and receiver, the ray angles to the local normal vector (θ_0 and θ), and the geometrical spreading factors g_0 (source-to-surface) and g (surface-to-receiver).
7. At each grid point, compute the reflection coefficient $R(\theta_0)$ and the factor $\cos\theta_0 + \cos\theta$.
8. At each grid point, compute the product $R(\theta_0)(\cos\theta_0 + \cos\theta)dA/(4\pi c g_0 g)$. Add the result to the digitized point of $J(t)$ that is closest to the total source-surface-receiver travel time, after first dividing the product by the digitization interval dt.
9. Repeat this for all grid points that produce travel times within the time interval of interest.
10. Convolve $J(t)$ with $f'(t)$, the derivative of the source-time function, to produce the final synthetic seismogram.
11. (very important) Repeat this procedure at a finer grid spacing, dx and dy, to verify that the same result is obtained. If not, the interface is undersampled and a finer grid must be used.

Generally the $J(t)$ function will contain high-frequency numerical "noise" that is removed through the convolution with $f'(t)$. It is computationally more efficient to compute $f'(t)$ and convolve with $J(t)$ than to compute $J'(t)$ and convolve with $f(t)$, particularly when multiple receiver positions are to be modeled.

7.7.3 Kirchhoff migration

Kirchhoff results can be used to implement migration methods for reflection seismic data that are consistent with wave propagation theory. For zero-offset data, $\theta_0 = \theta$

and $r_0 = r$ and (7.68) becomes

$$\phi_P(t) = \frac{1}{2\pi c} \int_S \delta \left(t - \frac{r + r_0}{c} \right) \frac{R(\theta_0)}{r^2} \cos \theta \, dS * f'(t). \qquad (7.69)$$

To perform the migration, the time derivative of the data is taken and the traces for each hypothetical scattering point are multiplied by the obliquity factor $\cos \theta$, scaled by the spherical spreading factor $1/r^2$ and then summed along the diffraction hyperbolas.

7.8 Exercises

1. (COMPUTER) Recall equation (7.17) for the vibroseis sweep function:

$$v(t) = A(t) \sin[2\pi(f_0 + bt)t].$$

   (a) Solve for f_0 and b in the case of a 20-s long sweep between 1 and 4 Hz. Hint: $b = 3/20$ is incorrect! Think about how rapidly the phase changes with time.
   (b) Compute and plot $v(t)$ for this sweep function. Assume that $A(t) = \sin^2(\pi t/20)$ (this is termed a Hanning taper; note that it goes smoothly to zero at $t = 0$ and $t = 20$ s). Check your results and make sure that you have the right period at each end of the sweep.
   (c) Compute and plot the autocorrelation of $v(t)$ between -2 and 2 s.
   (d) Repeat (b) and (c), but this time assume that $A(t)$ is only a short 2-s long taper at each end of the sweep, that is, $A(t) = \sin^2(\pi t/4)$ for $0 < t < 2$, $A(t) = 1$ for $2 \leq t \leq 18$, and $A(t) = \sin^2[\pi(20 - t)/4]$ for $18 < t < 20$. Note that this milder taper leads to more extended sidelobes in the autocorrelation function.
   (e) What happens to the pulse if autocorrelation is applied a second time to the autocorrelation of $v(t)$? To answer this, compute and plot $[v(t) \star v(t)] \star [v(t) \star v(t)]$ using $v(t)$ from part (b). Is this a way to produce a more impulsive wavelet?

2. A reflection seismic experiment produces the CMP gather plotted in Figure 7.21. Using the $t^2(x^2)$ method, determine the approximate rms velocity of the material overlying each reflector. Then compute an approximate depth to each reflector. Note: You should get approximately the same velocity for each reflector; do not attempt to solve for different velocities in the different layers.

3. Consider a simple homogeneous layer over half-space model (as plotted in Fig. 7.22) with P velocity α_1 and S velocity β_1 in the top layer and P velocity α_2 in the bottom layer.

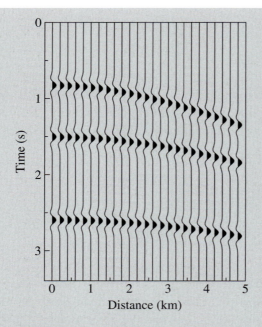

Figure 7.21 *P*-wave reflections from an individual CMP gather. Note the increase of travel time with source–receiver distance.

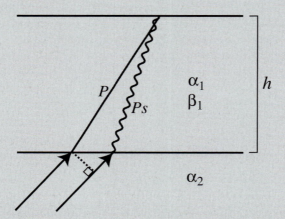

Figure 7.22 The *P* and *Ps* ray geometry for an upcoming plane wave incident on a constant velocity layer.

(a) Assuming a layer thickness of h and a ray parameter of p, derive an equation for the differential time $(T_{Ps} - T_P)$ between P and Ps at the surface. Hint: the travel time is equal along the wavefront shown by the dashed line in the bottom layer.

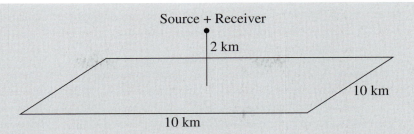

Figure 7.23 The geometry for Exercise 4, constructing a reflected pulse as a sum of secondary sources.

(b) What is $T_{P_s} - T_P$ for $h = 40$ km, $\alpha_1 = 6.3$ km/s, $\beta_1 = 3.6$ km/s, $\alpha_2 = 8$ km/s, and $p = 0.06$ s/km?

4. (COMPUTER) A common pulse shape used in reflection seismic modeling is the *Ricker wavelet*, defined as

$$s_R(t) = \left(1 - 2\pi^2 f_p^2 t^2\right) e^{-\pi^2 f_p^2 t^2}, \tag{7.70}$$

where f_p is the peak frequency in the spectrum of the wavelet.

(a) Using a digitization rate of 200 samples/s and assuming $f_p = 4$ Hz, make a plot of $s_R(t)$ and its time derivative between -0.5 and 0.5 s.

(b) Following Huygens' principle, model the plane reflector shown in Figure 7.23 as a large number of point sources. Use a velocity of 4 km/s, and a coincident source and receiver located 2 km from the center of a 10 km by 10 km plane, with secondary Huygens sources spaced every 0.1 km on the plane (i.e., 101 points in each direction, for a total of 10 201 sources). Construct and plot a synthetic seismogram representing the receiver response to the Ricker wavelet from part (a) by summing the contribution from each secondary source. At each point, compute the two-way travel time from/to the source/receiver and the geometric spreading factor $1/r^2$. Add the Ricker wavelet, centered on the two-way time and scaled by the geometric spreading factor, to your synthetic time series for each of the 10 201 points. Note that the resulting waveform for the reflected pulse is *not* the same shape as the Ricker wavelet.

(c) Repeat part (b), but this time use the Kirchhoff result of (7.68) and Section 7.7.2 that the secondary sources are given by the derivative of the Ricker wavelet. Assume that $R(\theta) = 1$ for this example. Show that the synthetic reflected pulse has the correct shape.

(d) Verify that the reflected pulse from part (c) has an amplitude of 0.25, the same as the predicted amplitude of a pulse 4 km away from a point source that has unit amplitude at $r = 1$.

(e) Note: Because convolution is a linear operation, parts (b) and (c) can be performed more efficiently by summing over the 10 201 points assuming a simple spike source ($s(t) = 1/r^2$ at the $t = 0$ point only) and then convolving the resulting time series with the Ricker wavelet or its time derivative to obtain the final synthetic seismogram. The intermediate time series will contain considerable high frequency noise but this is removed by the convolution.

(f) The main reflected pulse should arrive at $t = 1$ s. What is the origin of the small pulse at about 2.7 s?

8

Surface waves and normal modes

Our treatment to this point has been limited to body waves, solutions to the seismic wave equation that exist in whole spaces. However, when free surfaces exist in a medium, other solutions are possible and are given the name *surface waves*. There are two types of surface waves that propagate along Earth's surface: *Rayleigh waves* and *Love waves*. For laterally homogeneous models, Rayleigh waves are radially polarized (P/SV) and exist at any free surface, whereas Love waves are transversely polarized and require some velocity increase with depth (or a spherical geometry). Surface waves are generally the strongest arrivals recorded at teleseismic distances and they provide some of the best constraints on Earth's shallow structure and low-frequency source properties. They differ from body waves in many respects – they travel more slowly, their amplitude decay with range is generally much less, and their velocities are strongly frequency dependent. Surface waves from large earthquakes are observable for many hours, during which time they circle the Earth multiple times. Constructive interference among these orbiting surface waves, together with analogous reverberations of body waves, form the *normal modes*, or *free oscillations* of the Earth. Surface waves and normal modes are generally observed at periods longer than about 10 s, in contrast to the much shorter periods seen in many body wave observations.

8.1 Love waves

Love waves are formed through the constructive interference of high-order *SH* surface multiples (i.e., *SSS*, *SSSS*, *SSSSS*, etc.). Thus, it is possible to model Love waves as a sum of body waves. To see this, consider monochromatic plane-wave propagation for the case of a vertical velocity gradient in a laterally homogeneous model, a situation we previously examined in Section 6.4. In this case, a plane wave defined by ray parameter p will turn at the depth where $\beta = 1/p$. Along the

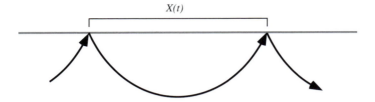

surface the plane wave will propagate with horizontal slowness defined by p. If the surface bouncepoints are separated by a distance $X(t)$, then the travel time along the surface between bouncepoints is given by $pX(p)$. This follows from our definition of a plane wave and does not depend upon the velocity model. In contrast, the travel time along the ray paths is given by $T(p)$ and is a function of the velocity–depth profile. Because these travel times are not the same, destructive interference will occur except at certain fixed frequencies. Along the surface, the phase (0 to 2π) of a harmonic wave will be delayed by $\omega pX(p)$, where ω is the angular frequency of the plane wave. The phase along the ray path is delayed by $\omega T(p) - \pi/2$, where the $-\pi/2$ comes from the WKBJ approximation for the phase advance at the plane-wave turning point (see Section 6.4). The requirement for constructive interference is thus

$$\omega pX(p) = \omega T(p) - \frac{\pi}{2} - n2\pi, \qquad (8.1)$$

where n is an integer. Rearranging, we obtain

$$\omega = \frac{n2\pi + \pi/2}{T(p) - pX(p)} = \frac{n2\pi + \pi/2}{\tau(p)}, \qquad (8.2)$$

where the delay time $\tau(p) = T(p) - pX(p)$ (see Section 4.3.2). The wave travels along the surface at velocity $c = 1/p$; thus (8.2) defines the $c(\omega)$ function for the Love waves, often termed the *dispersion curve*. The values of ω given for $n = 0$ are termed the fundamental modes; higher modes are defined by larger values of n. The frequency dispersion in the Love waves results from the ray geometry and does not require any frequency dependence in the body wave velocity β. Love wave dispersion is much stronger than the small amount of dispersion in S-wave velocities that results from intrinsic attenuation (see equation (6.91)).

The velocity defined by $c = 1/p$ is the velocity with which the peaks and troughs at a given frequency move along the surface and is termed the *phase velocity*. When the phase velocity varies as a function of frequency, as in (8.2), the wave is dispersed and the group velocity (the velocity that energy propagates) will be different from the phase velocity. In this example, the energy must move along the actual ray paths

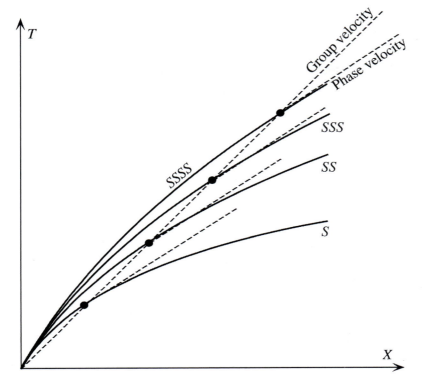

Figure 8.1 Love waves can be constructed as a sum of S surface multiples. The dashed lines show the group and phase velocities at a fixed value of the ray parameter p; the phase velocity is faster than the group velocity.

and thus the group velocity U is defined by

$$U = \frac{X(p)}{T(p)}. \tag{8.3}$$

For a prograde travel time curve (concave down), U will always be less than c. The relationship between phase and group velocity for Love waves is shown graphically in Figure 8.1 as a sum of SH surface multiples.

The group velocity is also often defined directly from the $c(\omega)$ dispersion relationship. To obtain this form, rewrite (8.1) in terms of the wavenumber $k = \omega p$, producing

$$\omega T - kX = \pi/2 + n2\pi. \tag{8.4}$$

Taking the derivative of this expression, we obtain

$$d\omega\, T + \omega\, dT - dk\, X - k\, dX = 0$$

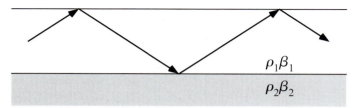

Figure 8.2 Love waves within a homogeneous layer can result from constructive interference between postcritically reflected body waves.

or

$$dω\, T - dk\, X + ω\, dX(dT/dX - k/ω) = 0. \tag{8.5}$$

Since $p = k/ω = dT/dX$, the rightmost term drops out, and we have

$$\frac{dω}{dk} = \frac{X}{T} = U, \tag{8.6}$$

and we see that the group velocity is also given by $dω/dk$.

Equation (8.2) is not very accurate at small values of n since a high frequency approximation was used to determine the phase shift at the turning point. However, it does give some understanding of how Love waves are formed through the positive interference of S surface multiples. More accurate Love wave calculations are generally performed using homogeneous layer techniques. In these methods, the plane wave response of a stack of layers is computed at a series of values of ray parameter; the frequencies of the different Love wave branches are then identified as the eigenvalues of the resulting set of equations.

8.1.1 Solution for a single layer

An exact equation may be derived for Love wave dispersion within a homogeneous layer. Consider a surface layer overlying a higher velocity half-space (Fig. 8.2). Equation (8.2) is still applicable, provided we replace the approximate $π/2$ phase shift at the turning point with the phase shift, $\phi_{\acute{S}\acute{S}}$, resulting from the SH reflection off the bottom of the layer:

$$ω = \frac{n2π - \phi_{\acute{S}\acute{S}}}{τ(p)}. \tag{8.7}$$

From (4.33), we may express the delay time $τ$ as

$$τ(p) = 2h\sqrt{1/β_1^2 - p^2}, \tag{8.8}$$

where h is the layer thickness and β_1 is the shear velocity in the top layer. For postcritical reflections, it can be shown from (6.46) and (6.59) that

$$\phi_{\breve{S}\acute{S}} = -2\tan^{-1}\left[\frac{\mu_2\sqrt{p^2 - 1/\beta_2^2}}{\mu_1\sqrt{1/\beta_1^2 - p^2}}\right]. \tag{8.9}$$

Substituting (8.8) and (8.9) into (8.7), we have

$$2h\omega\sqrt{1/\beta_1^2 - p^2} - n2\pi = 2\tan^{-1}\left[\frac{\mu_2\sqrt{p^2 - 1/\beta_2^2}}{\mu_1\sqrt{1/\beta_1^2 - p^2}}\right]$$

or

$$\tan\left[h\omega\sqrt{1/\beta_1^2 - p^2}\right] = \frac{\mu_2\sqrt{p^2 - 1/\beta_2^2}}{\mu_1\sqrt{1/\beta_1^2 - p^2}}. \tag{8.10}$$

This defines the dispersion curves for Love wave propagation within the layer. Note that the phase velocity, $c = 1/p$, varies between β_1 and β_2 ($c > \beta_2$ is not postcritical). For every value of c, there are multiple values of ω because of the periodicity in the tangent function. The smallest of the ω values defines the fundamental mode, the second smallest is the first higher mode, etc. There is no analytical solution to (8.10) for c; the $c(\omega)$ values must be determined numerically (see Exercise 8.1).

8.2 Rayleigh waves

For *SH* polarized waves, the reflection coefficient at the free surface is one, and the interference between the downgoing *SH* waves and those turned back toward the surface produces Love waves. The *P/SV* system is more complicated because the surface reflections involve both *P* and *SV* waves. In this case, the upgoing and downgoing body waves do not sum constructively to produce surface waves. However, a solution is possible for inhomogeneous waves trapped at the interface; the resulting surface waves are termed *Rayleigh waves*. The displacements of Love and Rayleigh waves are compared in Figure 8.3.

Let us begin by examining what occurs when *P* and *SV* waves interact with a free surface. For a laterally homogeneous medium, the displacements for harmonic

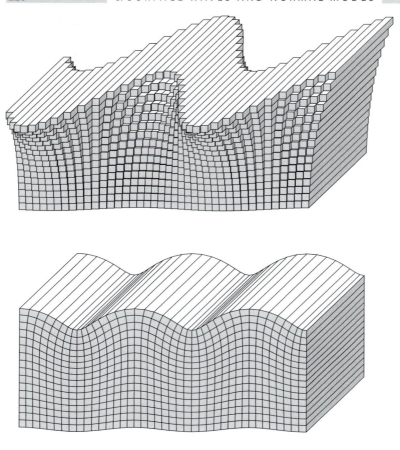

Figure 8.3 Fundamental Love (top) and Rayleigh (bottom) surface wave displacements (highly exaggerated) for horizontal propagation across the page. Love waves are purely transverse motion, whereas Rayleigh waves contain both vertical and radial motion. In both cases, the wave amplitude decays strongly with depth.

plane waves propagating in the $+x$ direction are given by

$$\mathbf{u} = \mathbf{A}e^{-i\omega(t-px-\eta z)}, \tag{8.11}$$

where p is the horizontal slowness and $\eta = \sqrt{1/c^2 - p^2}$ is the vertical slowness for wave velocity c. From Section 3.3.1, recall that we may express the displacement in terms of a P-wave scalar potential ϕ and a S-wave vector potential $\mathbf{\Psi}$, that is,

$$\mathbf{u} = \nabla\phi + \nabla \times \mathbf{\Psi}. \tag{8.12}$$

Now from (8.11), consider plane wave solutions for ϕ and Ψ_y (the only part of $\mathbf{\Psi}$ that produces SV motion for plane wave propagation in the x direction):

$$\phi = Ae^{-i\omega(t-px-\eta_\alpha z)}, \tag{8.13}$$

$$\Psi_y = Be^{-i\omega(t-px-\eta_\beta z)}, \tag{8.14}$$

where A and B are the amplitudes of the P and SV waves respectively, and the vertical slownesses are given by

$$\eta_\alpha = (1/\alpha^2 - p^2)^{1/2}, \tag{8.15}$$

$$\eta_\beta = (1/\beta^2 - p^2)^{1/2}. \tag{8.16}$$

The ray parameter p is constant; both P and SV are assumed to have the same horizontal slowness. Noting that ∂_y and u_y are zero for our P/SV plane wave geometry, the P-wave displacements are

$$u_x^P = \partial_x\phi = pAi\omega e^{-i\omega(t-px-\eta_\alpha z)}, \tag{8.17}$$

$$u_z^P = \partial_z\phi = \eta_\alpha Ai\omega e^{-i\omega(t-px-\eta_\alpha z)}, \tag{8.18}$$

and the SV-wave displacements are

$$u_x^S = -\partial_z\Psi_y = -\eta_\beta Bi\omega e^{-i\omega(t-px-\eta_\beta z)}, \tag{8.19}$$

$$u_z^S = \partial_x\Psi_y = pBi\omega e^{-i\omega(t-px-\eta_\beta z)}. \tag{8.20}$$

Now consider the boundary conditions at a free surface $z = 0$. Both the normal and shear tractions must vanish: $\tau_{xz} = \tau_{zz} = 0$. From (3.13), we have

$$\tau_{xz} = \mu(\partial_z u_x + \partial_x u_z), \tag{8.21}$$

$$\tau_{zz} = \lambda(\partial_x u_x + \partial_z u_z) + 2\mu\partial_z u_z. \tag{8.22}$$

Substituting (8.17)–(8.20) into (8.21) and (8.22), we obtain

$$\tau_{xz}^P = -A(2\mu p\eta_\alpha)\omega^2 e^{-i\omega(t-px-\eta_\alpha z)}, \tag{8.23}$$

$$\tau_{zz}^P = -A\left[(\lambda+2\mu)\eta_\alpha^2 + \lambda p^2\right]\omega^2 e^{-i\omega(t-px-\eta_\alpha z)}, \tag{8.24}$$

$$\tau_{xz}^S = -B\mu(p^2 - \eta_\beta^2)\omega^2 e^{-i\omega(t-px-\eta_\beta z)}, \tag{8.25}$$

$$\tau_{zz}^S = -B(2\mu\eta_\beta p)\omega^2 e^{-i\omega(t-px-\eta_\beta z)}. \tag{8.26}$$

At the free surface, we require

$$\tau_{xz} = \tau_{xz}^P + \tau_{xz}^S = 0, \tag{8.27}$$

$$\tau_{zz} = \tau_{zz}^P + \tau_{zz}^S = 0. \tag{8.28}$$

Substituting (8.23)–(8.26) into (8.27) and (8.28) at $z = 0$, and canceling the common terms, we obtain

$$A(2p\eta_\alpha) + B(p^2 - \eta_\beta^2) = 0, \tag{8.29}$$

$$A\left[(\lambda + 2\mu)\eta_\alpha^2 + \lambda p^2\right] + B(2\mu\eta_\beta p) = 0. \tag{8.30}$$

The equations for τ_{zz} can be written in terms of the P and S velocities by substituting $\lambda + 2\mu = \rho\alpha^2$, $\mu = \rho\beta^2$, and $\lambda = \rho(\alpha^2 - 2\beta^2)$ to give

$$A[2p\eta_\alpha] + B[p^2 - \eta_\beta^2] = 0, \tag{8.31}$$

$$A\left[\alpha^2\left(\eta_\alpha^2 + p^2\right) - 2\beta^2 p^2\right] + B[2\beta^2\eta_\beta p] = 0. \tag{8.32}$$

This coupled set of equations describes the free surface boundary condition for P- and SV-waves with horizontal slowness p. Recall that the vertical slownesses are given by $\eta_\alpha = (1/\alpha^2 - p^2)^{1/2}$ and $\eta_\beta = (1/\beta^2 - p^2)^{1/2}$. When $p < 1/\alpha$, there are two real solutions, a positive value of η_α for downgoing P waves and a negative value for upgoing P waves (assuming the z axis points downward). Similarly, when $p < 1/\beta$, then η_β is real and there exist both downgoing and upgoing SV waves. By defining different amplitude coefficients for the downgoing and upgoing waves, one could use (8.31) and (8.32) to solve for the P/SV reflection coefficients at the free surface.

However, our interest is in the case where $p > \beta^{-1} > \alpha^{-1}$ and both η_α and η_β are imaginary. From (8.11), if we factor out the depth dependence, we obtain

$$\mathbf{u} = \mathbf{A}e^{i\omega\eta z}e^{-i\omega(t-px)}, \tag{8.33}$$

and we see that imaginary values of η will lead to real values in the exponent. In this case we have the evanescent waves discussed in Chapter 6, for which amplitude grows or decays exponentially as a function of depth. The sign of η is chosen to give the solution that decays away from $z = 0$. For single imaginary values of η_α and η_β, the linear system of equations for A and B given in (8.31) and (8.32) has a non-trivial solution only when the determinant vanishes, that is, when

$$(p^2 - \eta_\beta^2)\left[\alpha^2(\eta_\alpha^2 + p^2) - 2\beta^2 p^2\right] - 4\beta^2 p^2\eta_\alpha\eta_\beta = 0. \tag{8.34}$$

Substituting for η_α and η_β, we can express this entirely in terms of p and the P and S velocities:

$$\left(2p^2 - \frac{1}{\beta^2}\right)^2 + 4p^2 \left(\frac{1}{\alpha^2} - p^2\right)^{1/2} \left(\frac{1}{\beta^2} - p^2\right)^{1/2} = 0, \tag{8.35}$$

after canceling a common factor of β^2. For imaginary η_α and η_β ($p > \beta^{-1} > \alpha^{-1}$), this can be rewritten as

$$\left(2p^2 - \frac{1}{\beta^2}\right)^2 - 4p^2 \left(p^2 - \frac{1}{\alpha^2}\right)^{1/2} \left(p^2 - \frac{1}{\beta^2}\right)^{1/2} = 0. \tag{8.36}$$

This is termed the *Rayleigh function* and has a single solution, with the exact value of p depending upon β and α. The corresponding phase velocity, $c = 1/p$, is always slightly less than the shear velocity, with $c = 0.92\beta$ for a Poisson solid. This result, obtained by Rayleigh over 100 years ago, shows that it is possible for two coupled inhomogeneous P and SV waves to propagate along the surface of a half-space.

By substituting the solution for p into (8.15), (8.16), (8.31), and (8.32), we may obtain the relative amplitudes of the P and SV components, and then substitution into (8.17)–(8.20) will give the vertical and horizontal displacements. Rayleigh wave particle motion for the fundamental mode is shown in Figure 8.4. The vertical

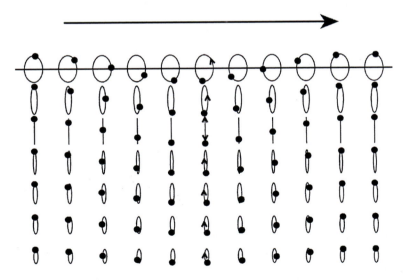

Figure 8.4 Particle motion for the fundamental Rayleigh mode in a uniform half-space, propagating from left to right. One horizontal wavelength (Λ) is shown; the dots are plotted at a fixed time point. Motion is counter clockwise (retrograde) at the surface, changing to purely vertical motion at a depth of about $\Lambda/5$, and becoming clockwise (prograde) at greater depths. Note that the time behavior at a fixed distance is given by looking from right to left in this plot.

and horizontal components are out of phase by $\pi/2$; the resulting elliptical motion changes from *retrograde* at the surface to *prograde* at depth, passing through a node at which there is no horizontal motion. For Rayleigh waves propagating along the surface of a uniform half-space there is no velocity dispersion (since there is no scale length in the model). However, in the Earth velocity dispersion results from the vertical velocity gradients in the crust and upper mantle; longer period waves travel faster since they sense the faster material at greater depths. As in the case of Love waves, Rayleigh wave dispersion curves for vertically stratified media may be computed using propagator matrix methods.

8.3 Dispersion

When different frequency components travel at different phase velocities, pulse shapes will not stay the same as they travel but will become dispersed as the frequencies separate. This leads to interference effects that cancel the wave energy except at particular times defined by the group velocity of the wave. This may be illustrated by considering the sum of two harmonic waves of slightly different frequency and wavenumber:

$$u(x, t) = \cos(\omega_1 t - k_1 x) + \cos(\omega_2 t - k_2 x). \tag{8.37}$$

Relative to an average frequency ω and wavenumber k, we have

$$\omega_1 = \omega - \delta\omega, \quad k_1 = k - \delta k, \tag{8.38}$$

$$\omega_2 = \omega + \delta\omega, \quad k_2 = k + \delta k. \tag{8.39}$$

Substituting into (8.37), we obtain

$$
\begin{aligned}
u(x, t) &= \cos(\omega t - \delta\omega t - kx + \delta kx) + \cos(\omega t + \delta\omega t - kx - \delta kx) \\
&= \cos\left[(\omega t - kx) - (\delta\omega t - \delta kx)\right] + \cos\left[(\omega t - kx) + (\delta\omega t - \delta kx)\right] \\
&= 2\cos(\omega t - kx)\cos(\delta kx - \delta\omega t), \tag{8.40}
\end{aligned}
$$

where we have used the identity $2\cos A \cos B = \cos(A + B) + \cos(A - B)$. The resulting waveform consists of a signal with the average frequency ω whose amplitude is *modulated* by a longer period wave of frequency $\delta\omega$ (Fig. 8.5).

In acoustics, this phenomenon is termed *beating* and may be observed when two musical notes are slightly out of tune. The short-period wave travels at velocity ω/k and the longer period envelope travels at velocity $\delta\omega/\delta k$. The former is the phase velocity c; the latter is the group velocity U. In the limit as $\delta\omega$ and δk approach

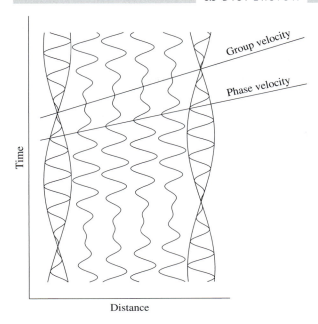

Group velocity

Phase velocity

Time

Distance

Figure 8.5 The sum of two waves of slightly different frequencies results in a modulated wave. The group velocity is the velocity of the wave packets; the phase velocity is the velocity of the individual peaks.

zero, we thus have

$$U = \frac{d\omega}{dk},\tag{8.41}$$

which agrees with our previous result in (8.6). Using the various relationships between the harmonic wave parameters (see Table 3.1), the group velocity may be alternatively expressed as

$$U = \frac{d\omega}{dk} = c + k\frac{dc}{dk} = c\left(1 - k\frac{dc}{d\omega}\right)^{-1}.\tag{8.42}$$

For Earth, the phase velocity c of both Love and Rayleigh waves generally increases with period; thus $dc/d\omega$ is negative and from (8.42) it follows that the group velocity is less than the phase velocity ($U < c$). Figure 8.6 plots Love and Rayleigh dispersion curves computed from the PREM model. A minimum or maximum point on the group velocity dispersion curve will result in energy from a range of periods arriving at nearly the same time. This is termed an *Airy phase* and occurs in Earth for Rayleigh waves at periods of about 50 and 240 s.

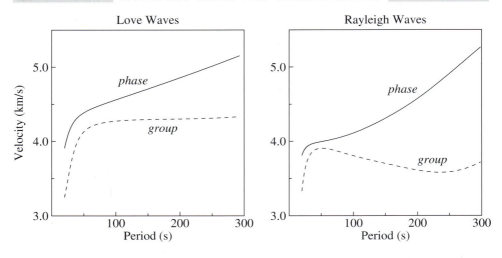

Figure 8.6 Fundamental Love and Rayleigh dispersion curves computed from the isotropic PREM model (courtesy of Gabi Laske).

8.4 Global surface waves

Love and Rayleigh waves in Earth travel along great circle paths radiating away from the source. Since they are confined to the surface of a sphere, geometrical spreading effects are reduced compared to body waves (which spread within a volume). At a given receiver location, the first surface wave arrival will travel along the minor (shorter) great circle arc and a later arrival will result from the major arc path on the opposite side of Earth (Fig. 8.7). The second arrival is due to surface waves that have passed through the *antipode*, the point directly opposite the source. The first and second arriving Love wave arrivals are termed $G1$ and $G2$, respectively, while the corresponding Rayleigh waves are called $R1$ and $R2$. The waves do not stop at the receiver, but continue traveling around the globe and these multiple orbits produce a series of later arrivals that can be observed for many hours following large earthquakes. The odd-numbered surface waves (e.g., $R1$, $R3$, $R5$, etc.) leave the source in the minor arc direction, while the even numbered waves depart in the major arc direction.

This is illustrated in Figure 8.8, which plots three components of motion from an earthquake at 230 km depth in the Tonga subduction zone recorded by the IRIS/IDA station NNA in Peru. Notice that the SH polarized Love wave arrivals appear most prominently on the transverse component, while the P/SV polarized Rayleigh waves are seen mostly on the vertical and radial components. Attenuation of surface waves can be seen in the decay of the amplitude of the arrivals with time.

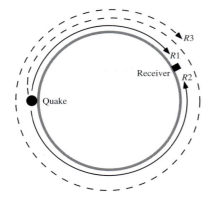

Figure 8.7 Ray paths for the first three Rayleigh wave arrivals.

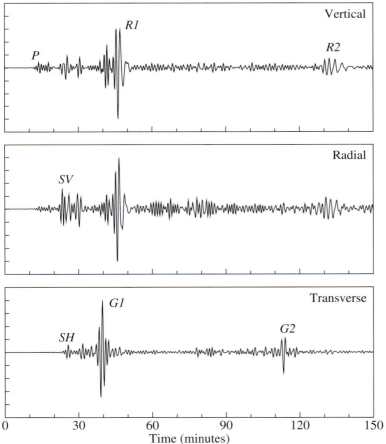

Figure 8.8 The vertical, radial, and transverse components of motion for a March 11, 1989, earthquake at 230 km depth in the Tonga trench recorded at IRIS/IDA station NNA in Peru. *P*, *SV*, and Rayleigh waves are most visible on the vertical and radial components; *SH* and Love waves appear on the transverse component.

At long periods, Rayleigh waves are sufficiently coherent that it is possible to stack records from many different events to produce a global picture (Fig. 8.9) of the seismic wavefield that images the surface wave arrivals (Shearer, 1994). This vertical-component image illustrates many of the concepts that we have developed in this chapter. The dispersion of the Rayleigh waves is clearly apparent, particularly in the later part of the image. Very long period (≥ 300 s) waves travel the fastest, arriving before the pronounced shorter-period banding in the Airy phase. The high amplitude of the Airy phase results from a local minimum in the group velocity dispersion curve near 240 s. The difference between phase and group velocity can be seen clearly in the image of the Airy phase. The lines of constant phase, defined by the peaks and troughs in the seismograms, are not parallel to the overall direction of energy transport. Rather, they cut across at a slightly more horizontal orientation, since the phase velocity is higher than the group velocity.

The major P and SV body-wave phases can also be seen in this image, in the triangular shaped region before the first Rayleigh wave ($R1$). Additional body-wave arrivals are visible between $R1$ and the second Rayleigh wave ($R2$). These include some P phases, but most prominent are the high-order S surface multiples and the families of S-to-P converted phases that they spawn upon each surface reflection. These can be traced to beyond $720°$ and are the major source of seismic energy between the Rayleigh wave arrivals. In the surface wave literature, these arrivals are termed overtone packets and are sometimes referred to as X phases (e.g., Tanimoto, 1987).

8.5 Observing surface waves

Surface waves are generally the strongest arrivals at teleseismic distances and contain a great deal of information about crust and upper mantle structure as well as the seismic source. Much of the power of surface wave observations comes from the fact that velocity can be measured at a number of different frequencies from a single seismogram, providing direct constraints on the velocity versus depth profile everywhere along the source–receiver path. In contrast, the corresponding body wave observations provide only a single travel time per phase, and recovering the complete velocity structure requires stations at a wide range of source–receiver distances.

A major goal in most surface wave studies is to determine the group or phase velocity at a number of periods. This can be done in several ways. If the location and origin time of the source are known, then the group velocity may be estimated from a surface wave record at a single station by measuring the travel time to the station for energy at a particular frequency. This can be done by applying narrow

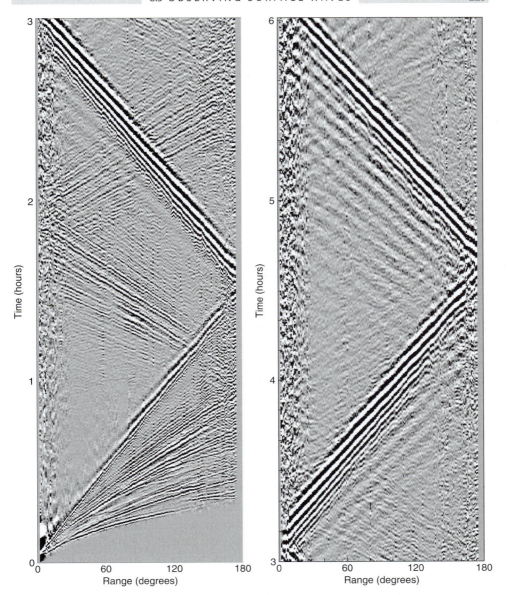

Figure 8.9 An image of Earth's long-period seismic response on vertical component seismographs as a function of time and distance to an earthquake. Positive amplitudes are shown as black, and negative amplitudes are shown as white. The Rayleigh wave arrivals $R1$ and $R2$ are visible in the left panel showing the first 3 hours of data, whereas $R3$ and $R4$ are seen on the right panel.

passband filters to the record to isolate the wave packet for a target frequency, or, more crudely, by measuring the time between successive peaks in a single dispersed seismogram. The same approach can be used to determine the group velocity between two stations along a great circle ray path through the source by measuring the difference in the arrival times at the stations. In this case (the *two-station method*), precise details of the source are not required, provided the location is approximately correct.

Many modern surface wave analyses measure the phase velocity rather than the group velocity. This is done by computing the Fourier spectrum of the record to determine the phase of each frequency component. If the phase is known at the source (this requires the focal mechanism or moment tensor for the event), then phase velocity measurements are possible from a single receiver; alternatively the two-station method can be used to determine the phase velocity between a pair of receivers. The tricky part of phase velocity measurements is that the observed phase ϕ at a particular frequency varies only between 0 and 2π and there will typically be many cycles between observation points, so that the total phase shift Φ is actually $2\pi n + \phi$, where n is an integer.

For example, consider measuring the Rayleigh wave phase velocity in Fig. 8.9 at a period of 240 s (close to the dominant period of the high-amplitude Airy phase) using stations at 90° and 120°. Phase measurements at these ranges alone do not tell us how many cycles, n, occurred between the stations; the phase velocity cannot be determined without independent knowledge of n. At long periods this is not a significant problem since n can be accurately estimated from standard one-dimensional Earth models. However, at shorter periods it becomes increasingly difficult to calculate n with confidence, since lateral velocity variations in the upper mantle cause n to vary with position as well as range. In this case, a useful approach is to measure the phase velocity at the longest periods first, and then gradually move to shorter periods, keeping track of the total accumulated phase shift Φ. This will work provided the phase velocity dispersion curve is a smooth and continuous function of frequency.

Comprehensive studies of surface wave phase velocities, using a global distribution of sources and receivers, can be used to invert for maps of phase velocity for both Love and Rayleigh waves. This is done separately for each period using techniques analogous to the body-wave velocity inversion problem discussed in Chapter 5. The structure seen in these maps is related to Earth's lateral velocity variations; the depth dependence in this heterogeneity is constrained by the results at different periods. Inverting surface-wave phase velocity observations is currently one of the best ways to resolve three-dimensional velocity variations in the upper few hundred kilometers of the mantle.

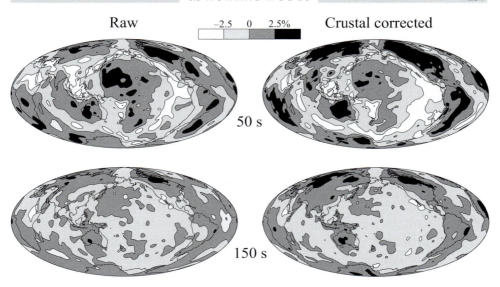

Figure 8.10 Rayleigh wave phase velocity at (top) 50 s and (bottom) 150 s period. The right panels have been corrected for crustal thickness variations using the model CRUST 2.0 (Laske *et al.*, http://mahi.ucsd.edu/Gabi/rem.dir/crust/crust2.html). Velocity perturbations are contoured at 2.5% intervals, with black indicating regions that are 2.5% faster than average, and white indicating velocities over 2.5% slower than average. Maps produced by Guy Masters (personal communication) using measurements from Ekström *et al.* (1997).

Figure 8.10 plots maps of Rayleigh wave phase velocity at 50 and 150 s period. Notice the ocean-continent signal is enhanced after corrections are applied for variations in crustal thickness. In general, the thicker crust beneath continents compared to the oceans causes slower surface-wave velocities, but this is counteracted by generally faster upper-mantle velocities beneath continents, which are especially strong in shield regions. When corrections for variations in crustal properties are applied (to obtain what the velocity would be for a globally uniform crust), the fast continental roots become even more prominent, particularly at shorter periods, which are more sensitive to shallow structure. Global mantle tomography models rely heavily on surface-wave analyses to constrain upper-mantle heterogeneity. Notice the similarity between the 50 s crustal-corrected phase velocity map and the velocity structure at 150 km depth in Figure 1.7.

8.6 Normal modes

Thus far we have considered the propagation of body and surface waves largely as if the Earth were of infinite extent. However, the Earth is a finite body in which

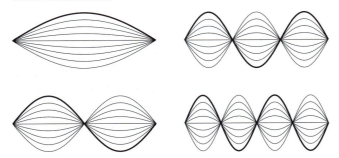

Figure 8.11 The first four modes of vibration of a string between fixed endpoints.

all wave motions must be confined. Body waves are reflected from the surface; surface waves orbit along great circle paths. At a particular point on the Earth's surface, there will be a series of arrivals of different seismic phases. The timing between these arrivals will result in constructive and destructive interference such that only certain frequencies will resonate over long time intervals. These resonant frequencies are termed Earth's *normal modes* and provide a way of representing wave propagation that is an alternative to the traveling wave approach.

The vibrations of a string fixed at both ends provide an analogy that may be familiar from your physics classes. The string will resonate only at certain frequencies (Fig. 8.11). These are termed the *standing waves* for the string and any motion of the string can be expressed as a weighted sum of the standing waves. This is an eigenvalue problem; the resonant frequencies are termed the *eigenfrequencies*; the string displacements are termed the *eigenfunctions*. In a musical instrument the lowest frequency is called the fundamental mode; the higher modes are the overtones or *harmonics*. For the vibrating string, the eigenfunctions are sines and cosines and it is natural to use a Fourier representation.

Normal modes for the Earth are also specified by their eigenfrequencies and eigenfunctions. A detailed treatment of normal mode theory for the Earth is beyond the scope of this book, and computation of eigensolutions for realistic Earth models is a formidable task. However, it is useful to remember some of the properties of the eigenfunctions of any vibrating system:

1. They are complete. Any wave motion within the Earth may be expressed as a sum of normal modes with different excitation factors.
2. They are orthogonal in the sense that the integral over the volume of the Earth of the product of any two eigenfunctions is zero. This implies that the normal mode representation of wave motion is unique.

What do Earth's normal modes look like? For a spherically symmetric solid, it can be shown that there are two distinctly different types of modes: *spheroidal modes*, which are analogous to P/SV and Rayleigh wave motion, and *toroidal modes*, which are analogous to SH and Love wave motion. The Earth's departures from spherical symmetry mean that this separation is not complete, but it is a very good first-order approximation. Toroidal modes involve no radial motion and are sensitive only to the shear velocity, whereas spheroidal modes have both radial and horizontal motion and are sensitive to both compressional and shear velocities. Spheroidal mode observations at long periods are also sensitive to gravity and provide the best direct seismic constraints on Earth's density structure.

The lateral variations in normal mode eigenfunctions are best described in terms of *spherical harmonics*, which provide an orthogonal set of basis functions on a spherical surface (spherical harmonics are useful in many areas of geophysics for representing functions on the surface of a sphere, and descriptions are available in many of the standard texts; see Aki and Richards (1980, 2002) or Lay and Wallace (1995) for treatments focusing on seismology). A common normalization for the spherical harmonics is

$$Y_l^m(\theta, \phi) = (-1)^m \left[\frac{2l+1}{4\pi} \frac{(l-m)!}{(l+m)!} \right]^{1/2} P_l^m(\cos\theta) e^{im\phi} \qquad (8.43)$$

where θ and ϕ are spherical polar coordinates (θ is the polar angle) and P_l^m is the *associated Legendre function*. The spherical harmonic function is written as Y_l^m, where l is termed the *angular order number* and m is the *azimuthal order number*. The index l is sometimes also termed the *spherical harmonic degree* and is zero or a positive integer up to any value. The angular order number, m, may take on $2l + 1$ integer values between $\pm l$. The order numbers determine the number of lines of zero crossings that are present in the function. The total number of zero crossings is given by l; the number of zero crossings through the pole is given by $|m|$. Figure 8.12 plots examples of Y_l^m for some of the lower harmonic degrees.

Note that the harmonics are defined relative to a particular coordinate system and depend upon the location of the poles. If the coordinate system is rotated, any spherical harmonic function in the old coordinate system may be expressed as a sum of spherical harmonics of the same l but differing m in the new coordinate system. A rotation of coordinates does not affect the angular order number but will change the relative weights of the azimuthal order numbers. For example, a rotation of 90° can change Y_1^0 to Y_1^1.

Expansions of global observations in terms of spherical harmonics are common in geophysics. Examples include Earth's surface geoid and seismic velocity perturbations at a particular depth. For Earth's normal modes, we are interested in displacement, which is a vector quantity and most conveniently expressed in terms

$l = 0$ $l = 1$ $l = 2$ $l = 3$

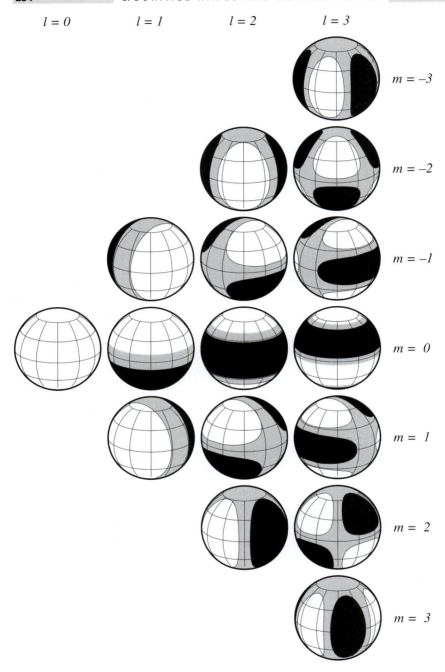

$m = -3$

$m = -2$

$m = -1$

$m = 0$

$m = 1$

$m = 2$

$m = 3$

Figure 8.12 Spherical harmonic functions Y_l^m up to degree $l = 3$. Positive values are shown as white, negative as black, with near-zero values as gray. There are $2l + 1$ values of m at each degree. Note that the negative m harmonics are rotated versions of the positive m harmonics.

of the *vector spherical harmonics*, which are defined as

$$\mathbf{R}_l^m(\theta, \phi) = Y_l^m \hat{\mathbf{r}} \tag{8.44}$$

$$\mathbf{S}_l^m(\theta, \phi) = \frac{1}{\sqrt{l(l+1)}} \left[\frac{\partial Y_l^m}{\partial \theta} \hat{\boldsymbol{\theta}} + \frac{1}{\sin \theta} \frac{\partial Y_l^m}{\partial \phi} \hat{\boldsymbol{\phi}} \right] \tag{8.45}$$

$$\mathbf{T}_l^m(\theta, \phi) = \frac{1}{\sqrt{l(l+1)}} \left[\frac{1}{\sin \theta} \frac{\partial Y_l^m}{\partial \phi} \hat{\boldsymbol{\theta}} - \frac{\partial Y_l^m}{\partial \theta} \hat{\boldsymbol{\phi}} \right] \tag{8.46}$$

where $\hat{\mathbf{r}}$, $\hat{\boldsymbol{\theta}}$, and $\hat{\boldsymbol{\phi}}$ are unit vectors in the r, θ, and ϕ directions, respectively. The vector fields associated with Earth's spheroidal motions can be expressed in terms of $\mathbf{R}$ and $\mathbf{S}$, while the toroidal motions are are expressed with $\mathbf{T}$.

Earth's normal modes are specified in terms of the spherical harmonic order numbers l and m and a *radial order number*, n, that describes the number of zero crossings in radius that are present. Toroidal modes are thus designated $_n T_l^m$ and spheroidal modes as $_n S_l^m$. The solutions for $n = 0$ are called the fundamental modes; the solutions for $n > 0$ are termed overtones. For a spherically symmetric Earth the eigenfrequencies at constant n and l are identical for all values of m and it is common to denote modes only by their radial and angular order numbers, that is, as $_n T_l$ and $_n S_l$ and the corresponding frequencies as $_n \omega_l$.

The fundamental spheroidal mode $_0 S_0$ is termed the "breathing" mode and represents a simple expansion and contraction of the Earth. It has a period of about 20 minutes. $_0 S_1$ is not used in seismology since it describes a shift in the center of mass of the Earth; this cannot result from purely internal forces. $_0 S_2$ has a period of about 54 minutes and represents an oscillation between an ellipsoid of horizontal and vertical orientation (Fig. 8.13). This is sometimes termed the "rugby" mode

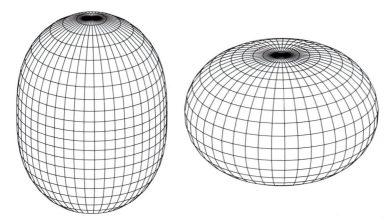

Figure 8.13 A highly exaggerated picture of the normal mode $_0 S_2$. This mode has a period of about 54 minutes; the two images are separated in time by 27 minutes.

for obvious reasons. The toroidal mode $_0T_1$ represents a change in Earth's rotation rate; this can happen but occurs at such long time intervals that it is unimportant in seismology. The toroidal mode $_0T_2$ has a period of about 44 minutes and describes a relative twisting motion between the northern and southern hemispheres. Because of the fluid outer core, toroidal modes do not penetrate below the mantle.

Although theoretical solutions for the normal modes of a solid sphere date back to Lamb in 1882, definitive observations for the Earth did not occur until the great 1960 earthquake in Chile. The enormous size of this event (the largest since seismographs began recording about a century ago), together with improvements in instrument design at long periods, made it possible to identify a few dozen normal modes. The next two decades were perhaps the golden age in normal mode seismology as over a thousand modes were identified (e.g., Gilbert and Dziewonski, 1975) and new methods were derived for inverting the observations for Earth structure (e.g., Backus and Gilbert, 1967, 1968, 1970).

The normal mode eigenfrequencies are identical for different azimuthal order number m only for a spherically symmetric solid (this is called *degeneracy* in the eigenfrequencies). Earth's small departures from spherical symmetry (e.g., ellipticity, rotation, general 3-D velocity variations) will cause the eigenfrequencies to separate. This is termed *splitting*; a single spectral peak will split into a *multiplet* composed of the separate peaks for each value of m. Earth's rotation rate and ellipticity are well known, but splitting due to 3-D structure is also observed, so measurements of mode splitting can be using to constrain three-dimensional velocity variations.

As an example of normal mode observations, Figure 8.14 plots the spectrum of 240 hours of data from the 2004 Sumatra-Andaman earthquake recorded on the vertical component of station ARU in Russia. The low-order spheroidal modes (labeled) are seen with excellent signal-to-noise because of the size of this earthquake ($M_W = 9.1$, the largest since the 1964 Alaskan earthquake and the first giant subduction zone earthquake to be recorded by modern broadband seismometers). Many of the modes are clearly split and $_0S_2$ is shown at an expanded scale to illustrate its splitting into five peaks, corresponding to its five m values, i.e., $_0S_2^{-2}$, $_0S_2^{-1}$, etc. The regular spacing of these peaks in the $_0S_2$ multiplet is characteristic of splitting due to Earth's rotation. Modes $_3S_1$ and $_1S_3$ have slightly different center frequencies but overlap so much that they cannot be separately resolved.

Since wave motion in the Earth can be described to equal precision with either traveling waves or normal modes, what is the advantage of the normal mode approach? Largely it comes from analysis of long time series from large events, where the multiplicity of different phase arrivals makes a traveling wave representation awkward. For example, it would be extremely difficult to attempt to model all of the

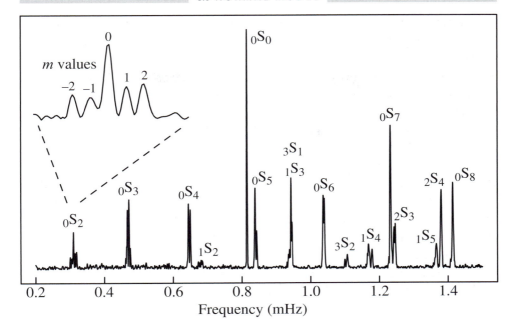

Figure 8.14 Low-order spheroidal modes visible in the spectrum of 240 hours of data from the 2004 Sumatra-Andaman earthquake ($M_W = 9.1$) recorded at station ARU at Arti, Russia (courtesy of Guy Masters). Mode $_0S_2$ is plotted at an expanded scale from 0.29 to 0.33 mHz to show its splitting into a five-peaked multiplet, corresponding to azimuthal order number (m) values from -2 to 2.

arrivals visible in Figure 8.9 with a time domain approach. However, by taking the Fourier transform of individual records and measuring the position of the spectral peaks (and any splitting that can be observed), it is possible to identify the various modes and use them to constrain Earth structure. This is the only practical way to examine records at very long periods (> 500 s) and provides information about the Earth's density structure that cannot be obtained any other way. Normal modes also are able to examine Earth properties, such as the shear response of the inner core, that are difficult to observe directly with body waves. Attenuation causes the mode amplitudes to decay with time, and so normal mode observations help to constrain Q at very long periods. Finally, normal modes provide a complete set of basis functions for the computation of synthetic seismograms that naturally account for Earth's sphericity. Computing synthetic seismograms by summing normal modes is standard practice in surface-wave and long-period body-wave seismology. The number of modes required increases rapidly at higher frequencies, but with modern computers, normal mode summation is an increasingly attractive alternative to other methods.

8.7 Exercises

1. (a) Show how (8.9) follows from (6.46) and (6.59) by proving that if

$$re^{i\theta} = \frac{a - ib}{a + ib} \tag{8.47}$$

then

$$r = 1, \tag{8.48}$$

$$\theta = -2\tan^{-1}\frac{b}{a}. \tag{8.49}$$

   (b) Model the crust as a simple layer over a half-space as in Figure 8.2, with $h = 40\,\mathrm{km}$, $\rho_1 = 2.7\,\mathrm{g/cm^3}$, $\rho_2 = 3.3\,\mathrm{g/cm^3}$, $\beta_1 = 3.5\,\mathrm{km/s}$, and $\beta_2 = 4.5\,\mathrm{km/s}$. Find the lowest value of ω (i.e., the fundamental mode) that satisfies equation (8.10) at values of phase velocity ($c = 1/p$) of 3.8, 4.0, 4.2, and 4.4 km/s. Convert ω to period, T, and list your results in a table with the c and T values. Hint: Make sure that you use consistent units in computing the μ values in (8.10).

   (c) (COMPUTER) Make a plot of the $c(\omega)$ dispersion curve for the fundamental and first two higher modes for values of c from β_1 to β_2. Include many more points than your result from (b).

2. (COMPUTER) Solve equation (8.36) numerically and plot the ratio of the Rayleigh wave phase velocity to the shear velocity (i.e., c/β) as a function of the P-to-S velocity ratio (α/β) for α/β values between 1.5 and 2.0.

3. Figure 8.15 shows a surface wave from the 2002 M_W 7.9 Denali earthquake in Alaska recorded near São Paulo in Brazil on a vertical-component station.

   (a) Is this most likely a Rayleigh or a Love wave? Why?

   (b) Measure the average time and the time separation between successive waveform troughs over the interval from 3200 s to 3600 s. Use your results to make a table of velocity as a function of the wave period (Hint: $1° = 111.19\,\mathrm{km}$). Is this a measure of group or phase velocity? How do your results compare to the PREM predictions in Figure 8.6?

4. (COMPUTER) Assume that the Rayleigh wave phase velocity at periods between 50 and 500 s can be approximated by the polynomial representation

$$c(T) = 4.020 - 1.839 \times 10^{-3}\,T + 3.071 \times 10^{-5}\,T^2 - 3.549 \times 10^{-8}\,T^3,$$

where c is the phase velocity in km/s and T is the period in seconds.

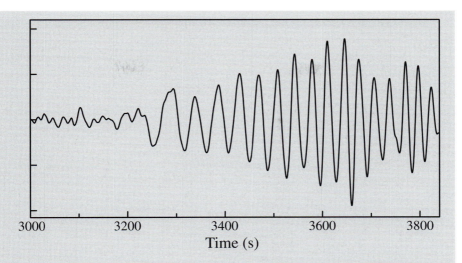

Figure 8.15 The 2002 Denali earthquake recorded at GEOSCOPE station SPB (vertical component) in Brazil at an epicentral distance of 115.4°. Time is from the earthquake origin time.

(a) Plot both the group and phase velocity dispersion curves as a function of period for $T = 50$ to 500 s, using (8.42) to generate the $U(T)$ curve (this can be done either analytically or through numerical differencing on the computer after converting $c(T)$ points to $\omega(k)$ points).

(b) Use this relationship to write a computer program to compute Rayleigh wave synthetic seismograms at source–receiver ranges of 0, 30, 90, and 150 degrees. Define your seismograms to be 1.5 hours long with a digitization interval of 5 s. Construct the synthetics as a sum of cosine functions at frequencies from 0.002 to 0.02 Hz with a frequency spacing of 0.0002 Hz. Apply a phase shift to each frequency component to account for the propagation distance. Make a plot of the synthetic seismogram at each range.

(c) Pick two adjacent peaks in your synthetic Rayleigh wave at 150° and measure their time separation. Using this as an approximation for the period T, show that the arrival time is in reasonable agreement with the group velocity curve plotted in part (a).

(d) Note: You may notice a "wrap-around" phase at 5000 s at zero distance. Note that $5000 = 1/df$ where $df = 0.0002$ Hz. To get rid of this phase, use a smaller value for df (which will push it back to later times) or simply window it out of your plot. Your synthetics will not be accurate past 180° unless a $\pi/2$ phase advance is added to correct for the effect of the focusing at the antipode (see

Aki and Richards, 2002, p. 351). An additional $\pi/2$ phase advance occurs for each additional epicentral or antipodal passage.

Hint: Here is the key part of a FORTRAN program to solve this problem:

```
a0= 4.02
a1=-1.839e-3
a2= 3.071e-5
a3=-3.549e-8
(initialize dt,npts,f1,f2,df,delta, set y array
    to zero)
    pi=3.1415927
    x=delta*111.19        !convert degrees to km
    do i=1,npts
        time(i)=float(i-1)*dt
        do f=f1,f2,df
            t=1./f
            pvel=a0+a1*t+a2*t**2+a3*t**3
            toff=x/pvel
            om=2.*pi*f
            hann=sin(pi*(f-f1)/(f2-f1))**2
            y(i)=y(i)+cos(om*(time(i)-toff))*hann
        enddo
    enddo
(output time(i),y(i) for i=1,npts)
```

In this case, we apply a Hanning taper ($\sin^2$ function) to smoothly reduce the amplitudes close to the frequency limits. This minimizes ringing and other artifacts in the final synthetic seismogram that are caused by the finite bandwidth. If you are curious why your synthetics are missing the fast waves at very long periods, try removing the Hanning taper. You will now see the early-arriving, long-period energy, but your synthetics will suffer ringing from the abrupt frequency limits in the calculation.

5. Comment on the validity of the following argument: "The earthquake source may be thought of as a delta function that generates energy at all frequencies. However, Earth's lowest normal mode has a period of 54 minutes. Thus, the normal mode representation for seismic displacements is incomplete because it cannot represent very long period energy generated by the source (i.e., periods longer than 54 minutes)."

9

Earthquakes and source theory

In the preceding chapters we have described methods for modeling the propagation of seismic waves, but we have largely neglected the question of where the waves come from and how the radiated seismic energy relates to the physical properties of the source. These topics can often be ignored if our interest is solely in learning about details of Earth structure outside of the source regions, such as travel time studies of velocity structure. However, in many cases resolving seismic structure requires some knowledge of the source characteristics, and, of course, resolving source properties is fundamental to any real understanding of earthquakes. Because seismic source theory can be very complex, we will not formally derive most of the equations in this chapter; instead we will summarize many of the important results that are of practical use in seismology and refer the reader to Aki and Richards (2002), Stein and Wysession (2002), or Kanamori and Brodsky (2004) for more details.

9.1 Green's functions and the moment tensor

A major goal in this chapter is to understand how the observed seismic displacements at some distance from a seismic event can be related to the source properties. Let us begin by recalling the momentum equation for an elastic continuum

$$\rho \frac{\partial^2 u_i}{\partial t^2} = \partial_j \tau_{ij} + f_i, \tag{9.1}$$

where ρ is the density, u_i is the displacement, τ_{ij} is the stress tensor, and f_i is the body force term. Now consider the displacement field in a volume V bounded by a surface S. The displacements within V must be a function solely of the initial conditions, the internal forces within V, and the tractions acting on S. A more formal statement of this fact is termed the *uniqueness* or *representation* theorem and is derived in Section 2.3 of Aki and Richards (2002). It turns out that specifying either the tractions or the displacement field on S, together with the body forces $\mathbf{f}$, is sufficient to uniquely determine $\mathbf{u}$ throughout V.

Solving (9.1) in general is quite difficult if we include the f_i term, and in Chapter 3 we quickly dropped it to concentrate on the homogeneous equation of motion. Let us now explore how the properties of the source can be modeled and related to the seismic displacements observed in the Earth. Consider a unit force vector $\mathbf{f}(\mathbf{x}_0, t_0)$ applied at point $\mathbf{x}_0$ at time t_0. By itself, this is not a realistic seismic source; rather, it is what would result if the hand of God could reach inside the Earth and apply a push to a particular point. Nonetheless, the unit force function is a useful concept because more realistic sources can be described as a sum of these force vectors. Consider the displacement $\mathbf{u}(\mathbf{x}, t)$ measured at a receiver at position $\mathbf{x}$ that results from this source. In general, $\mathbf{u}(\mathbf{x}, t)$ will be a complicated function of the Earth's seismic velocity and density structure and will include multiple seismic phases and reverberations. The $\mathbf{u}(t)$ function will vary for different source and receiver positions. However, for every $\mathbf{f}(\mathbf{x}_0, t_0)$ and $\mathbf{x}$, there is a unique $\mathbf{u}(t)$ that describes the Earth's response, which could be computed if we knew the Earth's structure to sufficient accuracy.

In considering this problem, it is helpful to develop a notation that separates the source terms from all the other details of the wave propagation. This is done by defining a Green's function $\mathbf{G}(\mathbf{x}, t)$ that gives the displacement at point $\mathbf{x}$ that results from the unit force function applied at point $\mathbf{x}_0$. In general we may write

$$u_i(\mathbf{x}, t) = G_{ij}(\mathbf{x}, t; \mathbf{x}_0, t_0) f_j(\mathbf{x}_0, t_0), \qquad (9.2)$$

where $\mathbf{u}$ is the displacement, $\mathbf{f}$ is the force vector, and $\mathbf{G}$ is termed the elastodynamic Green's function. The actual computation of $\mathbf{G}$ is quite complicated and involves taking into account all of the elastic properties of the material and the appropriate boundary conditions, and we defer discussion of specific forms for $\mathbf{G}$ until later. However, assuming that $\mathbf{G}$ can be computed, notice the power of this equation. Because it is linear, the displacement resulting from any body-force distribution can be computed as the sum or *superposition* of the solutions for the individual point sources. It also implies that knowledge of the displacement field may permit us to invert for the body-force distribution.

An earthquake is usually modeled as slip on a fault, a discontinuity in displacement across an internal surface in the elastic media. This parameterization cannot be used directly in (9.2) to model ground motion. Fortunately, however, it can be shown that there exists a distribution of body forces that produces exactly the same displacement field as slip on an internal fault. These are termed the *equivalent body*

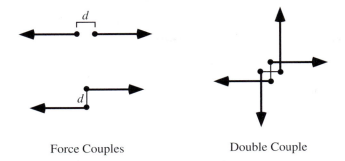

Force Couples   Double Couple

Figure 9.1 Force couples are opposing point forces separated by a small distance. A double couple is a pair of complementary couples that produce no net torque.

forces for the fault model. Before describing the relationship between these forces and the fault slip, let us first explore the different types of body forces that can occur within Earth.

For now, consider sources small enough compared to the wavelength of the radiated energy that they can be thought of as point sources. A single force acting at a point could only result from external forces; otherwise momentum would not be conserved. Internal forces resulting from an explosion or stress release on a fault must act in opposing directions so as to conserve momentum. For example, we could have two force vectors of magnitude f, pointing in opposite directions, separated by a distance d (Fig. 9.1). This is termed a *force couple* or *vector dipole*. Alternatively, the vectors could be separated in a direction perpendicular to the force orientation. In this case angular momentum is not conserved unless there also exists a complementary couple that balances the forces. The resulting pair of couples is termed a *double couple*.

We define the force couple M_{ij} in a Cartesian coordinate system as a pair of opposing forces pointing in the i direction, separated in the j direction. The nine different force couples are shown[1] in Figure 9.2.

The magnitude of M_{ij} is given by the product fd and is assumed constant as d goes to zero in the limit of a point source. It is then natural to define the *moment tensor* **M** as

$$\mathbf{M} = \begin{bmatrix} M_{11} & M_{12} & M_{13} \\ M_{21} & M_{22} & M_{23} \\ M_{31} & M_{32} & M_{33} \end{bmatrix}. \tag{9.3}$$

[1] Alert readers will notice that our coordinate system has flipped again, so that x_3 points upward as it did in Chapters 2 and 3, rather than downward as it did in Chapters 4–8. An upward pointing x_3 axis is the usual convention in source studies.

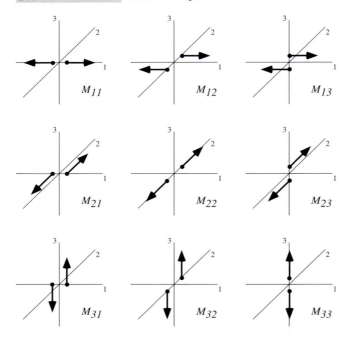

Figure 9.2 The nine different force couples that make up the components of the moment tensor.

The condition that angular momentum be conserved requires that $\mathbf{M}$ is symmetric (e.g., that $M_{ij} = M_{ji}$). Therefore $\mathbf{M}$ has only six independent elements. The moment tensor provides a general representation of the internally generated forces that can act at a point in an elastic medium. Although it is an idealization, it has proven to be a good approximation for modeling the distant seismic response for sources that are small compared with the observed seismic wavelengths. Larger, more complicated sources can also be modeled using the moment tensor representation by considering a sum of point forces at different positions.

Using (9.2), we may express the displacement resulting from a force couple at $\mathbf{x}_0$ in terms of the point-force Green's function as

$$u_i(\mathbf{x}, t) = G_{ij}(\mathbf{x}, t; \mathbf{x}_0, t_0) f_j(\mathbf{x}_0, t_0) - G_{ij}(\mathbf{x}, t; \mathbf{x}_0 - \hat{\mathbf{x}}_k d, t_0) f_j(\mathbf{x}_0, t_0)$$

$$= \frac{\partial G_{ij}(\mathbf{x}, t; \mathbf{x}_0, t_0)}{\partial (x_0)_k} f_j(\mathbf{x}_0, t_0) d, \tag{9.4}$$

where the force vectors f_j are separated by a distance d in the $\hat{\mathbf{x}}_k$ direction. The product $f_j d$ is the kth column of M_{jk} and thus

$$u_i(\mathbf{x}, t) = \frac{\partial G_{ij}(\mathbf{x}, t; \mathbf{x}_0, t_0)}{\partial (x_0)_k} M_{jk}(\mathbf{x}_0, t_0), \tag{9.5}$$

and we see that there is a linear relationship between the displacement and the components of the moment tensor that involves the spatial derivatives of the point-force Green's functions. The synthetic seismogram algorithms discussed in Chapter 3 can be used to calculate the Green's functions, which include all the body- and surface-wave phases connecting $\mathbf{x}_0$ and $\mathbf{x}$. Given a specified moment tensor, source location and Earth model, it is possible to compute displacement functions (i.e., seismograms) anywhere within the Earth. Because (9.5) is linear, once the Green's functions are computed for a reference Earth model, it is straightforward to use seismic observations, $\mathbf{u}(\mathbf{x}, t)$, to invert for the components of the moment tensor. This is now done routinely for globally recorded earthquakes by several groups. The most widely used moment tensor catalog is from the Global Centroid Moment Tensor (CMT) project (see http://www.globalcmt.org/CMTsearch.html). This project was started by Adam Dziewonski and for many years was called the Harvard CMT catalog (e.g., Dziewonski and Woodhouse, 1983). The CMT solution provides the moment tensor and also a *centroid* time and position that represents the average time/space origin of the long-period seismic radiation. The centroid location should not be confused with the earthquake hypocenter, which is usually determined from short-period P arrival times and which represents the starting point of the earthquake rupture.

9.2 Earthquake faults

Let us now consider models of slip on earthquake faults and how they relate to the moment tensor formalism that we have just discussed. Earthquakes may be idealized as movement across a planar fault of arbitrary orientation (Fig. 9.3). The fault plane is defined by its *strike* (ϕ, the azimuth of the fault from north where it intersects a horizontal surface) and *dip* (δ, the angle from the horizontal). For non-vertical faults, the lower block is termed the *foot wall*; the upper block is the *hanging wall*. The slip vector is defined by the movement of the hanging wall relative to the foot wall; the *rake*, λ, is the angle between the slip vector and the strike. Upward movement of the hanging wall is termed *reverse* faulting, whereas downward movement is called *normal* faulting. Reverse faulting on faults with dip angles less than 45° is also called *thrust* faulting; nearly horizontal thrust faults are termed *overthrust* faults. In general, reverse faults involve horizontal compression in the direction perpendicular to the fault strike whereas normal faults involve horizontal extension. Horizontal motion between the fault surfaces is termed strike–slip and vertical motion is called dip–slip. If an observer, standing on one side of a fault, sees the adjacent block move to the right, this is termed *right-lateral* strike–slip motion (with the reverse indicating *left-lateral* motion). To define the rake for

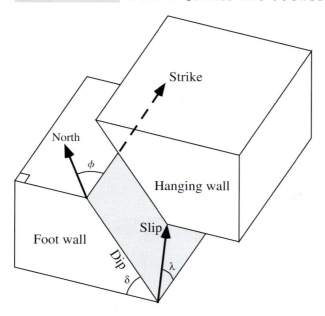

Figure 9.3 A planar fault is defined by the strike and dip of the fault surface and the direction of the slip vector.

vertical faults, the hanging wall is assumed to be on the right for an observer looking in the strike direction. In this case, $\lambda = 0°$ for a left-lateral fault and $\lambda = 180°$ for a right-lateral fault. The San Andreas Fault is a famous example of a right-lateral fault. Parts of California to the west of the fault are moving northward (right) relative to the rest of the United States.

The strike ($0 \le \phi < 360°$), the dip ($0 \le \delta \le 90°$), the rake ($0 \le \lambda < 360°$), and the magnitude of the slip vector, D, define the most basic seismic model of faulting or event *focal mechanism*. It can be shown that the seismic energy radiated from such a fault can be modeled with a double-couple source, the equivalent body-force representation of the displacement field. For example, right-lateral movement on a vertical fault oriented in the x_1 direction corresponds to the moment tensor representation

$$\mathbf{M} = \begin{bmatrix} 0 & M_0 & 0 \\ M_0 & 0 & 0 \\ 0 & 0 & 0 \end{bmatrix}, \tag{9.6}$$

where M_0 is defined as the *scalar seismic moment* and is given by

$$M_0 = \mu \overline{D} A, \tag{9.7}$$

where μ is the shear modulus, $\overline{D}$ is the average fault displacement, and A is the area of the fault. Scalar seismic moment was defined by Aki (1966) and is the most fundamental and widely used measure of earthquake strength. The reader should verify that the units for M_0 are N m, the same as for the force couples defined earlier.[2] More generally, M_0 can be computed from any moment tensor from

$$M_0 = \frac{1}{\sqrt{2}} \left(\sum_{ij} M_{ij}^2 \right)^{1/2}. \tag{9.8}$$

The connection between scalar moment as defined in (9.7) and the components of the moment tensor is complicated to prove (e.g., Aki and Richards, 2002, pp. 42–8), but is one of the most important results in seismology because it relates a real, physical property of the earthquake source to the double-couple model and ultimately to seismic observations. From the orientations of the different force couples, it is easy to see how any fault in which the strike, dip, and rake are multiples of 90° can be defined with a moment tensor representation. However, more generally, a fault plane and slip of any orientation can be described with a suitable rotation of the moment tensor in (9.6). Because $M_{ij} = M_{ji}$, there are two fault planes that correspond to a double-couple model. For example, (9.6) is also appropriate for a left-lateral strike–slip fault oriented in the x_2 direction (Fig. 9.4). Both faults have

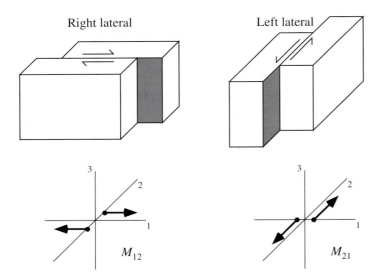

Figure 9.4 Owing to the symmetry of the moment tensor, these right-lateral and left-lateral faults have the same moment tensor represention and the same seismic radiation pattern.

[2] Older references sometimes express M_0 in dyne-cm. Note that $1\,\text{N} = 10^5$ dyne and thus $1\,\text{Nm} = 10^7$ dyne-cm.

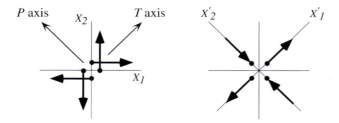

Figure 9.5 The double-couple pair on the left is represented by the off-diagonal terms in the moment tensor, M_{12} and M_{21}. By rotating the coordinate system to align with the P and T axes, the moment tensor in the new coordinate system is diagonal with opposing M_{11} and M_{22} terms.

the same moment tensor representation. This is a fundamental ambiguity in inverting seismic observations for fault models. In general, there are two fault planes that are consistent with distant seismic observations in the double-couple model. The real fault plane is termed the *primary fault plane*; the other is termed the *auxiliary fault plane*. This ambiguity is not a defect of the double-couple model (which has been shown to provide an excellent match to seismic observations) but reflects the fact that both faults produce exactly the same seismic displacements in the far field. Distinguishing between the primary and auxiliary fault planes requires examination of factors that go beyond a point source model (e.g., considering different parts of the rupture plane) or use of other information, such as aftershock locations or observed surface rupture.

Because the moment tensor is symmetric, it can be diagonalized by computing its eigenvalues and eigenvectors and rotating to a new coordinate system (just as we did for the stress and strain tensors in Chapter 2). For the example moment tensor given in (9.6), the principal axes are at $45°$ to the original x_1 and x_2 axes (Fig. 9.5), and the rotated moment tensor becomes

$$\mathbf{M}' = \begin{bmatrix} M_0 & 0 & 0 \\ 0 & -M_0 & 0 \\ 0 & 0 & 0 \end{bmatrix}. \tag{9.9}$$

The x_1' coordinate is termed the *tension axis*, T, and x_2' is called the *pressure axis*, P. The two sets of force couples plotted in Figure 9.5 are equivalent; they have the same moment tensor representation and they produce the same seismic radiation.

9.2.1 Non-double-couple sources

Double-couple sources arising from shear fracture have a specific moment tensor representation, in which both the trace and determinant of $\mathbf{M}$ are zero. However, the moment tensor is a more general description of possible sources than double-couple sources alone, and moment tensors computed from seismic data may include contributions from other types of events. The trace of the moment tensor is a measure of volume changes that accompany the event and is always zero for simple shear sources. In contrast, the moment tensor for an isotropic source (e.g., an explosion) has the form

$$\mathbf{M} = \begin{bmatrix} M_{11} & 0 & 0 \\ 0 & M_{22} & 0 \\ 0 & 0 & M_{33} \end{bmatrix}, \tag{9.10}$$

where $M_{11} = M_{22} = M_{33}$.

From a general moment tensor, we can extract the isotropic part as

$$\mathbf{M}^0 = \tfrac{1}{3}(\text{tr } \mathbf{M})\mathbf{I} \tag{9.11}$$

and decompose $\mathbf{M}$ into isotropic and deviatoric parts:

$$\mathbf{M} = \mathbf{M}^0 + \mathbf{M}' \tag{9.12}$$

where $\text{tr } \mathbf{M}' = 0$. The deviatoric moment tensor, $\mathbf{M}'$, is free of any isotropic sources but may contain additional non-double-couple components. We can diagonalize $\mathbf{M}'$ by computing its eigenvalues and eigenvectors and rotating to coordinates defined by its principal axes. We then have

$$\mathbf{M}' = \begin{bmatrix} \sigma_1 & 0 & 0 \\ 0 & \sigma_2 & 0 \\ 0 & 0 & \sigma_3 \end{bmatrix}, \tag{9.13}$$

where the eigenvalues are ordered such that $\sigma_1 > \sigma_2 > \sigma_3$. Because $\text{tr } \mathbf{M}' = 0$, we also have $\sigma_2 = -\sigma_1 - \sigma_3$. For a pure double-couple source, $\sigma_2 = 0$ and $\sigma_3 = -\sigma_1$. Following Knopoff and Randall (1970) we can further decompose $\mathbf{M}'$ into a best-fitting double-couple, $\mathbf{M}^{DC}$, and a second term called a *compensated linear vector dipole*, $\mathbf{M}^{CLVD}$

$$
\begin{aligned}
\mathbf{M}' &= \mathbf{M}^{DC} + \mathbf{M}^{CLVD} \\
&= \begin{bmatrix} \tfrac{1}{2}(\sigma_1 - \sigma_3) & 0 & 0 \\ 0 & 0 & 0 \\ 0 & 0 & -\tfrac{1}{2}(\sigma_1 - \sigma_3) \end{bmatrix} + \begin{bmatrix} -\sigma_2/2 & 0 & 0 \\ 0 & \sigma_2 & 0 \\ 0 & 0 & -\sigma_2/2 \end{bmatrix}.
\end{aligned} \tag{9.14}
$$

The complete decomposition of the original $\mathbf{M}$ is thus

$$\mathbf{M} = \mathbf{M}^0 + \mathbf{M}^{DC} + \mathbf{M}^{CLVD}. \tag{9.15}$$

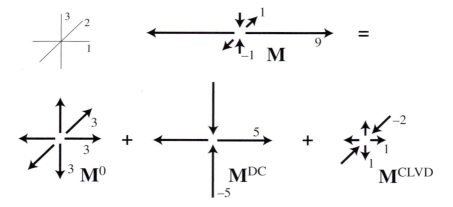

Figure 9.6 Example of the decomposition of a moment tensor into isotropic, best-fitting double couple, and compensated linear vector dipole terms.

Note that the decomposition of $\mathbf{M}'$ into $\mathbf{M}^{DC}$ and $\mathbf{M}^{CLVD}$ is unique only because we have defined $\mathbf{M}^{DC}$ as the *best-fitting* double-couple source, that is, we have minimized the CLVD part. There are alternative decompositions that will give a larger CLVD component and a correspondingly smaller double-couple moment tensor. Here is an example of the separation of a moment tensor into the components in (9.15), assuming it has already been rotated into its principal axes coordinates:

$$\mathbf{M} = \begin{bmatrix} 9 & 0 & 0 \\ 0 & 1 & 0 \\ 0 & 0 & -1 \end{bmatrix} = \begin{bmatrix} 3 & 0 & 0 \\ 0 & 3 & 0 \\ 0 & 0 & 3 \end{bmatrix} + \begin{bmatrix} 5 & 0 & 0 \\ 0 & 0 & 0 \\ 0 & 0 & -5 \end{bmatrix} + \begin{bmatrix} 1 & 0 & 0 \\ 0 & -2 & 0 \\ 0 & 0 & 1 \end{bmatrix}$$

as displayed in Figure 9.6. Alternatively, $\mathbf{M}'$ can be decomposed into two double-couple sources

$$\mathbf{M}' = \begin{bmatrix} \sigma_1 & 0 & 0 \\ 0 & \sigma_2 & 0 \\ 0 & 0 & \sigma_3 \end{bmatrix} = \begin{bmatrix} \sigma_1 & 0 & 0 \\ 0 & -\sigma_1 & 0 \\ 0 & 0 & 0 \end{bmatrix} + \begin{bmatrix} 0 & 0 & 0 \\ 0 & -\sigma_3 & 0 \\ 0 & 0 & \sigma_3 \end{bmatrix}, \qquad (9.16)$$

where we have used $\sigma_2 = -\sigma_1 - \sigma_3$. The larger and smaller of the two terms are called the *major* and *minor* double couples, respectively. This decomposition has the peculiar property that the major and minor double couples become nearly equal in size as σ_2 approaches zero; thus in most cases the decomposition of (9.15) is preferred.

Most earthquakes are well-described with double-couple sources, but the search for possible non-double-couple contributions has been a significant area of research (e.g., see reviews by Julian *et al.*, 1998, and Miller *et al.*, 1998). At one time it was hypothesized that deep focus earthquakes might involve volume changes caused by sudden implosive phase changes in minerals within the subducting slabs. However, results have generally indicated that these events do not have significant isotropic components (e.g., Kawakatsu, 1991). Moment tensor inversions will sometimes impose the constraint that $\mathbf{M}$ is purely deviatoric; this reduces the number of free parameters and can often lead to more stable results. A measure of the misfit between $\mathbf{M}'$ and a pure double-couple source is provided by the ratio of σ_2 to the remaining eigenvalue with the largest magnitude

$$\epsilon = \sigma_2/\max(|\sigma_1|, |\sigma_3|), \tag{9.17}$$

where $\epsilon = 0$ is obtained for a pure double-couple and $\epsilon = \pm 0.5$ is obtained for a pure compensated linear vector dipole.

Physically, non-double-couple components can arise from simultaneous faulting on faults of different orientations or on a curved fault surface. For example, CMT solutions for some Iceland earthquakes near Bardarbunga volcano suggest reverse faulting on outward dipping cone-shaped faults associated with caldera structures (Nettles and Ekström, 1998). Magma injection events can cause both isotropic and compensated linear vector dipole terms. For example, Kanamori *et al.* (1993) identified a dominant CLVD source for a 1984 earthquake near Tori Shima island in the Izu-Bonin arc. Perhaps the most exotic sources of all are volcanic eruptions (e.g., Kanamori *et al.*, 1984), landslides, such as Mt. St. Helens (Kanamori and Given, 1982), and glacial sliding events, such as the Greenland quakes recently discovered by Ekstöm *et al.* (2003). Seismic data from these events cannot be fit with standard force couples and moment tensor analysis and require single force models.

9.3 Radiation patterns and beach balls

To use the equivalent body-force representation to predict displacements, we need to know the elastodynamic Green's function, $\mathbf{G}$, in (9.2). In general, solving for $\mathbf{G}$ is rather complicated. However, some insight into the nature of the solutions that are obtained may be found by considering the simple case of a spherical wavefront from an isotropic source. In Chapter 3, we described how the solution for the P-wave potential in this case is given by

$$\phi(r, t) = \frac{-f(t - r/\alpha)}{r}, \tag{9.18}$$

where α is the P velocity, r is the distance from the point source, and $4\pi\delta(r)\,f(t)$ is the source-time function. Note that the amplitude of the potential diminishes as $1/r$, as we derived earlier from geometrical spreading considerations for a spherical wavefront. The displacement field is given by the gradient of the displacement potential

$$u(r, t) = \frac{\partial \phi(r, t)}{\partial r} = \left(\frac{1}{r^2}\right) f(t - r/\alpha) - \left(\frac{1}{r}\right) \frac{\partial f(t - r/\alpha)}{\partial r}. \tag{9.19}$$

Defining $\tau = t - r/\alpha$ as the *delay time*, where r/α is the time that it takes a P wave to travel the distance r from the source, we have

$$\frac{\partial f(t - r/\alpha)}{\partial r} = \frac{\partial f(t - r/\alpha)}{\partial \tau} \frac{\partial \tau}{\partial r} = -\frac{1}{\alpha} \frac{\partial f(t - r/\alpha)}{\partial \tau},$$

and so (9.19) can be expressed as

$$u(r, t) = \left(\frac{1}{r^2}\right) f(t - r/\alpha) + \left(\frac{1}{r\alpha}\right) \frac{\partial f(t - r/\alpha)}{\partial \tau}. \tag{9.20}$$

This equation is relatively simple because it applies only to P waves and involves no radiation pattern effects as the source is assumed to be spherically symmetric. The first term decays as $1/r^2$ and is called the *near-field term* since it is important only relatively close to the source. It represents the permanent static displacement due to the source. The second term decays as $1/r$ and is called the *far-field term* because it will become dominant at large distances from the source. It represents the dynamic response – the transient seismic waves that are radiated by the source that cause no permanent displacement. These waves have displacements that are given by the first time derivative of the source-time function.

More complicated expressions arise for point force and double-couple sources, but these also involve near- and far-field terms. Most seismic observations are made at sufficient distance from faults that only the far-field terms are important. The far-field P-wave displacement from the jk component of a moment tensor source at $\mathbf{x} = 0$ in a homogeneous whole space is given by

$$u_i^P(\mathbf{x}, t) = \frac{1}{4\pi\rho\alpha^3} \frac{x_i x_j x_k}{r^3} \frac{1}{r} \dot{M}_{jk}\left(t - \frac{r}{\alpha}\right), \tag{9.21}$$

where $r^2 = x_1^2 + x_2^2 + x_3^2$ is the squared distance to the receiver and $\dot{M}$ is the time derivative of the moment tensor. This is a general expression that gives the far-field P displacements for any moment tensor representation of the source.

Now let us consider the more specific example of a fault described by a double-couple source. Without loss of generality we may assume that the fault is in the $(x_1,$

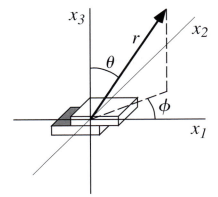

Figure 9.7 Spherical coordinates for a vector relative to a fault in the (x_1, x_2) plane with slip in the x_1 direction.

x_2) plane with motion in the x_1 direction (Fig. 9.7). We then have $M_{13} = M_{31} = M_0$ and

$$u_i^P(\mathbf{x}, t) = \frac{1}{2\pi\rho\alpha^3} \frac{x_i x_1 x_3}{r^3} \frac{1}{r} \dot{M}_0 \left(t - \frac{r}{\alpha}\right). \tag{9.22}$$

Note that the factor of two difference from (9.21) is due to the sum over M_{13} and M_{31}. If we define spherical coordinates relative to the fault as shown in Figure 9.7, we have

$$x_3/r = \cos\theta,$$

$$x_1/r = \sin\theta\cos\phi, \tag{9.23}$$

$$x_i/r = \hat{r}_i,$$

and thus, substituting for x in (9.22) and using $\cos\theta\sin\theta = \frac{1}{2}\sin 2\theta$, we have

$$\mathbf{u}^P = \frac{1}{4\pi\rho\alpha^3} \sin 2\theta \cos\phi \frac{1}{r} \dot{M}_0 \left(t - \frac{r}{\alpha}\right) \hat{\mathbf{r}}. \tag{9.24}$$

The P-wave radiation pattern is illustrated in Figure 9.8. Note that the fault plane and the auxiliary fault plane (the plane perpendicular to the fault plane and the slip vector) form nodal lines of zero motion that separate the P-wave polarities into four quadrants. The outward pointing vectors represent outward P-wave displacement in the far field (assuming $\dot{M}$ is positive); this is termed the compressional quadrant. The inward pointing vectors occur in the dilatational quadrant. The tension (T axis) is in the middle of the compressional quadrant; the pressure (P axis) is in the middle of the dilatational quadrant. (Yes, it's confusing! The *tension* axis is in the

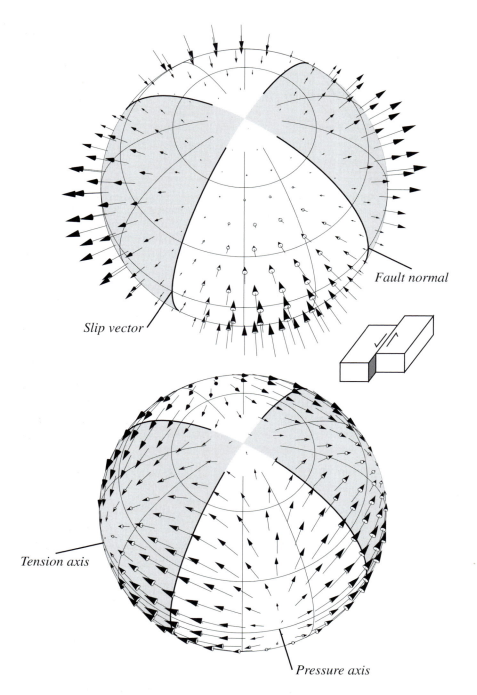

Figure 9.8 The far-field radiation pattern for *P* waves (top) and *S* waves (bottom) for a double-couple source. The orientation of the small arrows shows the direction of first motion; their length is proportional to the wave amplitude. The primary and auxiliary fault planes are shown as heavy lines; the compressional quadrants are shaded. *P*-wave first motions are outward in the compressional quadrant and inward in the dilatational quadrant with nodal lines in between. *S*-wave first motions are generally away from the pressure axis and toward the tension axis; there are six nodal points and no nodal lines in *S*. Because of the ambiguity between the primary and auxiliary fault planes, the positions of the slip and fault normal vectors in the top plot could be reversed.

compressional quadrant; compressional in this case refers to the outward direction of P first motion.)

For S waves the equations are only slightly more complicated. The far-field S displacements as a function of M_{jk} are given by

$$u_i^S(\mathbf{x}, t) = \frac{(\delta_{ij} - \gamma_i\gamma_j)\gamma_k}{4\pi\rho\beta^3} \frac{1}{r} \dot{M}_{jk}\left(t - \frac{r}{\beta}\right),$$

(9.25)

where β is the shear velocity and the direction cosines are $\gamma_i = x_i/r$. For a double-couple source with the geometry shown in Figure 9.7, we may rewrite this as

$$\mathbf{u}^S(\mathbf{x}, t) = \frac{1}{4\pi\rho\beta^3}(\cos 2\theta \cos\phi \,\hat{\boldsymbol{\theta}} - \cos\theta \sin\phi \,\hat{\boldsymbol{\phi}})\frac{1}{r}\dot{M}_0\left(t - \frac{r}{\beta}\right),$$

(9.26)

where $\hat{\boldsymbol{\theta}}$ and $\hat{\boldsymbol{\phi}}$ are unit Cartesian vectors in the θ and ϕ directions. The resulting S-wave radiation pattern is illustrated in Figure 9.8. There are no nodal planes, but there are nodal points. S-wave polarities generally point toward the T axis and away from the P axis.

The first motions of P waves have long been used to determine earthquake focal mechanisms using the double-couple model. The advantages of this approach, compared to more sophisticated methods of moment tensor inversion, are that only vertical component instruments are required, amplitude calibration is not needed, and the sense of the first P motion (i.e., up or down) can be easily noted from the seismogram at the same time the arrival time is picked, even on analog records. The initial motion of the P wave determines whether the ray left the source in a compressional (upward first motion at a surface receiver) or dilatational quadrant (downward first motion), regardless of sensor type (e.g., displacement, velocity, or acceleration). Ray theory is then used to project the rays from all of the observations back to the angle at which they left the source. The results are plotted on what is termed the *focal sphere*, an imaginary sphere surrounding the source that shows the takeoff angles of the rays. Usually only the lower hemisphere of the focal sphere is plotted, as most rays at teleseismic distances depart downward from the source. (*P* first motions for upward propagating rays may be plotted on the appropriate opposing point on the lower hemisphere, as the P-wave radiation pattern is symmetric about the origin.)

If enough polarity measurements are plotted, it is possible to divide the focal sphere into compressional and dilatational quadrants. The focal mechanism is then determined by finding two orthogonal planes and their great circle projections onto the focal sphere that separate these quadrants. As discussed above, there is no way to tell from these observations alone which of these planes is the true fault plane and which is the auxiliary fault plane. In the old days, this method was implemented

Strike Slip

Normal

Reverse

Oblique

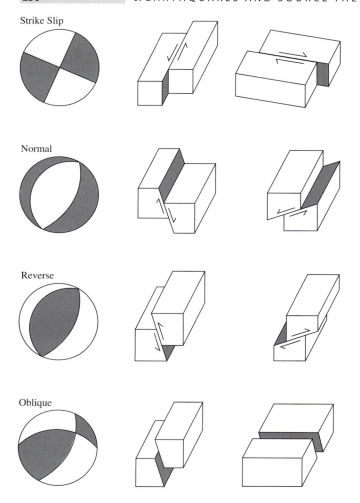

Figure 9.9 Examples of focal spheres and their corresponding fault geometries. The lower half of the focal sphere is plotted to the left, with the compressional quadrants shaded. The block diagrams on the right show the two fault geometries (the primary and auxiliary fault planes) that could have produced the observed radiation pattern.

by hand on special map projections. Today, it is fairly simple to find the focal mechanism using a computer to perform a grid search on the three parameters that define the focal mechanism (strike, dip, rake), directly identifying those solutions that fit the polarity observations most closely.

The focal sphere is also used as a means of displaying focal mechanisms. The lower hemisphere is plotted and the compressional quadrants are shaded to produce the traditional "beach ball" image. This is illustrated in Figure 9.9 for different types of focal mechanisms. In interpreting these plots, remember that the shaded regions

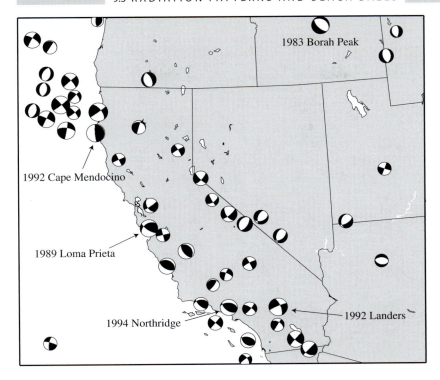

1983 Borah Peak

1992 Cape Mendocino

1989 Loma Prieta

1994 Northridge

1992 Landers

Figure 9.10 Selected focal mechanisms from the Global CMT catalog in the southwestern United States.

represent P waves leaving downward from the source with outward first motions that will produce upward first motions at the receivers, while the unshaded regions will result in downward first motions at the receivers. The tension axis is in the middle of the shaded region; the pressure axis is in the unshaded region. Normal and reverse faulting may be distinguished in beach ball plots by noting if the center of the plot is white or black. If it is white in the middle with black edges, then it represents a normal fault and a probable region of extension, whereas black in the center with white edges indicates a reverse or thrust fault and a likely compressional regime. Examples of strike–slip, normal, and reverse faulting earthquakes are shown in Figure 9.10, which plots global CMT results for the southwestern United States.

Note that the pressure and tension axes give the directions of maximum compression and tension in the Earth only if the fault surface corresponds to a plane of maximum shear. Because this is usually not true, the fault plane solution does not uniquely define the stress tensor orientation (although it does restrict the maximum compression direction to a range of possible angles). However, if multiple focal mechanisms are available at different orientations within a volume in which

Figure 9.11 Selected focal mechanisms from the Global CMT catalog.

the stress can be assumed homogeneous, then it is possible to estimate the stress tensor orientation (e.g., Gephart and Forsyth, 1984; Michael, 1987). This approach has been used to constrain the principal stress directions in many areas of active seismicity and address questions such as the possible rotation of the stress tensor near active faults.

Focal mechanisms began to be computed from first motions on a routine basis following the establishment of the WWSSN seismic network in the early 1960s. These results confirmed the double-couple theory for earthquake sources and showed that the earthquake mechanisms in different regions were consistent with the emerging theory of plate tectonics. Most earthquakes occur along the boundaries that separate the rigid plates. Strike–slip events are found along active transform faults, such as the San Andreas Fault in California, where the plates are sliding past each other. Reverse fault earthquakes are seen in subduction zones and normal faults in extensional regimes.

The radiation pattern equations presented here are for body waves. Analogous equations are used to describe the generation of surface waves and normal modes. Analyses of longer-period data for source characteristics typically involve a waveform fit and direct inversion for the components of the moment tensor as described earlier. Usually the moment tensor obtained is very close to a pure double couple, and, for convenience, the Global CMT catalog (see Section 9.1 or www.globalcmt.org/CMTsearch.html) provides the strike, dip, and rake for the best-fitting double-couple source. Examples of global focal mechanisms from this catalog between 1976 and 2005 are plotted in Figures 9.10 and 9.11. The complete catalog contains thousands of events; these figures plot the largest earthquakes in each region.

9.3.1 Example: Plotting a focal mechanism

Assume we are given that the strike, dip and rake of an earthquake are $30°$, $60°$, and $40°$ respectively and want to sketch the focal mechanism. Figure 9.12 shows how this is done. The left plot is an equal-area lower-hemisphere map of the focal sphere. The numbers around the outside circle show the fault strike in degrees. The circles show fault dip angles with $0°$ dip (a horizontal fault) along the outer circle and $90°$ dip (a vertical fault) at the center point. The fault strike of $30°$ defines point A at $30°$ and point C at $30 + 180 = 210°$. The fault dip of $60°$ defines point B, which is on the $60°$ dip contour on the right side of the line from C to A (recall that faults always dip to the right of the strike direction; see Figure 9.3). The curved line ABC shows the intersection of the fault with the focal sphere. But this is just one of the two fault planes that define the focal mechanism. With a bit of algebra, we can compute the

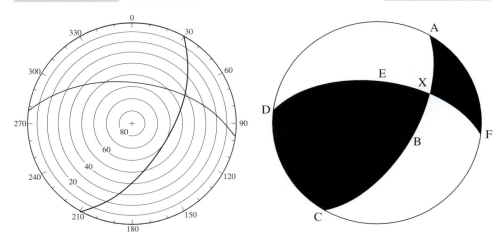

Figure 9.12 An example showing how a focal mechanism is plotted using a lower hemisphere projection.

orientation of the auxiliary plane from the strike, dip, and rake of the first plane. This is described on pp. 228–9 of Stein and Wysession (2002) and is also provided in the subroutine GETAUX in Appendix D. The strike, dip, and rake of this plane are 277.2°, 56.2° and 143.0° and are used to plot curve DEF.

The two curves divide the focal sphere into four parts and the final step is to shade in the correct two quadrants to create the beach ball plot on the right side of Figure 9.12. To get this right, refer again to Figure 9.3 and consider the first fault plane. The rake of 40° gives the angle along the fault plane from the strike direction that the hanging wall side of the fault moves. In lower-hemisphere projections, the hanging wall side is the side that does not include the center point of the plot, i.e., the side to the right of ABC in this example. Because the rake is less than 90°, this side is moving crudely in the direction of point A and thus the area AXF must be in the compressional quadrant and is shaded. The other side of the fault moves in the opposite direction and thus area DEXBC is also shaded. The bottom focal mechanism of Figure 9.9 is close to the orientation of this example and may help in visualizing the slip geometry.

9.4 Far-field pulse shapes

The displacement that occurs on opposite sides of a fault during an earthquake is permanent; the Earth does not return to its original state following the event. Thus, the equivalent body force representation of the displacement field must involve a

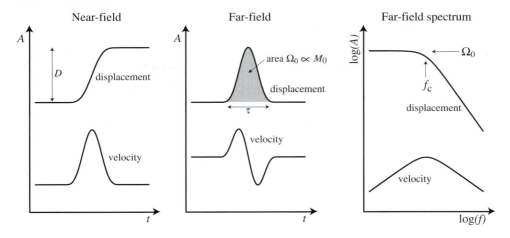

Figure 9.13 The relationships between near-field displacement and far-field displacement and velocity for time series (left two panels) and spectra (right panel).

permanent change in the applied forces. In addition, the displacement is not instantaneous but occurs over some finite duration of rupture. We can accommodate these properties by generalizing our moment tensor representation to be time dependent. For instance, one of the components of the moment tensor could be expressed as $M(t)$ and might have the form shown at the top left of Figure 9.13. This is what the near-field displacement would look like; for example, this might describe the path of a house near the San Andreas Fault during a large earthquake. These displacements are permanent and can be measured at some distance away from large earthquakes by geodetic means (such as surveying or GPS) after the shaking has subsided. The strain changes at Piñon Flat following the 1992 Landers earthquake that we examined in the exercises for Chapter 2 are an example of this.

The expressions for the far-field displacements from isotropic or double-couple sources (9.20–9.26) all involve the time derivative of the moment tensor. The time derivative of $M(t)$ is proportional to the far-field dynamic response (the middle panel of Figure 9.13), such as would be observed in a P- or S-wave arrival. Note that this is a displacement pulse and that there is no permanent displacement after the wave passes. Most seismometers measure velocity $\dot{u}(t)$ rather than displacement $u(t)$, in which case what is actually recorded will have an additional time derivative. In problems of Earth structure, it generally matters little whether we use velocity rather than displacement provided we assume an extra derivative for the source when we are modeling the waveforms. However, when studying seismic sources, velocity is almost always converted to displacement. This is done by integrating the velocity record and normally also involves a correction for the instrument response.

The aim is to recover an unbiased record of $\dot{M}(t)$ at the source. We will assume for most of this section that we are measuring far-field displacement.

The spectrum of the far-field displacement pulse (see top right of Figure 9.13) at low frequencies will be flat at a level, Ω_0, equal to the area beneath the pulse. The displacement spectrum will then roll off at higher frequencies, with the *corner frequency*, f_c, inversely proportional to the pulse width, τ. In the frequency domain the effect of the time derivatives is to multiply the spectrum by f. Thus velocity records are enhanced in high frequencies relative to displacement records.

The long-period spectral level, Ω_0, is proportional to the scalar seismic moment, M_0. In the case of body waves, integrating (9.24) and (9.26) over time, we obtain

$$M_0 = \frac{4\pi\rho c^3 r \Omega_0}{U_{\phi\theta}} \tag{9.27}$$

where ρ is the density, c is the wave velocity, r is the distance from the source, and $U_{\phi\theta}$ is the radiation pattern term. This equation is for spherical wavefronts expanding in a whole space but can be applied to more complicated velocity models using ray theory if the r factor is replaced with the appropriate term for geometrical spreading. If Ω_0 is measured from a station at the Earth's surface, then corrections must be applied to account for the wave amplification that occurs from the surface reflections. There are analogous expressions for computing M_0 from surface waves. These equations are important because they show how a fundamental property of the earthquake source – scalar moment – can be obtained directly from seismic wave observations at great distances. Because Ω_0 is measured at the lowest possible frequency, it is relatively insensitive to the effects of scattering and attenuation, making scalar moment estimates more reliable than measurements of source properties that require higher frequency parts of the spectrum. However, note that (9.27) does require knowledge of the focal mechanism owing to the $U_{\phi\theta}$ term. If a focal mechanism is not available, sometimes M_0 is estimated by averaging results from many stations and replacing $U_{\phi\theta}$ with the mean radiation term over the focal sphere (0.52 and 0.63 for P and S waves, respectively). Of course, if the complete moment tensor is computed using (9.5), then the scalar moment can be obtained directly using (9.8).

9.4.1 Directivity

Consider a point source characterized by a ramp function in $M(t)$. The corresponding $\dot{M}(t)$ function and the far-field displacement pulse $u(t)$ will be a boxcar (Fig. 9.14). This is sometimes termed the *Haskell source* and is the simplest realistic representation of a source. The duration of the boxcar pulse τ_r is called the *rise*

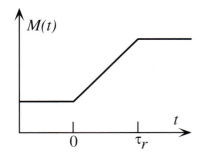

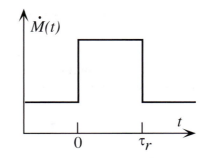

Figure 9.14 The Haskell source model.

time. For earthquakes that can be approximated as point sources, the Haskell source often provides a good description of the far-field response. Note that in this case the amplitude of the observed boxcar pulse will vary with azimuth as a function of the radiation pattern, but the duration of the pulse remains constant.

However, for larger events in which the rupture is extended in time and space we generally must include directivity effects. An important principle in seismology is that of *linear superposition*, which states that the response of a large fault can be described as the sum of the response from a number of individual pieces of the fault. Thus we can model a fault by integrating over individual point-source displacements on the fault surface. Let us examine how this works in the case of a long, narrow fault. We assume that the rupture propagates along the fault of length L from left to right at a rupture velocity v_r. In the far field we will observe the rupture from each point on the fault at a different time. For example, in the case where the fault is rupturing directly toward us, the apparent rupture duration time τ_d for P waves is given by

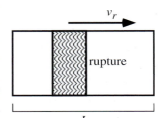

$$\tau_d(\text{toward}) = L\left(\frac{1}{v_r} - \frac{1}{\alpha}\right), \tag{9.28}$$

whereas the observed time for rupture directly away from us would be

$$\tau_d(\text{away}) = L\left(\frac{1}{v_r} + \frac{1}{\alpha}\right). \tag{9.29}$$

In general τ_d is a function of the orientation of the fault relative to the receiver and the direction and velocity of the rupture. The changes in τ_d as a function of receiver location are termed *directivity* effects. For example, for a vertical, strike–slip fault such as the San Andreas we would expect to see an azimuthal dependence in τ_d. The rupture velocity v_r is generally observed to be somewhat less than the shear-wave velocity, although sometimes much slower ruptures occur and the possibility of occasional ruptures faster than the shear velocity (termed *super shear*) is a current research topic. Because *P* waves travel faster than the rupture, the hypocenter, which represents the point of rupture initiation, can be unambiguously located using *P*-wave arrivals.

Now imagine that the actual slip at a point on the fault can be described by a ramp function in displacement as the rupture passes by. The shape of the far-field displacement pulse will be given by the convolution of two boxcar functions, one of width τ_r, the rise time, and the other of width τ_d, the apparent rupture duration time (see Appendix E for a review of time series analysis and convolution). The resulting pulse will be a trapezoid (Fig. 9.15). This is termed the *Haskell fault model* and is valid for a simple model of a line source. The width of the trapezoid will vary with azimuth to the fault as the apparent rupture duration time τ_d changes. However, the shorter pulses are larger in amplitude, such that the area of the trapezoid, Ω_0, is preserved. Thus we should expect to see higher amplitudes and shorter durations for pulses propagating in the direction of rupture propagation and weaker amplitudes and longer durations for pulses radiating in the opposite direction (Fig. 9.16). Because Ω_0 is constant, however, these differences do not affect estimates of the scalar seismic moment, M_0, derived from these pulses (this

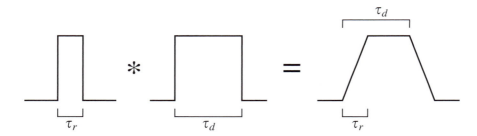

Figure 9.15 The Haskell fault model consists of the convolution of two boxcar functions with widths given by the rise time and rupture duration time.

Figure 9.16 Displacement pulses radiated in the direction of rupture propagation will be higher in amplitude, but shorter in duration, than pulses radiated in the opposite direction.

comparison assumes that factors that affect Ω_0, such as geometrical spreading and the radiation pattern are constant between the two paths).

9.4.2 Source spectra

Let us now consider the characteristics of far-field pulses in the frequency domain. Those readers who require a review of Fourier analysis of time series should consult Appendix E. The Fourier transform of a boxcar of unit height and width is given by

$$\mathcal{F}[B(t)] = \int_{-1/2}^{1/2} e^{i\omega t} dt = \frac{1}{i\omega} \left(e^{i\omega/2} - e^{-i\omega/2} \right)$$

$$= \frac{1}{i\omega} [i \sin(\omega/2) - i \sin(-\omega/2) + \cos(\omega/2) - \cos(-\omega/2)]$$

$$= \frac{1}{i\omega} 2i \sin(\omega/2) = \frac{\sin(\omega/2)}{\omega/2}. \tag{9.30}$$

The function $\sin x / x$ is commonly referred to as sinc x. Using the scale rule for Fourier transforms we can express the Fourier transform of a boxcar of unit height and width τ_r as

$$\mathcal{F}[B(t/\tau_r)] = \tau_r \text{sinc}(\omega \tau_r / 2). \tag{9.31}$$

This is illustrated in Figure 9.17. Note that the first zero crossing occurs at $\omega = 2\pi/\tau_r$, corresponding to the frequency $f = 1/\tau_r$. The Haskell fault model, given by the convolution of two boxcars of widths τ_r and τ_d, may be expressed as a product of two sinc functions in the frequency domain:

$$\mathcal{F}[B(t/\tau_r) * B(t/\tau_d)] = \tau_r \tau_d \text{ sinc}(\omega \tau_r / 2) \text{sinc}(\omega \tau_d / 2). \tag{9.32}$$

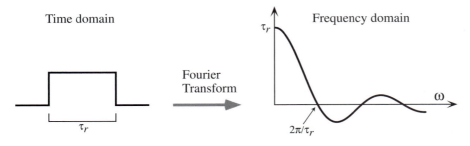

Figure 9.17 A boxcar pulse in the time domain produces a sinc function in the frequency domain.

Thus, the far-field amplitude spectrum $|A(\omega)|$ for the Haskell fault model may be expressed as

$$|A(\omega)| = gM_0 \, |\text{sinc}(\omega\tau_r/2)| \, |\text{sinc}(\omega\tau_d/2)| \,, \qquad (9.33)$$

where g is a scaling term that includes geometrical spreading, etc. Source spectra are usually plotted using a log–log scale. Taking the logarithm of (9.33) we have

$$\log|A(\omega)| = G + \log(M_0) + \log|\text{sinc}(\omega\tau_r/2)| + \log|\text{sinc}(\omega\tau_d/2)| \,, \qquad (9.34)$$

where $G = \log g$. We can approximate $|\text{sinc}\, x|$ as 1 for $x < 1$ and $1/x$ for $x > 1$. We then obtain

$$
\begin{aligned}
\log|A(\omega)| - G &= \log M_0, & \omega &< 2/\tau_d \\
&= \log M_0 - \log\frac{\tau_d}{2} - \log\omega, & 2/\tau_d &< \omega < 2/\tau_r \\
&= \log M_0 - \log\frac{\tau_d\tau_r}{4} - 2\log\omega, & 2/\tau_r &< \omega \qquad (9.35)
\end{aligned}
$$

where we have assumed $\tau_d > \tau_r$. Thus we see that in the case of the spectrum of a trapezoidal source-time function, we should expect to see a low-frequency part that is flat at a level proportional to M_0, a ω^{-1} segment at intermediate frequencies, and a ω^{-2} fall off at high frequencies (Fig. 9.18).

The frequencies corresponding to $\omega = 2/\tau_r$ and $\omega = 2/\tau_d$ are called the *corner frequencies* and divide the spectrum into the three different parts. By studying the spectra of real earthquakes we can, in principle, recover M_0, τ_r, and τ_d for this model. However, we often can only identify a single corner, defined by the intersection of the ω^0 and ω^{-2} asymptotes. Caution should be applied in any interpretation of an observed spectrum directly in terms of source properties, since attenuation and near-surface effects can distort the spectrum, particularly at higher frequencies.

Many different theoretical earthquake source models have been proposed and they predict different shapes for the body-wave spectra. Brune (1970) described one of the most influential models, in which the displacement amplitude spectrum is given by

$$A(f) = \frac{\Omega_0}{1 + (f/f_c)^2} \qquad (9.36)$$

where f_c is the corner frequency. Note that the high-frequency falloff rate agrees with the Haskell fault model. A more general model is

$$A(f) = \frac{\Omega_0}{[1 + (f/f_c)^{\gamma n}]^{1/\gamma}} \qquad (9.37)$$

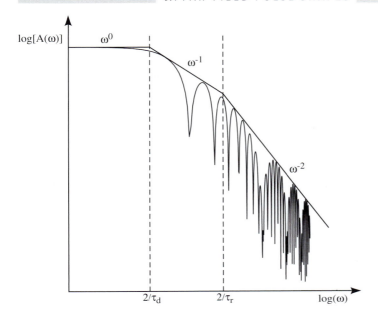

Figure 9.18 The amplitude spectrum for the Haskell fault model. The spectrum is the product of two sinc functions, corresponding in the time domain to the convolution of two boxcar functions of durations τ_d and τ_r. The spectral amplitudes fall off as ω^{-1} for $2/\tau_d < \omega < 2/\tau_r$ and as ω^{-2} for $\omega > 2/\tau_r$. For the spectrum plotted in this figure, $\tau_d = 8\tau_r$.

which was found by Boatwright (1980) with $\gamma = 2$ to provide a better fit to the sharper corners that he found in his data. Equations (9.36) and (9.37) with $n = 2$ are often called ω^{-2} source models. Some theoretical source models, particularly those which consider elongated fault geometries, predict ω^{-3} fall off at high frequencies. However, studies of both globally and locally recorded earthquakes over a wide range of sizes have generally shown that their average high-frequency falloff rate is close to ω^{-2}, although individual earthquakes often have quite different spectral behavior.

9.4.3 Empirical Green's functions

One of the challenging aspects of studying seismic spectra is separating out what originates from the source and what is caused by attenuation or other path effects. For example, for a simple constant Q model the spectra will drop off exponentially at high frequencies

$$A(f) = A_0(f)e^{-\pi ft/Q}. \tag{9.38}$$

In principle, this falloff has different curvature than the power law decay with frequency of theoretical source models and one approach has been to use (9.38) together with (9.36) or (9.37) to simultaneously solve for Q and f_c (and sometimes n and γ as well). However, with the irregular spectra and limited bandwidth of real data it can be difficult to separately resolve the source and attenuation contributions and there is often a tradeoff between them.

Another approach is to use records from a smaller earthquake near the target earthquake to compute an empirical path and attenuation correction. The assumption is that the second quake is small enough that its corner frequency is above the observation band and its spectrum is nearly flat, i.e., it is effectively a delta-function source. In this case one can either deconvolve its waveform from the target earthquake record in the time domain or simply correct the observed spectrum in the frequency domain. This is called the *empirical Green's function* or EGF method (e.g., Müeller, 1985; Hough, 1997) and is widely used in source studies. It does, however, require that there be a suitable event close enough to the target earthquake that the path effects will be approximately the same.

9.5 Stress drop

The seismic moment, $M_0 = \mu \overline{D} A$, does not distinguish between an earthquake involving small slip on a large fault and one with large slip on a small fault, provided the product of the average slip $(\overline{D})$ and fault area (A) remains constant. However, these earthquakes would change the stress on the fault by very different amounts. This change may be defined as the *stress drop*, which is the average difference between the stress[3] on a fault before an earthquake to the stress after the earthquake:

$$\Delta\sigma = \frac{1}{A} \int_S [\sigma(t_2) - \sigma(t_1)] \, dS \,, \tag{9.39}$$

where the integral is performed over the surface of the fault and A is the fault area. Analytical solutions for the stress drop have been derived for a few specialized cases of faults embedded within homogeneous material. For a circular fault in a whole space, Eshelby (1957) obtained

$$\Delta\sigma = \frac{7\pi\mu\overline{D}}{16r} = \frac{7M_0}{16r^3} \,, \tag{9.40}$$

---

[3] In this section "stress" refers specifically to the shear stress across the fault plane.

where r is the fault radius, μ is the shear modulus, and $\overline{D}$ is the average displacement. For strike–slip motion on a shallow, rectangular fault of length L and width w ($L \gg w$), Knopoff (1958) obtained

$$\Delta\sigma = \frac{2\mu\overline{D}}{\pi w} = \frac{2M_0}{\pi w^2 L}. \tag{9.41}$$

More generally, we may write

$$\Delta\sigma = C\mu \left[\frac{\overline{D}}{\tilde{L}}\right], \tag{9.42}$$

where $\tilde{L}$ is a *characteristic rupture dimension* (r in the case of the circular fault, w for the long rectangular fault) and C is a non-dimensional constant that depends upon the geometry of the rupture. Notice that physically it makes sense that the shear stress change on the fault will be proportional to the ratio of the displacement to the size of the fault. Large slip on a small fault will cause more stress than small slip on a large fault. It should be noted that these solutions assume smooth forms for the slip function on the fault surface and thus represent only approximations to the spatially averaged stress drop on real faults, for which the displacement and corresponding stress drop may vary in complicated ways owing to non-uniform elastic properties and initial stresses. A widely used result to obtain results for faults made up of arbitrary rectangular slip patches is the half-space solution of Okada (1992).

For large earthquakes for which the fault geometry can be constrained from surface rupture or aftershock studies, the stress drop can then be estimated from the moment. For large, shallow earthquakes, $\Delta\sigma$ varies from about 1 to 10 MPa (10 to 100 bars in the units often used in older studies) with no observed dependence on moment for M_0 variations from 10^{18} to 10^{23} N m (Kanamori and Anderson, 1975; Kanamori and Brodsky, 2004). Earthquakes near plate boundaries (*interplate* events) generally have been observed to have somewhat lower stress drops than those that occur in the interior of plates (*intraplate* events) (e.g., Kanamori and Anderson, 1975; Kanamori and Allen, 1986). Average $\Delta\sigma$ for interplate quakes is about 3 MPa compared to about 6 MPa for intraplate events (Allmann and Shearer, 2009). This implies that intraplate faults are "stronger" in some sense than interplate faults and have smaller fault dimensions for the same moment release.

For small earthquakes, direct observations of the rupture geometry are not possible so the fault dimensions must be estimated from far-field observations of the radiated seismic waves. In this case it is necessary to make certain assumptions about the source properties. In particular, these methods generally assume that the source dimension is proportional to the observed body-wave pulse width (after correcting for attenuation). The first quantitative model for estimating stress drop in this way was derived by Brune (1970), who assumed a simple kinematic model for a circular fault with effectively infinite rupture velocity and showed that the expected high-frequency spectral falloff rate is ω^{-2} and that the corner frequency is inversely proportional to the source radius. This result, together with a number of other proposed rupture models, predicts that the fault radius varies as

$$r = \frac{k\beta}{f_c}, \tag{9.43}$$

where r is the fault radius, f_c is the observed corner frequency (see Figure 9.13) and k is a constant that depends upon the specific theoretical model. Currently, perhaps the most widely used result is from Madariaga (1976), who performed dynamic calculations for a circular fault using finite differences. Assuming that the rupture velocity is 90% of the shear-wave velocity ($v_r = 0.9\beta$), he obtained $k = 0.32$ and 0.21 for the P- and S-wave corner frequencies, respectively, with an ω^{-2} high-frequency falloff rate. His model predicts a P-wave corner frequency about 50% higher than the S-wave corner frequency ($f_c^P \simeq 1.5 f_c^S$). Figure 9.19 plots predicted P-wave spectra for the Madariaga (1976) model for a wide range of M_0, assuming a constant stress drop of 3 MPa. Note that the corner frequency varies as $M_0^{-1/3}$, with higher corner frequencies for smaller earthquakes.

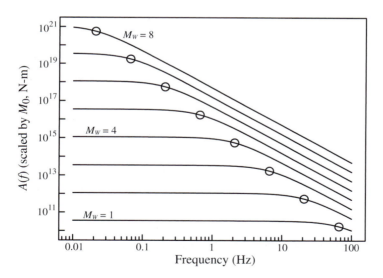

Figure 9.19 Predicted P-wave spectra from the Madariaga (1976) source model, assuming a constant stress drop of 3 MPa. The spectra have been scaled such that their amplitudes at low frequency are equal to their moments, M_0. The circles show the corner frequencies (f_c). Individual spectra are for moment magnitudes, M_W, from 1 to 8 (see (9.73) for the definition of M_W).

From (9.40) and (9.43), we have

$$\Delta\sigma = \frac{7}{16}\left(\frac{f_c}{k\beta}\right)^3 M_0 \,. \tag{9.44}$$

This is how stress drop can be estimated directly from far-field body-wave spectra using corner-frequency measurements, together with measurements of M_0 (which can be computed from the low-frequency part of the spectrum; see Ω_0 in Figure 9.13). Because this equation involves the cube of the $(f_c/k\beta)$ term, the computed $\Delta\sigma$ is extremely sensitive to differences in the assumed theoretical model (which determines the value of k and in general depends upon the assumed rupture velocity) and to variations in the estimated corner frequency f_c. The Brune (1970) model has a k value about 1.7 times larger than the Madariaga (1976) model, which translates to stress drop estimates about 5 times smaller. The corner frequency, f_c, can be tricky to measure from individual spectra, which are rarely as smooth as the theoretical models predict, and are sensitive to corrections for attenuation effects. Published stress drop values exhibit considerable scatter and it can be difficult to determine what part of these variations are real and what part may be attributed to differences in the modeling assumptions and analysis methods. However, there are large variations in individual earthquake stress drops even within single studies, suggesting that much of the observed scatter is real. For example, Shearer *et al.* (2006) analyzed *P*-wave spectra from over 60 000 small earthquakes in southern California using the Madariaga (1976) model and obtained $\Delta\sigma$ values from 0.2 to 20 MPa, with the bulk of the events between 0.5 to 5 MPa.

In principle, stress drop, like moment, is essentially a static measurement of permanent changes caused by an earthquake. However, the methods for estimating stress drops for small earthquakes are derived from body-wave pulse shapes and assumptions about the dynamics of the source. Because these are not direct measurements of static stress drop, they are sometimes termed *Brune-type* stress drops, although they may not be computed exactly as in Brune (1970). It is important to remember that these measurements involve a number of modeling assumptions that may not be true for individual earthquakes. For example, variations in rupture speed will cause a change in corner frequency even if the stress drop remains constant. Finally, note that measurements of the stress drop do not constrain the absolute level of stress. The absolute level of stress in the crust near faults is currently a topic of controversy that we will discuss later in this chapter.

9.5.1 Self-similar earthquake scaling

The fact that earthquake stress drops appear to be at least approximately constant over a wide range of earthquake sizes has implications for earthquake scaling relationships. Aki (1967) proposed that the physics of large and small earthquakes may be fundamentally similar, in which case we should expect *scale-invariance* or *self-similarity* of the rupture process. This implies that regardless of which theoretical earthquake source model is correct, the properties of the source will change in predictable ways as a function of earthquake size.

This is illustrated in Figure 9.20, which shows the expected change in pulse shape and spectrum when an earthquake rupture plane is increased in size by a factor b. Assuming the dimensions of the larger rupture are scaled proportionally, then the fault area, A, will increase by a factor b^2, the displacement, D, will increase by b, and the moment, $M_0 = \mu D A$, will increase by a factor of b^3. Stress drop remains constant because it is proportional to $DA^{-1/2}$. It follows that moment will scale with fault area as

$$M_0 \propto A^{3/2} \tag{9.45}$$

and such a scaling is observed to be approximately correct for large earthquakes (e.g., Kanamori and Anderson, 1975; Kanamori and Brodsky, 2004).

For an identical source–receiver geometry, no attenuation, and constant rupture velocity (predicted from self-similarity), the far-field displacement pulse will increase in duration by a factor of b and in amplitude by a factor of b^2. Note that the area under the pulse, Ω_0, also increases by b^3, as expected since Ω_0 is proportional

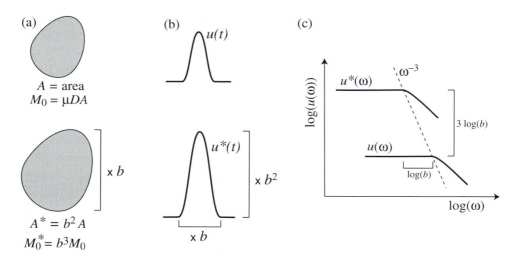

Figure 9.20 Illustration of the effects of self-similarity when an earthquake is increased in size by a factor b, showing the behavior of (a) rupture area and moment, (b) far-field displacement pulses, and (c) displacement spectra. Figure adapted from Prieto *et al.* (2004).

to M_0. It follows that the displacement pulse, u^*, recorded by the second earthquake can be expressed as

$$u^*(t) = b^2 u(t/b), \tag{9.46}$$

where $u(t)$ is the recorded displacement pulse of the first earthquake. The radiated seismic energy, E_R, in the recorded pulse will be proportional to $\int \dot{u}^2(t)\, dt$ (the integrated square of the slope of the pulse), so the second pulse will contain a factor b^3 more energy than the first pulse. Thus the radiated seismic energy to moment ratio (E_R/M_0) remains constant.

Using the similarity theorem for the Fourier transform, it follows that the spectrum of the second earthquake is given by

$$u^*(\omega) = b^3 u(b\omega) \tag{9.47}$$

where $u(\omega)$ is the spectrum of the first earthquake. This relationship predicts that the shape of all spectra on a log-log plot will be identical, but offset along a line of ω^{-3} (Fig. 9.20c). This means that corner frequency will vary as

$$f_c \propto M_0^{-1/3} \tag{9.48}$$

as is seen in Figure 9.19.

Self-similarity appears to be at least roughly true for average earthquake properties, although this has been a subject of considerable debate and there are large variations among individual earthquakes. It should be noted that self-similarity may break down for very large earthquakes that rupture through the entire seismogenic zone. In this case, ruptures are much longer than they are wide, with aspect ratios of 10 or more, which might make them behave differently than the less elongated rupture planes expected of smaller earthquakes (e.g., Scholz, 1982, 1997; Heaton, 1990). For example, the 1906 San Francisco earthquake ruptured for about 450 km to a depth of no more than 10 km (Thatcher, 1975).

9.6 Radiated seismic energy

Seismic moment and static stress drop are fundamental properties of the slip geometry of an earthquake, but they say nothing directly about the dynamics of the event, such as how fast the rupture propagated or how fast the two sides of the fault moved. This is why it is possible to estimate M_0 and $\Delta\sigma$ from geodetic measures of Earth deformation long after an earthquake; they are measures of the permanent static displacements across faults. Fault creep events that are too slow

to radiate seismic energy at observable frequencies can nonetheless have significant moments and stress drops (although as noted in Section 9.5, some methods
of actually computing stress drops require seismic wave observations and make
assumptions about source dynamics).

In contrast, one of the most fundamental measures of earthquake dynamics is the
total radiated energy, E_R, which represents the seismic energy that would propagate
to the far field in a whole space with no attenuation. Recalling the expressions for
seismic energy flux in Section 6.1, we have (e.g., Venkataraman *et al.*, 2006)

$$E_R = \rho \int_S \int_{-\infty}^{\infty} \left[\alpha \dot{u}_\alpha^2(t, \theta, \phi) + \beta \dot{u}_\beta^2(t, \theta, \phi) \right] dt \, dS, \tag{9.49}$$

where $\dot{u}_\alpha$ and $\dot{u}_\beta$ are velocity seismograms for P and S waves, respectively, and S
is a spherical surface at a large distance around the source. Of course, we cannot
integrate over the entire focal sphere; we must use seismic observations from a
discrete number of seismic stations. Using ray theory, we can correct the observed
amplitudes for varying amounts of geometrical spreading and determine the ray
takeoff angles, θ and ϕ, at the source. Because of radiation pattern effects, $\dot{u}_\alpha$ and $\dot{u}_\beta$
vary greatly over the surface of the sphere and thus a large number of observations
from different seismic stations would be necessary to estimate E_R reliably from
(9.49) directly. However, if the focal mechanism and thus the radiation pattern is
known, then single station estimates are possible, i.e.,

$$E_R = E_R^P + E_R^S = 4\pi\rho\alpha r^2 \frac{\langle {}_P U_{\phi\theta}{}^2 \rangle}{{}_P U_{\phi\theta}{}^2} I_P + 4\pi\rho\beta r^2 \frac{\langle {}_S U_{\phi\theta}{}^2 \rangle}{{}_S U_{\phi\theta}{}^2} I_S \tag{9.50}$$

where ${}_P U_{\phi\theta}$ and ${}_S U_{\phi\theta}$ are the P and S radiation pattern terms and $\langle U_{\phi\theta}{}^2 \rangle$ is the
mean over the focal sphere of $(U_{\phi\theta})^2$ ($\langle {}_P U_{\phi\theta}{}^2 \rangle = 4/15$ for P waves and $\langle {}_S U_{\phi\theta}{}^2 \rangle = 2/5$ for S waves), and I_P and I_S are the time-integrated values of $\dot{u}_\alpha^2$ and $\dot{u}_\beta^2$, as
corrected for geometrical spreading and any near-receiver effects (e.g., free-surface
reflections or amplifications from slow velocities in shallow layers) to what they
would be at a uniform distance r in the absence of attenuation.

I_P and I_S are usually computed in the frequency domain from body-wave spectra
because it is easier to correct for attenuation and instrument response effects, as
well as to check for adequate signal-to-noise properties. From Parseval's theorem,
we have

$$I = \int_{-\infty}^{\infty} |v(t)|^2 \, dt = \int_{-\infty}^{\infty} |v(f)|^2 \, df. \tag{9.51}$$

In principle, the integration is performed to infinite frequency. However, the velocity
spectrum peaks near the corner frequency (see Figure 9.13), and this peak becomes

even stronger when the velocity is squared. For the ω^{-2} model, calculations have shown that 90% of the total energy is obtained if the integration is performed out to 10 times the corner frequency (Ide and Beroza, 2001). Often data do not have this much bandwidth, which can lead to underestimation of the energy. To correct for this, the integration can be extrapolated beyond the observed bandwidth of the data by assuming that the spectral fall off continues at a fixed rate. However, in this case the result is no longer a direct measurement from the data because it relies on assumptions about the nature of the source.

The ratio of S-wave energy to P-wave energy is defined as

$$q = E_R^S / E_R^P. \tag{9.52}$$

For a point-source model in which the P- and S-wave pulses have identical shapes (and thus identical corner frequencies f_c^P and f_c^S), we have from (9.24), (9.26), (9.50), and the values of $\langle _p U_{\phi\theta}{}^2 \rangle$ and $\langle _s U_{\phi\theta}{}^2 \rangle$, that $q = 1.5(\alpha/\beta)^5 \simeq 23.4$ for a Poisson solid. However, many theoretical finite source models predict that the P-wave pulse will be shorter in duration than the S-wave pulse (i.e., $f_c^P > f_c^S$), which will result in lower values for q. For example, the Madariaga (1976) model has $f_c^P \simeq 1.5 f_c^S$, from which one can compute that q is about 7 (Boatwright and Fletcher, 1984). Observations have generally suggested average q values between 9 and 25, with a large amount of scatter for individual earthquakes.

Measuring E_R is much more difficult than measuring M_0 and results among different groups for the same earthquakes often differ by factors of 2 or more. This is because E_R is derived from high-frequency parts of the source spectrum where corrections for attenuation are critically important. Most of the energy is radiated as S waves, which are particularly sensitive to attenuation. If only E_R^P measurements are available, E_R can still be estimated if a fixed value of q is assumed, but once again this detracts from the directness of the observation. Because energy is proportional to the square of the wave amplitudes, the effects of the radiation pattern are more severe for E_R calculations compared to M_0 calculations. The $U_{\phi\theta}$ terms in the denominators of (9.50) go to zero at the nodes in the radiation pattern. This can lead to artificially high energy estimates if measurable wave amplitudes are seen near the nodes, which can happen due to scattering, 3-D structure, or inaccuracies in the focal mechanism. Finally, rupture directivity does not affect M_0 estimates (because Ω_0 is preserved despite changes in the pulse amplitudes) but produces large variations in I_P and I_S (e.g., Ma and Archuleta, 2006). If directivity effects are important, then (9.50) is incomplete and can produce biased results, depending upon whether or not the critical takeoff angles with the highest amplitudes are included in the available data.

The ratio of the radiated energy to the moment

$$\tilde{e} = \frac{E_R}{M_0} = \frac{1}{\mu} \frac{E_R}{\overline{D}A} \tag{9.53}$$

is called the *scaled energy* and is dimensionless (note that 1 joule $= 1\,\mathrm{N\,m}$). The parameter $\mu\tilde{e} = E_R/\overline{D}A$ has units of stress and has traditionally been called *apparent stress* but this term can be confusing because it is not directly related to either absolute stress or stress drop. The scaled energy, $\tilde{e}$, is proportional to the energy radiated per unit fault area and per unit slip. As noted in the previous section, if earthquakes are self-similar then $\tilde{e}$ should be constant as a function of moment. Whether this is indeed the case has been the subject of some controversy (e.g., see recent review by Walter *et al.*, 2006). Some have argued that average $\tilde{e}$ grows with moment approximately as $M_0^{1/4}$ (e.g., Mayeda and Walter, 1996) while others have maintained that average $\tilde{e}$ is seen to be nearly constant with M_0 when one carefully corrects for possible biases in the data analysis (e.g., Ide and Beroza, 2001). Figure 9.21 plots $\tilde{e}$ versus M_0, showing results from a number of different studies. Note that there is a great deal of scatter in the $\tilde{e}$ estimates, which span over an order of magnitude even at the same moment. However, there is some evidence for an increase in $\tilde{e}$ with moment, particularly for the smaller earthquakes. Ide and Beroza (2001) have argued, however, that this may be an artifact of the data selection method in the Abercrombie (1995) study. An important issue is the fact that energy

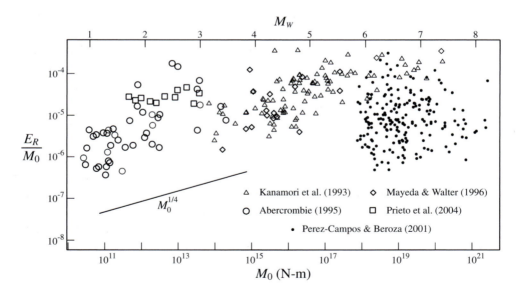

Figure 9.21 The observed radiated seismic energy to moment ratio, $\tilde{e} = E_R/M_0$, plotted as a function of moment. The $M_0^{1/4}$ trend noted in some studies is plotted for reference.

estimates derived from teleseismic data tend to be about 10 times smaller than those obtained from local records (Singh and Ordaz, 1994; Mayeda and Walter, 1996). This can be seen in Figure 9.21, noting that Perez-Campos and Beroza (2001) is the only teleseismic study plotted. If these points are excluded, the $M_0^{1/4}$ trend becomes much clearer.

9.6.1 Earthquake energy partitioning

The total strain and gravitational energy released during an earthquake is given by

$$E = \tfrac{1}{2}\overline{(\sigma_1 + \sigma_2)D}A, \tag{9.54}$$

where σ_1 is the initial stress, σ_2 is the final stress, D is displacement, A is the fault area, and the overbar means the spatial average. Note that $\tfrac{1}{2}(\sigma_1 + \sigma_2) = \overline{\sigma}$ is the average shear stress on the fault so this is analogous to "work = force × distance" from basic physics. As discussed in Kanamori and Brodsky (2004) and Kanamori and Rivera (2006), this is usually approximated as

$$E = \overline{\sigma}\overline{D}A = \tfrac{1}{2}\Delta\sigma\overline{D}A + \sigma_2\overline{D}A, \tag{9.55}$$

where the average stress drop $\Delta\sigma = \sigma_1 - \sigma_2$. The total energy can be partitioned into three parts:

$$E = E_R + E_F + E_G \tag{9.56}$$

where E_R is the radiated seismic energy, E_F is the frictional energy (often released as heat), and E_G is the energy used to fracture the rock, although the separation between E_F and E_G is not always clear cut. In principle, E_R and E_G can be estimated from seismic data. However, E_F cannot be measured from direct seismic wave observations and depends upon the absolute level of stress on the fault, which is difficult to determine (see Section 9.9).

This energy balance is shown graphically in Figure 9.22 for two idealized earthquakes on faults of unit area and total displacement D. In the first example, the *Orowan fault model* (e.g., Orowan, 1960; Kostrov, 1974), the stress on the fault, σ_f, drops abruptly to σ_2 as soon as the fault starts moving. In this case, there is no fracture energy, E_G, and σ_2 represents the dynamic frictional stress on the fault. The total energy released is the shaded trapezoid, which is the sum of E_R and E_F. Generalizing to a fault of area A, we have

$$E_R = \tfrac{1}{2}(\sigma_1 - \sigma_2)DA = \tfrac{1}{2}\Delta\sigma DA, \tag{9.57}$$

$$E_F = \sigma_2 DA. \tag{9.58}$$

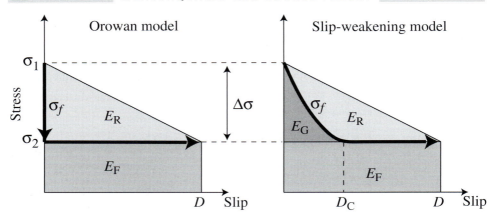

Figure 9.22 The shear stress, σ_f, on a point on a fault as a function of slip for the Orowan fault model and a simple example of a slip-weakening fault model. σ_1 and σ_2 are the initial and final stresses, D is the total slip, D_C is the critical slip, E_R is the radiated seismic energy, E_F is the frictional energy dissipated, and E_G is the fracture energy released.

In this case, the stress drop can be expressed as

$$\Delta\sigma(\text{Orowan}) = \frac{2E_R}{DA} = \frac{2\mu E_R}{M_0} = 2\mu\tilde{e} \qquad (9.59)$$

and we see that this model predicts a very simple relationship between stress drop and scaled energy, $\tilde{e}$. This is sometimes termed the *Orowan stress drop* to make clear that it only represents the true stress drop if the earthquake obeys this simple model. Assuming $\Delta\sigma = 3$ MPa and $\mu = 30$ GPa (typical values for crustal earthquakes), the Orowan model predicts $\tilde{e} = 5 \times 10^{-5}$, which is in rough agreement with direct observations of $\tilde{e}$ for large earthquakes (see Fig. 9.21).

In general, however, we expect the rupture process to be more involved than the Orowan model and the σ_f function may follow a complicated trajectory. In some models, σ_f rises above σ_1 at the onset of rupture to what is termed the yield stress before dropping as slip begins. It is also possible for σ_f to fall below σ_2 during part of the rupture and for σ_f to end at a value above or below the final stress state once the earthquake is completely over (the latter phenomena are called overshoot and undershoot, respectively, and are predicted by some theoretical models).

The right part of Figure 9.22 shows an example of a *slip-weakening model* in which the stress drops from σ_1 to σ_2 over a distance D_C (sometimes called the *critical slip*) and then continues at a constant stress $\sigma_f = \sigma_2$. The radiated seismic energy, E_R, is reduced by the area to the left of the curve, which represents the fracture energy E_G. In this case we have

$$E_G = E - E_F - E_R = \tfrac{1}{2}\Delta\sigma DA - E_R = \frac{\Delta\sigma}{2\mu}M_0 - E_R \qquad (9.60)$$

and

$$\Delta\sigma = \frac{2\mu(E_R + E_G)}{M_0} \geq \Delta\sigma(\text{Orowan}) \qquad (9.61)$$

and we see that in principle we can estimate the fracture energy E_G if we are able to separately measure M_0, $\Delta\sigma$ and E_R, and that the Orowan stress drop represents the minimum possible stress drop, given values of E_R and M_0, at least for simple models in which $\sigma_f \geq \sigma_2$. It should be noted that σ_f for real earthquakes may follow more complicated trajectories than those plotted in Figure 9.22, in which case E_F is not determined by the final stress and the partitioning in (9.60) and (9.61) between E_F and E_G does not necessarily have physical significance in the faulting process.

The *radiation efficiency* is defined as the ratio

$$\eta_R = \frac{E_R}{E_R + E_G} \qquad (9.62)$$

and is an important measure of the dynamic properties of earthquakes. Note that $\eta_R = 1$ for the Orowan fault model. For our simple slip-weakening model, it can be expressed as

$$\eta_R = \frac{E_R}{\tfrac{1}{2}\Delta\sigma DA} = \frac{2\mu}{\Delta\sigma}\frac{E_R}{M_0} = 2\mu\frac{\tilde{e}}{\Delta\sigma}, \qquad (9.63)$$

and thus is proportional to the ratio between the scaled energy and the stress drop. As discussed in Kanamori and Brodsky (2004), the radiation efficiency can be related to the rupture velocity, v_r, in theoretical crack models:

$$\eta_R = 1 - g(v_r) \qquad (9.64)$$

where $g(v_r)$ is a function that depends upon the specific crack model and the ratio of v_r to the Rayleigh or shear wave velocity. For example, for Mode III (transverse shear) cracks,

$$g(v) = \sqrt{\frac{1 - v_r/\beta}{1 + v_r/\beta}}, \qquad (9.65)$$

in which case η_R approaches 1 and the fracture energy, E_G, goes to zero as the rupture velocity approaches the shear wave velocity. For about 30 earthquakes of $6.6 < M_W < 8.3$, Venkataraman and Kanamori (2004) obtained radiation efficiency estimates generally between 0.25 and 1.0. One class of earthquakes that appear to have $\eta_R < 0.25$ are *tsunami earthquakes*, which involve slow rupture and generate large tsunamis relative to their moment.

The radiation efficiency should not be confused with the *seismic efficiency*, η, defined as the fraction of the total energy that is radiated into seismic waves:

$$\eta = \frac{E_R}{E} = \frac{E_R}{\overline{\sigma} \overline{D} A} = \frac{\mu E_R}{\overline{\sigma} M_0} = \frac{\mu \tilde{e}}{\overline{\sigma}}. \tag{9.66}$$

The seismic efficiency is more difficult to estimate than the radiation efficiency because it depends upon the poorly constrained absolute stress level on the fault.

In the extreme case where we assume that the earthquake relieves all of the stress on the fault, then $\sigma_2 = 0$ and we say that the stress drop is total. In this case, $E_F = 0$ and we have

$$E_{\min} = \tfrac{1}{2} \overline{\Delta \sigma D} A = \frac{\Delta \sigma}{2\mu} M_0. \tag{9.67}$$

This represents the minimum amount of energy release for an earthquake with a given stress drop and moment.

The theories that describe how slip on a fault initiates, propagates, and comes to a halt can be very complicated, even for idealized models with uniform pre-stress and elastic properties. Much of the recent work in this area has involved theory and observations of *rate and state friction* (e.g., Dieterich, 1994) in which the frictional properties are time and slip dependent. Because these models vary in their behavior and it is likely that real earthquakes span a range of different rupture properties, it is important to keep in mind the distinction between parameters that are more-or-less directly estimated (e.g., moment, geodetically-determined static stress drop, and radiated energy) and those that depend upon modeling assumptions (e.g., Brune-type and Orowan stress drops) and thus are not truly independent measurements. For example, it would make little sense to use (9.60) to estimate E_G if both $\Delta \sigma$ and E_R are derived from fitting the observed body-wave spectra to the same theoretical model.

9.7 Earthquake magnitude

For historical reasons the most well-known measure of earthquake size is the earthquake *magnitude*. There are now several different types of magnitude scales, but all are related to the largest amplitude that is recorded on a seismogram. This is one of the easiest things to measure and is one reason for the continued popularity of magnitude scales. A recent comprehensive review of magnitude scales is contained in Utsu (2002b).

In the 1930s, Charles Richter introduced what is now called the *local magnitude* M_L. This was determined by measuring the largest amplitude A recorded on a standard instrument, the Wood–Anderson seismograph. Richter noticed that plots of log A versus epicentral distance for different earthquakes generally exhibited a similar decay rate (Fig. 9.23). This suggested that a distance-independent measure of earthquake size could be provided by the offset in log A from a reference event at the same range,

$$M_L = \log_{10} A(X) - \log_{10} A_0(X), \qquad (9.68)$$

where A_0 is the amplitude of the reference event and X is the epicentral distance. At each seismic station, a value of M_L may be obtained from the measured amplitude A and the value of $\log_{10} A_0$ at the appropriate source–receiver distance (Richter made a table of $\log_{10} A_0$ at different ranges). From the table of values of $A_0(X)$, an approximate empirical formula has been derived (e.g., Bullen and Bolt, 1985):

$$M_L = \log_{10} A + 2.56 \log_{10} X - 1.67, \qquad (9.69)$$

where A is the displacement amplitude in microns (10^{-6} m) and X is in kilometers. The formula is valid for $10 < X < 600$ km. For the Wood–Anderson torsion instrument the largest amplitude generally comes from the S-wave arrival. Individual

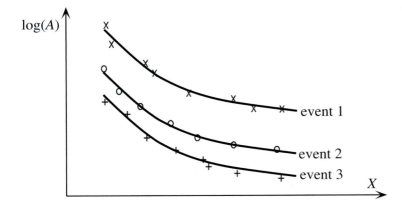

Figure 9.23 Different earthquakes are observed to have a similar falloff in log(amplitude) with distance.

estimates of M_L will exhibit some scatter owing to directivity, radiation pattern, focusing, and other effects. However, a stable estimate can generally be obtained by averaging the results from different stations.

Richter defined a fairly small reference event so that the magnitudes of all but the tiniest earthquakes are positive. Events below about $M_L = 3$ are generally not felt. Significant damage to structures in California begins to occur at about $M_L = 5.5$. The magnitude scale is logarithmic to account for the wide range in observed amplitudes. A $M_L = 6.0$ event implies a recorded amplitude 100 times greater than a $M_L = 4.0$ event.

The Richter magnitude scale provided a practical method of quickly determining the relative size of different events in California. Since the dominant period of the Wood–Anderson intrument (0.8 s) is close to that of many structures, the M_L scale has proven especially useful in engineering seismology. The local magnitude scale is also important because all subsequent magnitude scales have been tied to it. However, the portability of M_L is limited since it is based upon an amplitude versus range relationship that was defined specifically for southern California, and it depends on an intrument that is now rarely used. Caltech and Berkeley kept some Wood–Anderson seismographs operating into the 1990s just to maintain continuity of the magnitude scale. However, these venerable instruments have now been retired since the Wood–Anderson response can be simulated through suitable filtering of modern broadband data. Related to M_L for local earthquakes is the *coda magnitude* (e.g., Suteau and Whitcomb, 1979), which is derived from the amplitude of the scattered waves or coda that follow the direct P and S arrivals, and which has been calibrated to agree with the local magnitude scale. In many cases, coda magnitudes are more stable than M_L because the scattered energy that makes up the coda waves averages out spatial variations and provides a more uniform coverage of the radiation pattern.

A general magnitude scale used for global seismology is the *body-wave magnitude*, which is defined as

$$m_b = \log_{10}(A/T) + Q(h, \Delta), \tag{9.70}$$

where A is the ground displacement in microns, T is the dominant period of the measured waves, Δ is the epicentral distance in degrees, and Q is an empirical function of range and event depth h (e.g., Veith and Clawson, 1972). The Q function includes the details of the average amplitude versus epicentral distance and source depth behavior of the Earth. The measurement is normally made on the first few cycles of the P-wave arrival on short-period vertical-component intruments, for which the dominant wave period is usually about 1 s. As with the local magnitude scale, m_b estimates for the same event will vary between stations, with scatter of up to about ± 0.3. This is due to radiation pattern, directivity, and local station effects. A station correction term is often used to account for stations that consistently give higher or lower m_b values.

Another global seismology scale is the *surface wave magnitude*, which may be defined as

$$M_S = \log_{10}(A/T) + 1.66 \log_{10} \Delta + 3.3 \tag{9.71}$$

for Rayleigh wave measurements on vertical instruments. Since the strongest Rayleigh wave arrivals are generally at a period of 20 s, this expression is often written as

$$M_S = \log_{10} A_{20} + 1.66 \log_{10} \Delta + 2.0. \tag{9.72}$$

Note that this equation is applicable only to shallow events; surface wave amplitudes are greatly reduced for deep events.

The m_b and M_S scales were designed to agree with the M_L scale for local events in California. However, it is not possible to align the scales for all size events. This is because the magnitude scales are obtained at different periods and the frequency content of events changes as a function of event size. Consider the examples in the previous sections, in which the source spectrum falls as f^{-2} above a certain corner frequency. The corner frequency f_c generally moves to lower frequencies for larger events. If we assume stress drop is constant, then the fault dimension and corresponding rupture duration will scale approximately as $M_0^{1/3}$. The corner frequency is inversely proportional to the rupture duration and will scale as $M_0^{-1/3}$. In this case the position of the corner will fall off as f^{-3} (Fig. 9.24).

At frequencies below f_c there is a linear relationship between magnitude ($\log_{10}$ of the measured amplitude) and moment. However, at higher frequencies this linearity breaks down and the magnitude scale does not fully keep up with the increasing size of the events. This phenomenon is called magnitude *saturation*. At a given measurement frequency (e.g., 1 Hz for m_b) this begins to occur when the event size becomes large enough to move the corner frequency below the measurement frequency. Of course, not all earthquake spectra fall off exactly at f^{-2} but any degree of falloff will lead to some saturation of the magnitude scale. Another contributing factor to magnitude saturation can be the fixed window length used to measure the amplitudes, which may not be long enough to capture the true amplitude of larger events (e.g., Houston and Kanamori, 1986). Observed m_b values begin to saturate at about $m_b = 5.5$ and M_S values (measured at longer period) at about $M_S = 8$. For this reason it is rare for m_b to exceed 7 or for M_S to exceed 8.5, even for extremely large events.

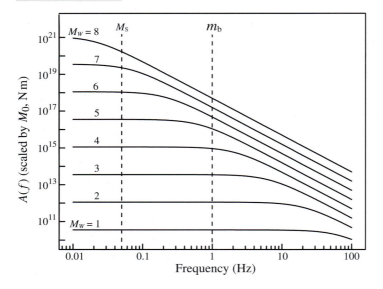

Figure 9.24 For larger events, the corner in the source spectrum moves to lower frequencies, reducing the observed amplitude increase at the fixed frequencies used to estimate M_S and m_b.

The saturation of the m_b and M_L scales for large events helped motivate development of the *moment magnitude M_W* by Kanamori (1977) and Hanks and Kanamori (1979). The moment magnitude is defined as

$$M_W = \tfrac{2}{3}\left[\log_{10} M_0 - 9.1\right], \tag{9.73}$$

where M_0 is the moment measured in N m (for M_0 in dyne-cm, replace the 9.1 with 16.1)[2]. The moment magnitude is derived entirely from the moment, with a scaling such that M_W is in approximate agreement with M_S for many events. The advantage of the M_W scale is that it is clearly related to a physical property of the source and it does not saturate for even the largest earthquakes. It is simply another way to express the moment, which provides units that are easier to quickly comprehend than M_0 numbers like 8.2×10^{19} N m.

However, (9.73) can be misused if it is naively applied to estimate M_0 from an earthquake magnitude. This is because the M_L, m_b, and M_S magnitudes exhibit considerable scatter among events of the same moment and even their average values do not agree with M_W over the full range of event sizes. To see this, consider M_S measurements and the self-similar ω^{-2} source spectra plotted in Figure 9.24.

---

[2] A minor source of confusion has existed in definitions of M_W, arising from a lack of precision in the final term. The original Hanks and Kanamori paper defined moment magnitude as $2/3 \log M_0 - 10.7$ (dyne-cm). However, many authors, including Aki and Richards (2002), use $\log M_0 = 1.5 M_W + 16.1$, which is slightly different (note that $1.5 \times 10.7 = 16.05$). Additional slight precision loss can occur in translating from dyne-cm to N m because of the 2/3 factor. Here we use the Aki and Richards definition of M_W.

For small earthquakes ($M_W < 6$), the measured amplitude at 20 s will scale linearly with moment and thus $M_S \propto \log_{10} M_0$. For larger earthquakes ($M_W > 8$), the result of the f^{-3} corner frequency falloff and f^{-2} high-frequency spectral falloff is that $M_S \propto \frac{1}{3} \log_{10} M_0$. M_W was defined to agree with M_S mainly for events between M 6 and 8, where a slope of 2/3 is approximately correct. Thus, we should expect M_S to underpredict M_W at both small and large magnitudes. A similar phenomenon should occur for m_b measurements, but shifted to smaller earthquakes because of the higher frequency of the m_b observations.

This is illustrated in Figure 9.25, which plots as gray corridors the distribution of M_S and m_b measurements from the US Geological Survey's Preliminary Determination of Epicenter (PDE) catalog as a function of M_0 values for the same earthquakes from the Global CMT catalog between 1976 and 2005. For reference, the straight line shows M_W values from (9.73), which has a slope of 2/3 on the log plot. M_S agrees approximately with M_W at magnitudes between about 6.5 and 7.5, but underpredicts M_W outside of this interval. The m_b values agree with M_W near magnitude 5, but increasingly underpredict M_W at larger magnitudes. The average m_b and M_S values agree exactly only near magnitude 5.5, where they slightly

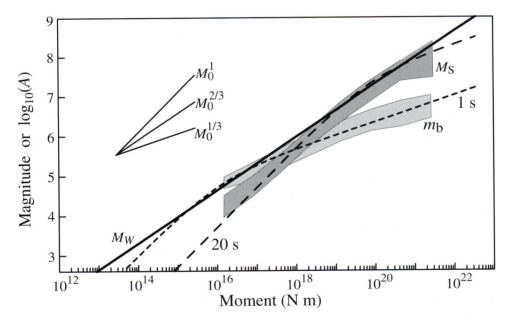

Figure 9.25 Magnitude as a function of moment, M_0, for m_b, M_S, and M_W, compared to predictions of log(amplitude) for an ω^{-2} source model at periods of 1 and 20 s. The gray corridors show m_b and M_S values ($\pm$ one standard deviation) from the USGS PDE catalog versus M_0 from the Global CMT catalog, compared to the definition of M_W (straight solid line). The dashed lines are predicted P-wave amplitudes for the Madariaga (1976) source model, assuming a stress drop of 3 MPa.

underpredict the M_W value of 5.8. This behavior can be explained nicely with an ω^{-2} source model, assuming constant stress drop. This model predicts that $\log_{10}(A)$ will vary as M_0 for small events, as $M_0^{2/3}$ for events with corner frequencies near the observation frequency, and as $M_0^{1/3}$ for larger events. Predicted $\log_{10}$ amplitudes at the m_b measurement period of 1 s and the M_S measurement period of 20 s exhibit similar behavior to the data.

Figure 9.25 makes clear how the 2/3 factor in the definition of M_W serves to make M_W values comparable on average to standard magnitudes over a fairly wide range of event sizes, provided m_b is used for earthquakes less than about 5.5 and M_S is used for larger events. However, this correspondence increasingly fails outside of $4 < M_W < 8$. Moment is not routinely estimated for earthquakes below about $M_W = 4$, which are only recorded locally. However, because M_L is computed at a similar frequency to m_b, we should not expect the M_W and M_L scales to agree very well for small earthquakes. Results for the slope of M_L versus $\log_{10}(M_0)$ have varied among different studies, but a systematic analysis of over 60 000 southern California earthquakes of $1 \leq M_L \leq 3$ by Shearer *et al.* (2006) gave a best-fitting slope of 0.96, close to the unit slope predicted from Figure 9.25.

The various seismic magnitude scales are important because they are ingrained in the history and practice of seismology. But if earthquake size is to be quantified with a single number, it is far better to use the moment because it is directly related to a fundamental physical property of the source, which can also be used in geodetic studies of earthquakes and comparisons to long-term geological slip rates. Although it is possible to understand the average behavior of the various magnitude scales with respect to moment, it should be remembered that M_L, m_b, and M_S measurements exhibit large scatter among individual events of the same moment. These differences are presumably not random and may reflect variations in stress drop or other source properties, and appear in some cases to have systematic regional variations (e.g., Ekstrom and Dziewonski, 1988). This is why it is better whenever possible to compute moment directly from the lowest frequency part of the seismic spectra rather than to use magnitude measurements as a proxy for moment.

From (9.73), a one unit increase in M_W corresponds to a $10^{3/2} \simeq 32$ times increase in moment. As we saw in Section 9.6, the average radiated seismic energy is approximately proportional to moment so this means that seismic energy also goes up by a factor of 32. Thus, on average a $M_S = 7$ earthquake releases about 32 times more energy than a $M_S = 6$ event and 1000 times more than a $M_S = 5$ event. This is consistent with the classic empirical Gutenberg-Richter relation between E_R and M_S

$$\log_{10} E_R \text{ (joules)} \simeq 4.8 + 1.5 M_S. \qquad (9.74)$$

Table 9.1: Some big earthquakes (M_0 in 10^{20} N m)					
Date	Region	m_b	M_S	M_W	M_0
1960 May 22	Chile		8.3	9.5	2000
1964 March 28	Alaska		8.4	9.2	820
2004 Dec 26	Sumatra-Andaman	6.2	8.5	9.1	680
1957 March 9	Aleutian Islands		8.2	9.1	585
1965 Feb 4	Aleutian Islands			8.7	140
2005 March 28	Sumatra	7.2	8.4	8.6	105
1977 Aug 19	Indonesia	7.0	7.9	8.3	36
2003 Sept 25	Hokkaido, Japan	6.9	8.1	8.3	31
1994 Oct 4	Shitokan, Kuriles	7.4	8.1	8.2	30
1994 June 9	Bolivia (deep)	6.9		8.2	26
2004 Dec 23	Macquarie Ridge	6.5	7.7	8.1	16
1989 May 23	Macquarie Ridge	6.4	8.2	8.2	20
1985 Sept 19	Michoacan, Mexico	6.5	8.3	8.0	14
1906 April 18	San Francisco		8.2	7.9	10
2008 May 12	Eastern Sichuan	6.9	8.0	7.9	9
2002 Nov 3	Denali, Alaska	7.0	8.5	7.8	7
2001 Nov 14	Kokoxili, Kunlun	6.1	8.0	7.8	6
1992 June 28	Landers, California	6.2	7.6	7.5	2

However, this agreement is not really a coincidence because this equation was one of the contributing reasons for the 2/3 factor in the definition of M_W (Kanamori, 1977; Hanks and Kanamori, 1979). If we substitute M_W from (9.73) into this equation, we can obtain $E_R/M_0 \simeq 6 \times 10^{-5}$, in rough agreement with the average $\tilde{e}$ values plotted in Figure 9.21, at least over the $6 < M_W < 8$ interval in which $M_W \simeq M_S$. However, it should be emphasized that the radiated energy E_R is best obtained through direct observations; the Gutenberg–Richter relation (9.74) provides a crude estimate, but it can be in error by more than an order of magnitude for individual events.

Table 9.1 lists some of the biggest earthquakes that have been recorded seismically, as well as some smaller strike–slip events for comparison purposes, and Figure 9.26 plots global CMT results between 1976 and 2005, scaled by moment. The listed m_b and M_S values show some of the effects of magnitude saturation. The largest earthquakes of M_W 8.5 and greater occur in subduction zones where the fault area (= length × width) can be very large. Examples include the 1960 Chile earthquake, the 1964 Alaska earthquake, and the 2004 Sumatra-Andaman earthquake.

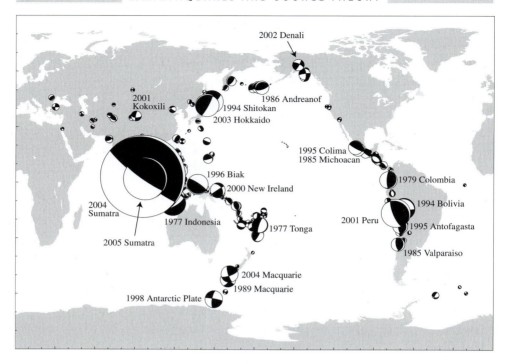

Figure 9.26 The largest earthquakes from 1976 to 2005, with the focal mechanism area proportional to seismic moment, as estimated from the Global CMT catalog.

In contrast, crustal strike–slip earthquakes generally do not exceed about $M_W = 8$ because their fault widths are limited to the upper crust. Recent examples of large strike–slip earthquakes include the 1989 and 2004 Macquarie Ridge earthquakes, the 1998 Antarctic Plate earthquake, the 2001 Kokoxili (China) earthquake, and the 2002 Denali (Alaska) earthquake. Most very large earthquakes are shallow; a notable exception was the 1994 Bolivian earthquake at 630 km depth.

9.7.1 The *b* value

Smaller earthquakes occur much more frequently than large earthquakes. This trend may be quantified in terms of a magnitude–frequency relationship. Gutenberg and Richter noted that this relationship appears to obey a power law[3] and obtained the empirical formula

$$\log_{10} N = a - bM, \tag{9.75}$$

---

[3] A power-law distribution for earthquake energy had earlier been suggested by K. Wadati.

where N is the number of events with magnitudes greater than or equal to M. In this equation, a describes the total number of earthquakes, while the parameter b is called the *b-value* and measures the relative number of large quakes compared to small quakes. The b-value is generally found to lie between 0.8 and 1.2 for a wide variety of regions and different magnitude scales (for a review, see Utsu, 2002a). As we will discuss in Chapter 10, the b value is often used to estimate the fractal dimension of fault systems (e.g., Aki, 1981; Turcotte, 1997). At $b = 1$ the number of earthquakes increases by a factor of 10 for every unit drop in magnitude. For example, if there is 1 $M = 6$ events per year in a region then we should expect about 10 $M = 5$ events per year, 100 $M = 4$ events, etc.

Figure 9.27 shows $N(M_W)$ computed for the global CMT catalog from 1976 to 2005. Between M_W values 5.5 and 7.5, the distribution is well fit with $b = 1$. At smaller magnitudes the increase in N drops off because earthquakes below about $M_W = 5$ are too small to be well recorded by the global seismic networks. Plots like this are often used to evaluate *catalog completeness* – the lowest magnitude to which a network or catalog includes all of the earthquakes. For $M_W > 7.5$, the numbers also drop below the $b = 1$ line, which may represent a change in the power law or a temporary deficit in the number of very large global earthquakes. Because moment increases by ~ 30 for every unit increase in M_W, while the number of events only decreases by a factor ~ 10, the total moment release from all of the seismicity in a region is dominated by the largest events, rather than the accumulated sum of many smaller events. Fortunately for humanity, (9.75) cannot remain valid for

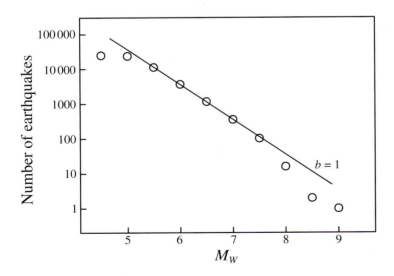

Figure 9.27 $N(M_W)$ for the global CMT catalog from 1976 to 2005, compared to the predictions of a power law decay with b value of 1.

	Table 9.2: The modified Mercalli scale, adapted from the abridged version in Bolt (1993).
I	Not felt except by a few under especially favorable circumstances.
II	Felt only by a few persons at rest, especially on upper floors of buildings. Delicately suspended objects may swing.
III	Felt quite noticeably indoors, especially on upper floors of buildings, but many people do not recognize it as an earthquake. Parked cars may rock slightly. Vibration like passing of truck. Duration can be estimated.
IV	During the day felt indoors by many, outdoors by few. At night some awakened. Dishes, windows, doors disturbed, walls make creaking noise. Parked cars rocked noticeably. (0.015–0.02 g)
V	Felt by nearly everyone, many awakened. Some dishes, windows, etc., broken; a few instances of cracked plaster; unstable objects overturned. Disturbance of trees, poles and other tall objects sometimes noticed. (0.03–0.04 g)
VI	Felt by all; many frightened and run outdoors. Some heavy furniture moved, a few instances of fallen plaster or damaged chimneys. Damage slight. (0.06–0.07 g)
VII	Everybody runs outdoors. Damage negligible in buildings of good design and construction; slight-to-moderate damage in well-built ordinary structures; considerable in poorly built or badly designed structures; some chimneys broken. Noticed by people driving cars. (0.10–0.15 g)
VIII	Damage slight in specially designed structures; considerable in ordinary substantial buildings, with partial collapse; great in poorly built structures. Panel walls thrown out of frame structures. Fall of chimneys, factory stacks, columns, monuments, walls. Heavy furniture overturned. Sand and mud ejected in small amounts. Changes in well water. Disturbs people driving cars. (0.25–0.30 g)
IX	Damage considerable in specially designed structures; well-designed frame structures thrown out of plumb; great in substantial buildings, with partial collapse. Buildings shifted off foundations. Ground cracked conspicuously. Underground pipes broken. (0.5–0.55 g)
X	Some well-built wooden structures destroyed; with foundations; ground badly cracked. Rails bent. Landslides considerable from river banks and steep slopes. Shifted sand and mud. Water splashed over banks. (> 0.6 g)
XI	Few, if any, masonry structures remain standing. Bridges destroyed. Broad fissures in ground. Underground pipelines completely out of service. Earth slumps and land slips in soft ground. Rails bent greatly.
XII	Damage total. Waves seen on ground surfaces. Lines of sight and level distorted. Objects thrown upward into the air.

arbitrarily large earthquakes because the finite extent of Earth's faults means there is a maximum possible earthquake size. We need not fear a $M_W = 11$ earthquake every 1000 years.

9.7.2 The intensity scale

Another measure of the earthquake strength is the seismic *intensity*, which describes the local strength of ground shaking as determined by damage to structures and the perceptions of people who experienced the earthquake. The intensity scale most often used today in the United States is the modified Mercalli scale, in which intensity ranges from I to XII (Roman numerals). As shown in Table 9.2, a value of I indicates shaking that is felt only by a few people, V is felt by almost everyone, VIII causes great damage in poorly built structures, and XII indicates total destruction. Although approximate peak accelerations can be assigned to these levels, the great advantage of the Mercalli scale is that it can be used to examine historic earthquakes that were not recorded by modern instruments. This is often done by interviewing witnesses and studying old newspaper accounts. Once the intensity has been estimated at a number of different sites, a contour map of the intensities can be constructed. The earthquake location is then identified from this map as the spot of maximum intensity, and an approximate magnitude can be estimated from the area surrounded by the different intensity contours. This technique provides the only practical way to obtain probable locations and magnitudes for many older events. The importance of these estimates is illustrated by the seismicity in the eastern United States, where no large earthquake was recorded in the twentieth century. However, several large events occurred in the nineteenth century, including a series of three large earthquakes that struck in 1811–1812 near New Madrid along the Mississippi River in Missouri. By studying and mapping accounts of these events, which were felt across much of the eastern United States, it is possible to constrain their sizes and locations and make estimates of the impact of a future earthquake in the same area. For many years the New Madrid events were thought to be about magnitude 8, but a recent reanalysis of intensity data by Hough *et al.* (2000) indicates that they were more likely M_W 7 to 7.5.

A modern way to quickly estimate earthquake intensities is the "Did You Feel It?" website maintained by the United States Geological Survey (USGS), in which users are asked to answer a short series of questions about how strongly they felt the shaking and to identify their location (http://earthquake.usgs.gov/eqcenter/dyfi.php). These responses are then compiled into an intensity map that can provide surprising detail, as shown in Figure 9.28 for the 2003 M 6.5 San Simeon earthquake in central California. Spatial resolution in the United States is currently limited to zip code boundaries, but in the future it is likely that more precise user locations will make possible more detailed maps.

9.8 Finite slip modeling

Up to this point, we have described seismic sources with a limited number of parameters, such as the moment tensor or focal mechanism, the scalar moment, the duration or corner frequency, the rupture velocity and direction, the stress drop, and the radiated energy. These represent averages over the source region and for small earthquakes this is most of what we can hope to learn because of the limited bandwidth of far-field seismic records. However, for large earthquakes it is often possible to invert for more detailed source properties because it it possible to separately resolve the seismic radiation from different parts of the fault. A common approach is to discretize the fault into a series of rectangular cells, solve for the Green's function that gives the response at each of the available seismic stations,

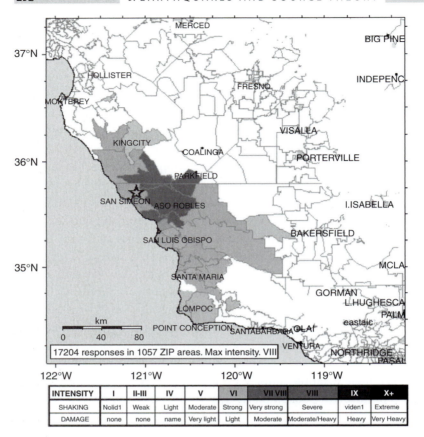

INTENSITY	I	II-III	IV	V	VI	VII VIII	VIII	IX	X+
SHAKING	Nolid1	Weak	Light	Moderate	Strong	Very strong	Severe	viden1	Extreme
DAMAGE	none	none	name	Very light	Light	Moderate	Moderate/Heavy	Heavy	Very Heavy

Figure 9.28 Ground shaking intensities for the 2003 San Simeon earthquake in California, as measured from 17 204 responses to the USGS "Did You Feel It?" website.

and then set up an inverse problem using the superposition principle to solve for the time-dependent slip distribution that best fits all of the data. If geodetic data are also available, these can also be included in the inversion to provide constraints on the total slip. As in the tomographic inversions for 3-D velocity structure discussed in Chapter 5, the inverse problem often is stabilized by applying regularization constraints, such as only permitting slip in one direction and solving for the smoothest or the minimum slip solution.

The resulting *finite slip models* typically show that large earthquakes have irregular slip distributions and that their total moment is dominated by one or two large slip regions where the stress drop is much higher than surrounding parts of the fault. Rupture velocity is also not always constant during earthquakes. As an example, Figure 9.29 shows the Custodio *et al.* (2005) model of the 2004 Parkfield earthquake obtained from an inversion of the strong motion data. From the hypocenter

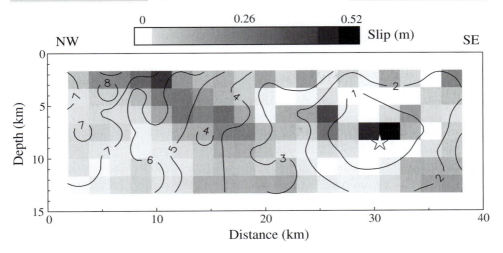

Figure 9.29 The Custodio *et al.* (2005) slip model for the 2004 Parkfield earthquake. The star shows the hypocenter location. The contours show the slip onset times at 1 s intervals.

the rupture propagated northwest along the San Andreas Fault for about 10 s. There were two major areas of slip, one near the hypocenter and the other 10 to 20 km to the northwest at a depth between 2 and 8 km.

Finite slip models provide detailed constraints on rupture dynamics that are valuable for understanding earthquake physics. Results so far point to highly heterogeneous faulting processes with large variations in moment release and stress drop. But there are problems with non-uniqueness in many of the inversions and results from different groups for the same earthquake do not always agree very well. As inversions move to higher frequencies, propagation path effects on strong motion data become increasingly important and link the rupture inversion problem to the 3-D velocity inversion problem. This is an active area of research and eventually improved modeling methods and denser seismic and geodetic networks will help to provide clearer images of fault rupture.

9.9 The heat flow paradox

As discussed in Section 9.6.1, the frictional and fracture energy release during faulting is given by

$$E_F = \sigma_f \overline{D} A \qquad\qquad (9.76)$$

where σ_f is the resisting shear stress on the fault, $\overline{D}$ is the average displacement, and A is the fault area. We can estimate σ_f from Byerlee's law

$$\sigma_f = \begin{cases} 0.85\sigma_n, & \sigma_n < 200\,\text{MPa}, \\ 50 + 0.6\sigma_n, & \sigma_n > 200\,\text{MPa}, \end{cases} \tag{9.77}$$

where σ_n is the normal stress on the fault. Byerlee's law has been found to be valid for a wide variety of rock types in laboratory experiments.

In general, we expect σ_n to grow rapidly with depth in the crust to support the weight of the overlying rock. The *lithostatic stress* for a crust of uniform density is given by

$$\sigma_{\text{litho}} = \rho g z \tag{9.78}$$

where ρ is the density, g is the acceleration due to gravity (about $10\,\text{m s}^{-2}$ near Earth's surface), and z is the depth. In the absence of topography or other lateral density gradients, the lithostatic stress is equal to the vertical normal force, σ_v, on a horizontal plane (the τ_{zz} component of the stress tensor). Assuming $\rho = 2.7\,\text{Mg/m}^3$, then $\sigma_v = 270\,\text{MPa}$ at $10\,\text{km}$ depth. However, the effective vertical stress, σ_v' in the crust can be reduced by the ambient pore pressure, P, such that

$$\sigma_v' = \sigma_v - P. \tag{9.79}$$

Assuming that the crust is fluid-saturated within an interconnected network of cracks (this is termed hydrostatic pore pressure conditions), then $P = \rho_f g z$, where the fluid density ρ_f is typically assumed to be $1.0\,\text{Mg/m}^3$ (i.e., the value for water). For crustal rocks of $\rho = 2.7\,\text{Mg/m}^3$, this results in effective vertical stresses for wet, saturated rocks that are 63% of their values under dry conditions. For a crust without shear stresses (i.e., behaving as a fluid), this would also give the normal force, or pressure, across planes of any orientation. However, for an elastic crust with tectonic forces producing deviatoric (non-hydrostatic) stresses, the situation is more complicated. In this case, one can compute the orientation of planes that, for a given coefficient of friction, will slip at the minimum value of deviatoric stress. Andersonian faulting theory (which assumes that one of the three principal stresses is vertical owing to the free-surface boundary condition) can be used to compute the shear stress on such optimally oriented faults. The result for vertical, strike–slip faults at $10\,\text{km}$ depth is that $\sigma_f = 80\,\text{MPa}$ for water-saturated (hydrostatic pore pressure) faults and $\sigma_f = 130\,\text{MPa}$ for dry faults (e.g., Brune and Thatcher, 2002). At the same depth the shear stress is about 40% lower for normal faults and about 250% larger for thrust faults.

Using these predicted values of shear stress in (9.76) together with appropriate values for the long-term slip rate, one can predict the average frictional heat generated on a fault as a function of time. This has been done for the San Andreas Fault in California and the predicted heat flow anomaly near the fault greatly exceeds the observed heat flow. Heat flow measurements above the thrust fault of the Cascadia subduction zone also are much less than standard laboratory faulting theory would predict. This discrepancy has been termed the *heat flow paradox* and has been a source of controversy for many years (for a recent review, see Brune and Thatcher, 2002).

Note that the large values of shear stress on the fault inferred from Byerlee's law would require that the seismic efficiency, η, of most observed earthquakes be very low. From (9.57) and (9.66), we have

$$\eta = \frac{E_R}{E} = \frac{E_R}{\overline{\sigma}DA} = \frac{\frac{1}{2}\Delta\sigma\overline{D}A - E_G}{\overline{\sigma}DA} = \frac{\Delta\sigma}{2\overline{\sigma}} - \frac{E_G}{\overline{\sigma}DA}, \qquad (9.80)$$

where E is the total energy release, E_R is the radiated seismic energy, E_G is the fracture energy, and $\overline{\sigma} = \frac{1}{2}(\sigma_1 + \sigma_2)$ is the average of the starting and ending stresses. Assuming the stress drop $\Delta\sigma = 3\,\text{MPa}$ (a typically observed value) and the absolute stress level is $100\,\text{MPa}$, then from (9.80) the inferred seismic efficiency is 1.5% or less.

The lack of observed heat flow anomalies near faults, together with the relatively low stress drops observed for earthquakes, has led most researchers to conclude that the shear stress on faults is lower than Byerlee's law suggests. The laboratory rock friction experiments are simply not applicable to large-scale faulting in the real Earth (although for an opposing view, see Scholz, 2000). This conclusion has been supported by measurements of stress levels near the San Andreas fault that suggest average shear stress levels of 10 to $20\,\text{MPa}$ (e.g., Zoback *et al.*, 1987).

These low values of shear stress imply that the fault zones are relatively weak, with effective coefficients of friction less than those obtained in laboratory experiments. The puzzle has now become to understand how slip on faults can occur so easily despite the low values of shear stress compared to the normal stresses pushing the two sides of the fault together. One explanation is that fluids, perhaps released during the faulting, are overpressured enough to "lubricate" the fault (e.g., Lachenbruch, 1980). Another suggested mechanism is elastohydrodynamic lubrication (Brodsky and Kanamori, 2001), in which material in a narrow and slightly rough fault zone acts as a viscous fluid and reduces friction for large events. In other models, friction-induced melting is predicted along the fault surface and rocks called *pseudotachylytes*, which are occasionally found in exhumed faults, are thought to be a result of this melting. However, the rarity of pseudotachylyte observations suggests that this is not a widespread phenomenon. Exotic properties have also been suggested for the fault gouge material, but these have not been seen in most laboratory experiments. However, laboratory experiments are generally performed at much lower slip rates than occur during earthquakes. Recent experiments on quartz rocks using higher slip rates have found a dramatic decrease in friction associated with the formation of a thin layer of silicon gel (Di Toro *et al.*, 2004).

The dynamics of earthquake rupture are also important and are not necessarily captured in traditional laboratory experiments. Heaton (1990) has found observational evidence that rupture during large earthquakes occurs in a narrow self-healing pulse of slip and that the effective friction on the fault surface is inversely related to the local slip velocity. A possible mechanism for reducing shear friction during rupture is "chatter" or movement normal to the fault plane. This process can dynamically reduce the normal stress on pieces of the fault (or even cause separation of the fault surfaces), an idea developed by Jim Brune based on his experiments that simulated earthquakes as sliding between two large blocks of foam rubber (e.g., Brune *et al.*, 1993). Fault chatter might be expected from collisions between local bumps and asperities on the fault surfaces, resulting in interface vibrations normal to the fault plane.

The problem of reconciling crustal faulting with the absence of heat flow anomalies and apparently low shear stresses has similarities to the mystery regarding the origin of subduction zone earthquakes at depths of 400 km and more. Seismic observations from these events are well fit by simple double-couple models, suggesting that the sources involve slip along a fault surface. However, at such great depths, the normal stresses due to the high pressures should prevent any simple form of frictional sliding at the expected levels of shear stress. Thus more exotic mechanisms have been invoked, such as a sudden changes in crystal structure (e.g., Kirby *et al.*, 1991; Green, 1994) or lubrication of the fault surface through melting of the rock (Kanamori *et al.*, 1998). However, these issues are far from settled and fundamental questions remain regarding exactly how and why earthquakes, at a wide range of depths, appear to occur at relatively low shear stresses.

9.10 Exercises

1. Assuming the following moment tensor,

$$\mathbf{M} = \begin{bmatrix} 6 & 0 & 0 \\ 0 & -6 & 0 \\ 0 & 0 & 3 \end{bmatrix},$$

(a) compute M_0, the scalar seismic moment, (b) give the decomposition of $\mathbf{M}$ into $\mathbf{M}^0$, $\mathbf{M}^{DC}$, and $\mathbf{M}^{CLVD}$ and compute the scalar moment of each part, and (c) compute ϵ, the measure of how well the double-couple model fits the deviatoric part of $\mathbf{M}$.

2. An interesting seismic event occurred near Tori Shima, Japan, on June 13, 1984, which generated an unusually large tsunami for its size. Kanamori *et al.* (1993) argued that it was probably caused by magma injection and obtained the following components for the moment tensor: $M_{11} = -1.8$, $M_{22} = -1.9$, $M_{33} = 3.7$, $M_{12} = -0.38$, $M_{31} = -0.96$, $M_{32} = 0.62$ (all numbers in $10^{17}\,\mathrm{N\,m}$). To obtain stable results, they constrained the isotropic component to be zero.

    (a) Compute M_0 and M_W for this event.

    (b) Compute the eigenvalues σ_1, σ_2, and σ_3 (sorted such that $\sigma_1 > \sigma_2 > \sigma_3$) and express $\mathbf{M}$ in its principal axes coordinates.

    (c) Compute the parameter ϵ, the measure of the misfit with a double-couple source. Is its value close to that expected for a pure double-couple (DC) source or a pure compensated linear vector dipole (CLVD) source?

    (d) Decompose $\mathbf{M}$ into $\mathbf{M}^{DC}$ and $\mathbf{M}^{CLVD}$ using equation (9.14) and compute the scalar moment of each part.

    (e) Devise an alternative decomposition of $\mathbf{M}$ into $\mathbf{M}^{DC}$ and $\mathbf{M}^{CLVD}$ that maximizes the CLVD part and compute the scalar seismic moment of each part.

    (f) Explain your results in (a), (d), and (e) in terms of your computed ϵ parameter in (c).

3. For the fault geometry shown in Figure 2.7, what is the expected P first motion (up or down) at PFO from the Landers earthquake?

4. Using (9.24) and (9.26) for a double-couple source, what is the ratio of maximum expected S-wave amplitude to maximum P-wave amplitude? Note: these need not be at the same takeoff angle, just compare the maximum in each case.

5. Make a photocopy of Figure 9.11. Using different colors for the different faulting types, circle five examples each of (a) normal faults, (b) reverse faults, and (c) strike–slip faults.

6. Table 9.3 is a list of earthquakes with double-couple fault plane solutions from the Global CMT catalog. Following the moment magnitude M_W for each event, the first three numbers give the strike, dip, and rake for one of the nodal planes; the final three numbers repeat this for the second nodal plane. Sketch lower-hemisphere focal spheres for each event, showing the orientation of the two possible fault planes. Shade in the compressional quadrants. Identify the dominant type of faulting for

Table 9.3: Some focal mechanisms from the Global CMT catalog								
Date	Region	M_W	ϕ_1	δ_1	λ_1	ϕ_2	δ_2	λ_2
10/28/83	Borah Peak, Idaho	7.0	304	29	-103	138	62	-83
09/19/85	Michoacan, Mexico	8.0	301	18	105	106	73	85
10/18/89	Loma Prieta, Calif.	6.9	235	41	29	123	71	128
06/20/90	Western Iran	7.4	200	59	160	300	73	32
07/16/90	Luzon, Phillipines	7.7	243	86	178	333	88	4
06/28/92	Landers, Calif.	7.3	318	88	178	48	88	2
01/17/94	Northridge, Calif.	6.7	278	42	65	130	53	111
06/09/94	Northern Bolivia	8.2	302	10	-60	92	81	-95
01/16/95	Kobe, Japan	6.9	324	70	12	230	79	160

each event (i.e., strike–slip, normal, reverse, or oblique). For the strike–slip faults, indicate with arrows the type of slip expected on each of the two possible fault planes (i.e., left-lateral or right-lateral).

7. Your borehole seismic experiment obtains the P-wave spectra plotted in Figure 9.30 at a distance of 10 km from an earthquake. Using a ruler, crudely estimate Ω_0 and f_c from the plot. Assuming that the density is $2700\,\text{kg/m}^3$, the P velocity is 6 km/s, and the S velocity is 3.46 km/s, compute the moment, M_0, the moment magnitude, M_W, the source radius, r, and the stress drop, $\Delta\sigma$, for this event. State any modeling assumptions that you make. Do not confuse the source radius r in Section 9.5 with the source–receiver distance r in equation (9.27). The recording station is deep enough that you may assume that the effect of the free surface can be ignored and that the rock properties are uniform between the earthquake and the station. You may also assume that attenuation is negligible (or that the spectrum has already been corrected for attenuation) and that the radiation pattern term is simply its average P-wave value of 0.52.

8. Estimate the radius and average displacement of a circular fault for M_W values ranging from -2 to 8 (make a table at unit increments). Use (9.40) and (9.73), a constant stress drop of 3 MPa, and a shear modulus, μ, of 30 GPa.

9. Suppose a fault existed completely around Earth's equator to a depth of 30 km and ruptured in a single earthquake with 20 m of slip. Estimate M_W for this event.

10. From Figure 9.21, estimate the average seismic energy radiated by a earthquake with $M_W = 8$. How does this compare with: (1) the US annual electricity production, and (2) the energy released during a major hurricane?

11. The Republic of Temblovia commisions you to estimate how often they should expect a $M_W > 8$ earthquake to occur. You study the available seismic data and

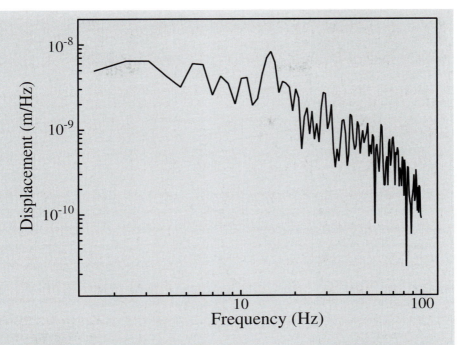

Figure 9.30 A *P*-wave spectrum from a borehole seismic experiment.

find that Temblovia has 100 events per year at $3.5 \le M_W \le 4$ (data from larger events are classified!). What do you tell them? Please state any assumptions that you make.

12. An earthquake occurs during a dinner party in Los Angeles. When the shaking subsides, Joe attempts to impress the other guests by proclaiming, "It felt like a 5.5 to me." Even assuming that Joe can reliably estimate the local ground accelerations, what vital piece of information does he lack that limits the accuracy of his magnitude estimate?

10

Earthquake prediction

Despite their usefulness as a research tool for illuminating Earth structure, earthquakes are generally considered harmful because of their potential for causing death and destruction. It is therefore unfortunate that the most useful thing that seismologists could do – predict earthquakes – is what they are least able to do. Although many ideas for earthquake prediction have been explored, the sad truth is that reliable prediction of damaging earthquakes is not currently possible on any time scale. In this section we will present some of the terminology and concepts in earthquake occurrence modeling and discuss possible reasons why major earthquakes are so difficult to predict.

10.1 The earthquake cycle

The idea that earthquakes represent a sudden release of accumulated stress in the crust was first documented by H. F. Reid, who examined survey lines taken before and after the 1906 earthquake in San Francisco. His results led to the *elastic rebound* theory of earthquake occurrence, in which stress and strain increase gradually and are then released during an earthquake by sudden movement along a fault (Fig. 10.1). This mechanism is now recognized to be the primary cause of earthquakes in the crust. Earthquakes occur mostly along the boundaries between Earth's surface plates (see Fig. 1.2), releasing the stress that results from the relative tectonic motion between the different plates. Observations of surface deformations, using ground- and satellite-based surveying techniques, can be used to monitor the slow strain changes that are seen in seismically active regions (often termed the *secular* strain rate) and the sudden change that occurs in the deformation field during earthquakes (these are termed *co-seismic* changes). Co-seismic changes in the strain field are often observed even at considerable distances away from earthquakes and can be used to constrain the distribution of slip on subsurface faults.

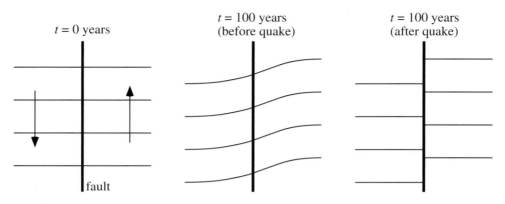

$t = 0$ years

$t = 100$ years
(before quake)

$t = 100$ years
(after quake)

fault

Figure 10.1 The elastic rebound model of earthquakes. In this example, a strike–slip earthquake releases the stress and strain that had built up slowly along the fault.

What causes faults to fail when they do? It is natural to imagine that there might be some threshold value of stress that represents the maximum shear stress that the fault can withstand. When the long-term stress accumulation reaches this level, the earthquake occurs.

A simple model of this kind of behavior is a block pulled by a spring. In this case we assume that the static friction coefficient μ_s is greater than the dynamic friction coefficient μ_d and that the spring is pulled at a rate v. The block will exhibit *stick–slip* behavior – when the force exerted by the spring exceeds the static friction, the block will slide until the dynamic friction balances the reduced level of stress.

If μ_s, μ_d, and v are all constant, then the "earthquakes" will repeat at regular intervals (commonly called *recurrence intervals*) and the system is completely predictable. This can be illustrated in plots of stress and accumulated slip on the block versus time (Fig. 10.2). During the time between events, no slip occurs and

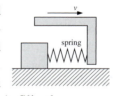

stress steadily builds up to a maximum value σ_2 (related to μ_s). Slip then occurs and the stress drops to a lower value σ_1 (related to μ_d).

We can add some additional complexity to this model by considering the case where μ_s or μ_d is not constant. For a model in which the dynamic friction μ_d randomly varies between events, the "size" of each event is not predictable but the time of occurrence can be predicted; thus this model is called *time predictable*. The time until the next event is proportional to the amount of slip in the last event (Fig. 10.2). Alternatively, the static friction μ_s might randomly vary between events, in which case the occurrence times cannot be predicted, but the amount of slip for an event at any time can be predicted; this is called a *slip predictable* model. The

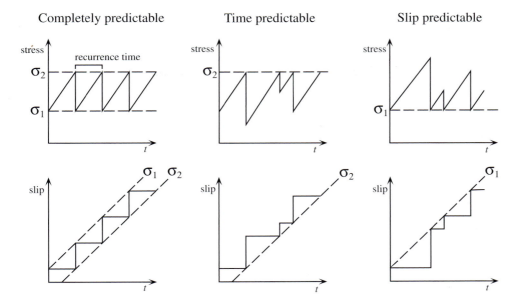

Figure 10.2 Simple models of recurring earthquakes parameterized by a threshold stress level σ_2 (related to the static friction on the fault) and a post earthquake stress level σ_1 (related to the dynamic friction on the fault). This diagram is based on Shimazaki and Nakata (1980).

amount of slip on the current event is proportional to the time since the last event (Fig. 10.2). Finally, both μ_s and μ_d might randomly vary between events, in which case neither the time nor the amount of slip is predictable.

A fundamental assumption in this type of model is that individual fault segments can be treated in isolation and a *characteristic earthquake* will occur at fairly regular intervals. In this case, long-term earthquake prediction might be feasible, although the exact time and date of future events would remain unknown. Unfortunately the real Earth is typically much more complicated than these simple models. A humbling example of this is provided by a sequence of earthquakes on the San Andreas Fault at Parkfield, California, where $m_b > 5.5$ events occurred in 1857, 1881, 1901, 1922, 1934, and 1966 (Fig. 10.3). The waveforms from many of these events are almost identical, suggesting that the same segment of fault ruptured each time. This pattern led the National Earthquake Prediction Evaluation Council (NEPEC) to "predict" in 1984 that a $m_b \sim 6$ event would occur before 1993 (the anomalous time of the 1934 quake was explained as a premature triggering of an expected 1944 quake). But the expected earthquake did not occur until 2004. What went wrong?

Some researchers have questioned whether the earliest events in the sequence plotted in Figure 10.3 truly were at Parkfield because they were before the instru-

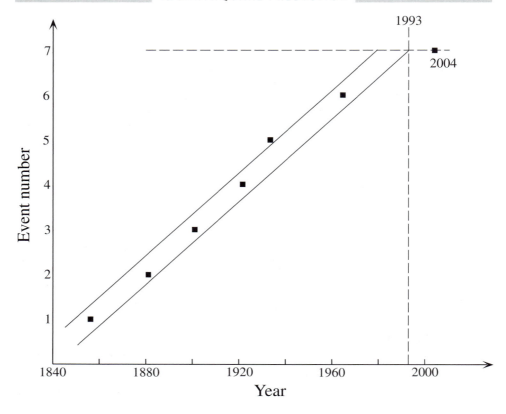

Figure 10.3 Significant earthquakes at Parkfield, California, have repeated at fairly regular intervals since 1850, leading to predictions of another event before 1993. However the earthquake did not occur until 2004.

mental era and their exact locations are unknown. For example, the 1983 M 6.7 Coalinga earthquake, 40 km to the north, might have been counted as a Parkfield earthquake had it occurred before 1900. More generally, however, a problem with the characteristic earthquake hypothesis is that it ignores the interactions with adjacent segments on the same fault, as well as interactions with other faults. Some of the effects of these interactions can be modeled using an expanded version of the block-slider model, in which a series of blocks are connected by springs to each other and a bar pulled at a constant rate (Fig. 10.4).

This type of model was first explored in a classic paper by Burridge and Knopoff (1967) and now has been studied in many different variations. Slip on one block can trigger slip on adjacent blocks and lead to larger events. For certain values of the spring constants and the individual friction coefficients, this type of model will produce a wide range of event sizes and a Gutenberg-Richter b-value close to

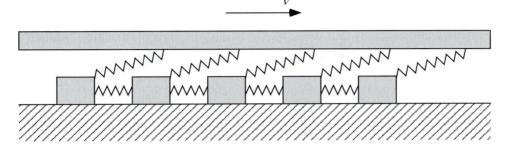

v

Figure 10.4 The sliding block model for simulating earthquakes.

that observed for real seismicity.[1] Under some conditions, the larger events occur at fairly regular intervals but it is not difficult to devise models in which the seismicity is *chaotic* with no clearly defined characteristic event size or recurrence time. This behavior is typical of non-linear dynamical systems. Indeed, even a two-block model has been shown to exhibit chaotic behavior under some conditions (e.g., Narkounskaia and Turcotte, 1992).

Real fault systems involve not just single faults but entire systems of faults. For example, in California the San Andreas Fault (SAF) is the dominant fault. But since the SAF is not straight, it is clear that other somewhat smaller faults must be involved to account for features such as the "Big Bend," a kink in the fault beneath the Transverse Ranges in southern California. These faults in turn create complications in the strain field that require the existence of smaller faults, and so on. The complex systems of faults of varying sizes in many regions appear to be *self-similar* and obey a *fractal* scaling relationship. Self-similarity is apparent in the power-law distribution of seismicity rates (the *b*-value relationship) and in the nearly constant value of stress drop over a wide range of earthquake sizes (implying that the displacement scales linearly with the size of the fault; see equation (9.42)). The fractal dimension D of a seismicity distribution can be shown to be approximately twice the b value (see the discussion in Turcotte, 1997, p. 59).

The events in real earthquake catalogs occur at apparently random times, with the exception of aftershock sequences. Earthquake times over large regions can be modeled reasonably well as a Poisson process, that is, the probability of an earthquake at any given time is constant and independent of the time of the last

---

[1] However, Rice and Ben-Zion (1996) make the important point that the observed Gutenberg-Richter (G-R) relation obtained for block slider models occurs for discretized systems with cell sizes much larger than the expected slip weakening distance for real faults. They were unable to obtain G-R behavior for smooth faults.

event.[2] However, at small scales there is a noticeable clustering of earthquakes in time and space that violates the simple Poisson model. Much of this can be explained as aftershock triggering, but some features, such as swarms without a clear mainshock, appear to require other mechanisms. In general, it is difficult to completely separate aftershocks from other earthquakes because even small events increase the probability of future events to some extent.

For small earthquakes, it is comparatively easy to test statistical models of earthquake occurrence since lengthy catalogs of events are available. However, larger earthquakes occur much less frequently, and some researchers have argued that great earthquakes ($M_W > 8$) may have more regular recurrence intervals. Since the average time spacing between these events on a particular fault may be hundreds of years or more, it is difficult to evaluate this hypothesis using modern seismicity catalogs. However, in some cases it is possible to detect ancient earthquakes by careful examination of the local geology along faults. Strike–slip faults commonly produce offset stream channels and other geomorphological features that, when dated, can yield earthquake occurrence times and long-term slip rates. Figure 10.5 shows a well-known example of offset stream channels along the San Andreas Fault in California. In other cases, examining fault scarps or trenches dug across the fault in areas of high sedimentation rate provides a means to measure and date observed offsets. These techniques, termed *paleoseismology*, only work at selected sites and cannot be applied everywhere along a fault. Statistical analysis of paleoseismic results is challenging because of the substantial error bars in the data for many of the events and uncertainties in correlating events among different sites. However, results for the southern San Andreas fault indicate that great earthquakes occur at irregular intervals with some tendency to cluster in time, arguing against a characteristic repeat time for these events (e.g., Sieh, 1996; Grant, 2002).

Since fairly simple block-slider models exhibit chaotic behavior, it seems reasonable that the much more complicated systems of faults in the real world will also lead to chaos. This would imply that very long term earthquake prediction is fundamentally impossible. But it is important to recognize that this limitation does not necessarily prevent earthquake prediction on shorter time scales. A useful analogy is weather forecasting, in which chaotic behavior is also important. Meteorologists are unlikely to ever be able to predict the time of an individual storm a year in advance. But storms can be predicted up to a week in advance and forecasting becomes quite precise at the level of hours in advance. Indeed there are numerous short-term precursors to storms (the barometer drops, the wind increases, etc.).

---

[2] For example, see Gardner and Knopoff, 1974, "Is the sequence of earthquakes in southern California, with aftershocks removed, Poissonian?" This paper is famous for having the shortest abstract in the geophysical literature – it simply says "Yes."

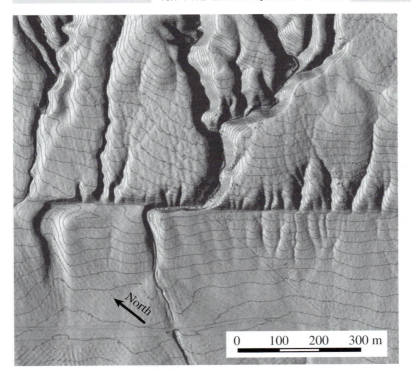

Figure 10.5 A LIDAR (Light Detection and Ranging) image of the San Andreas Fault at Wallace Creek, California, showing offset stream channels caused by the right-lateral movement along the fault. Contour interval is 5 feet (~1.5 m). Image courtesy of David Raleigh and Ken Hudnut.

Unfortunately, observations to date indicate that earthquakes very rarely produce definitive short-term precursors (see Section 10.3).

Chaotic behavior alone does not necessarily impose a fundamental limitation on predicting the next major earthquake or series of earthquakes in a region on time scales of years to tens of years. Given sufficient knowledge of the crustal stress and strain fields and the strengths of the existing faults, it is possible that reasonable predictions could be made as to where and when the next fault segment is likely to break. One assumption that seems intuitively reasonable and is often made is that the occurrence of a major earthquake on a fault temporarily relieves the stress on that fault, so that one should not expect another large quake on the same fault until the stress has had a chance to build up again. However, an earthquake will tend to increase the stress on adjacent segments of the fault and make them more likely to fail.

Reasoning along these lines has gone into the *seismic gap hypothesis*, which proposes that the probability of a large quake on an individual fault segment is greater for those segments that have not slipped in a long time. Since the continuous long-term slip rate is often known from far-field geodetic measurements, some estimates can be made of likely recurrence intervals for events with a given value of slip. The verdict on the usefulness of seismic gap reasoning is currently uncertain. In 1988, the Working Group on California Earthquake Probabilities (WGCEP) produced a conditional probability map for some of the major faults in California. Probabilities were ranked high for the Parkfield segment of the SAF, a section of the SAF near Santa Cruz just south of where the 1906 rupture from the San Franciso earthquake ended, and a section of the San Jacinto Fault. The 1989 M 7.1 Loma Prieta earthquake occurred near the location of the WGCEP "prediction" for the SAF near Santa Cruz, but it had a large thrust component, which was not expected for the predicted "characteristic" strike–slip event. The anticipated Parkfield earthquake did not occur until 2004 and no large earthquakes have occurred yet on the San Jacinto fault. In the meantime, the 1992 M 7.3 Landers, the 1994 M 6.7 Northridge, the 1999 M 7.1 Hector Mine, and the 2003 M 6.5 San Simeon earthquakes have occurred on "minor" faults not considered in the WGCEP report.

The seismic gap idea has also been applied to global seismology and the likelihood of subduction zone events related to the historical record of fault breaks. McCann *et al.* (1979) used this approach to map the probabilities of large earthquakes along the major plate boundaries. Kagan and Jackson (1991; see also Rong *et al.*, 2003) analyzed the McCann *et al.* paper and claimed that the seismic gap hypothesis was not supported by the subsequent record of events. Instead, Kagan and Jackson advocated a contrarian hypothesis – that the probability of an event is highest in the vicinity of recent earthquakes and that the event probability drops as the interval without an earthquake increases. This model is consistent with the earthquake triggering ideas that we will develop in the next section.

Although the seismic gap model in its original form is not supported by statistical tests of its global significance, many seismologists continue to believe that at least part of it remains true for the very largest earthquakes. The idea is that these earthquakes relieve so much of the stress that they create a *stress shadow*, in which major earthquakes are not likely to occur for some time. The 1906 San Francisco earthquake on the San Andreas fault is often cited as supporting the stress shadow concept because there were many more $M \geq 5.5$ earthquakes in the San Francisco Bay Area in the 50 years before 1906 than the 50 years after (e.g., Bakun, 1999). However, Felzer and Brodsky (2005) argue that similar changes in seismicity rate occurred outside of the Bay Area at the same time, casting doubt as to whether the drop in seismicity really supports the stress shadow hypothesis.

One problem with making reliable earthquake probability estimates for individual fault segments is that stress and strain fields in the crust are likely to be heterogeneous at many length scales and we currently can only make estimates as to their values at points so widely separated that spatial aliasing is a problem. This situation will improve as GPS (Global Positioning System, a method for determining precise locations using satellites) measurements become more common. Another promising technology is synthetic aperture radar (SAR), in which interferometer techniques are used to measure displacement changes between images taken at two different times. The ultimate goal is to obtain a complete picture of the surface strain field with high temporal and spatial resolution. Such an image might reveal subtleties in both the secular and co-seismic strain evolution that could provide clues regarding where future earthquakes are likely to occur.

10.2 Earthquake triggering

The most obvious example of non-random earthquake occurrence is the existence of aftershock sequences after large earthquakes. Although the exact timing of individual events is still random, an increased rate of activity is observed that is temporally and spatially correlated with the mainshock. The seismicity rate decays with time, following a power law relationship, called *Omori's law* after Omori (1894),

$$n(t) = \frac{K}{t+c}, \tag{10.1}$$

where $n(t)$ is the number of aftershocks per unit time above a given magnitude, t is the time measured from the mainshock, and K and c are constants. This is often generalized to the modified Omori's law

$$n(t) = K(t+c)^{-p}, \tag{10.2}$$

which permits a more general power law relation, but in which the exponent p is typically close to 1.

As an example, Figure 10.6 plots the aftershock rate following the 1994 Northridge, California, earthquake ($M_W = 6.7$), which is well fitted by Omori's law with $c = 3.3$ days and $p = 1$. The parameter c is related to a relative deficit of aftershocks immediately following the earthquake compared with a simple uniform power law. For the Northridge example, this is mainly caused by the inability of the seismic network to detect and locate the large number of events occurring in the first few days after the earthquake (e.g., Kagan and Houston, 2005). When care is taken to obtain a more complete catalog, the deficit in early aftershocks lasts only a few minutes after the mainshock (Peng *et al.*, 2007).

Earthquakes are thought to trigger aftershocks either from the dynamic effects of their radiated seismic waves or the resulting permanent static stress changes (for reviews, see Harris, 2002; Freed, 2005). A common assumption based on

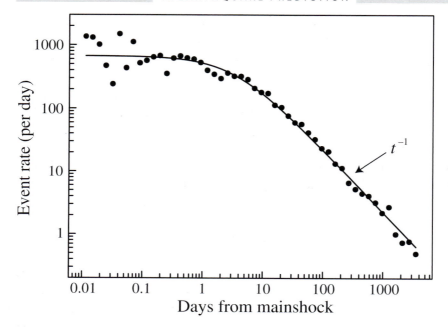

Figure 10.6 Aftershock rate for the 1994 Northridge, California, earthquake as a function of time after the mainshock. The line shows the Omori's law prediction for $K = 2230$, $c = 3.3$ days, and $p = 1$. Data are from the Southern California Seismic Network catalog within a lat/lon window of $(34.2°, 34.45°, -118.75°, -118.3°)$.

rock behavior in laboratory experiments is that earthquake occurrence on a fault is promoted by increases in the *Coulomb failure function* (CFF)

$$\text{CFF} = |\tau_s| + \mu(\tau_n + P), \tag{10.3}$$

where τ_s is the shear traction on the fault, τ_n is the normal traction (positive for tension), P is the pore fluid pressure, and μ is the coefficient of static friction (don't confuse this with the shear modulus!). The second term is negative because in our sign convention τ_n is negative for the hydrostatic compression forces at depth. Thus, increases in shear stress or decreases in fault normal compression (which "unclamp" the fault) will encourage failure, and the opposite changes will discourage failure. Numerous studies have searched for possible spatial correlations between aftershock occurrence and the sign of the CFF change predicted by mainshock slip models, and many have found that there tend to be more aftershocks in regions where the static stress changes should promote earthquakes (e.g., Reasenberg and Simpson, 1992; Harris and Simpson, 1992; Stein *et al.*, 1992; Stein, 1999). However, these correlations are not perfect and some aftershocks occur even in areas where the CFF changes are negative. The relative importance of static and dynamic triggering for aftershocks is also not yet firmly established.

Dynamic stress changes from seismic waves often trigger earthquakes at large distances from mainshocks and some have argued that dynamic effects could be the dominant triggering mechanism for near-field aftershocks as well (e.g., Kilb *et al.*, 2000; Felzer and Brodsky, 2006).

An obvious and important aspect of aftershocks is that they don't all occur instantly at the time of the mainshock – they have a time dependence that is described by Omori's law. This indicates that whatever their triggering mechanism, it must initiate a time-dependent failure process that causes events to occur at a wide range of times following the mainshock. There cannot simply be a precise threshold stress that, when exceeded, immediately triggers earthquakes (i.e., as shown in the simple models of Fig. 10.2). Additional evidence for this come from the lack of an obvious correlation between earthquake occurrence time and the solid Earth tides. Daily variations in crustal stresses caused by the tides greatly exceed the daily accumulation of stress from tectonic loading. Thus, any threshold level of stress will be first exceeded only at certain times in the tidal cycle, which might be expected to produce strong periodicities in earthquake occurrence times. Many researchers have searched for tidal signals in earthquake catalogs, but the most careful studies (e.g., Vidale *et al.*, 1998) have found little or no correlation between earthquakes and tidal stresses.

Omori's law does not say anything about the magnitude distribution of the aftershocks or their spatial relationship to the mainshock. However, by combining Omori's law with the Gutenberg–Richter magnitude–frequency law (9.75) and other empirical relationships, one can develop general models that predict the probability of future events based on the record of previous seismicity. The most well-known of these is called the *Epidemic Type Aftershock-Sequences* or *ETAS* model (for reviews, see Ogata, 1999; Helmstetter and Sornette, 2002). In the ETAS model, every earthquake, no matter how small, increases the probability of future nearby events. The increased probability is greatest immediately after an earthquake and then decreases following Omori's law until it reaches a background level of seismicity. These models do not require that aftershocks always be smaller than the triggering event. Sometimes mainshocks can be considered really big aftershocks of a *foreshock*, a smaller preceding earthquake that is spatially and temporally near the mainshock. Thus when any earthquake occurs, the possibility that it might be a foreshock increases the probability that a larger earthquake will soon follow. In California, for example, it has been estimated that an M 5.3 earthquake on the San Gorgonio Pass segment of the San Andreas Fault would produce a 1% chance of a much larger earthquake occurring within the next 3 days (Agnew and Jones, 1991).

The ETAS model in its original form does not include any spatial constraints on aftershock probabilities, that is the observed decay in aftershock density with distance from the mainshock. Felzer and Brodsky (2006) have explored this decay rate, and incorporating their result into the ETAS model of Ogata (1999), a general equation for estimated earthquake probability is

$$\lambda(\mathbf{x}, t) = \lambda_0 + \sum_i \kappa 10^{\alpha(m_i - m_0)} (t_i + c)^{-p} r_i^{-q} \tag{10.4}$$

where $\lambda(\mathbf{x}, t)$ is the predicted event density (events per unit volume and unit time) at position $\mathbf{x}$ and time t, λ_0 is a background rate (untriggered), which in general may be spatially varying, κ is a triggering productivity parameter, the summation is taken over all events in the catalog prior to t, m_i is the magnitude of each earthquake, m_0 is the minimum magnitude of the counted events, α (≈ 1) accounts for the fact that larger earthquakes trigger more events, t_i is the time from the ith event to t, c and p (≈ 1) are the Omori decay constants, r_i is the distance from the ith event to $\mathbf{x}$, and q defines the decay with distance. This type of model is an attempt to quantify the clustering in time and space of seismicity, i.e., the common observation that earthquakes are most likely to occur near recent earthquake activity. Note that this equation approximates earthquakes as point sources and would require modification to accurately predict aftershock density around the extended rupture of a large earthquake. By including the Gutenberg–Richter b-value relation (9.75), these models can also be used to estimate the probability as a function of earthquake size. Most of these models are purely empirical, but there have also been attempts to create physical models based on time-dependent failure mechanisms, such as the rate-and-state friction laws of Dieterich (1994). Although this work is unlikely to lead to deterministic predictions of individual events, understanding how earthquake occurrence relates to prior events is important for developing more accurate earthquake probability forecasts.

An example of this is the US Geological Survey sponsored effort to provide realtime estimates of the probability of significant ground shaking in California. As shown in Figure 10.7, the probability of earthquake occurrence increases following every earthquake, but then decays back to the background rate. Large earthquakes increase the risk of future events more than small earthquakes. Using seismicity catalogs it is possible to use such a model to predict the instantaneous probability of earthquake occurrence of a given magnitude as a function of location. This can be combined with the known relationship between earthquake magnitude and shaking intensity in California, as measured, for example, by the Mercalli scale (see Table 9.2). Integrating over all locations, the result can be plotted as the probability of ground motion exceeding a specified Mercalli intensity within a one-day period. An example of the results of this calculation is shown in Figure 10.8, comparing just before to just after the September 28, 2004 Parkfield earthquake (M 6). The probability of damaging shaking is greatly increased in a large region around Parkfield,

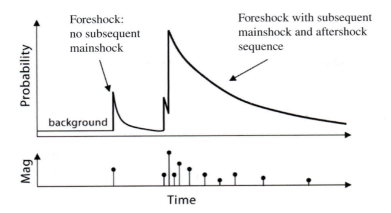

Figure 10.7 A cartoon illustrating how earthquake probability increases immediately after prior events, and then decays back to the background seismicity rate. Figure adapted from web material at: http://pasadena.wr.usgs.gov/step/.

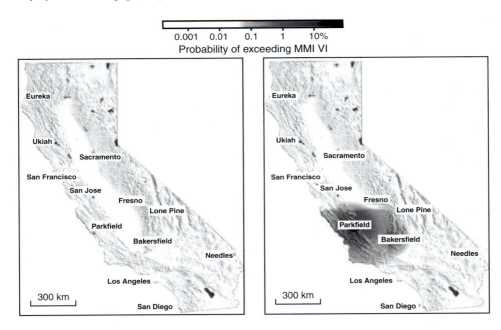

Figure 10.8 The probability of local ground motions of modified Mercalli intensity 6 or greater within a 24-hour period, immediately before and after the 2004 Parkfield earthquake in California. Source: http://pasadena.wr.usgs.gov/step/

reflecting the chance that the Parkfield earthquake might be a foreshock of a larger earthquake.

Although space–time clustering of earthquakes is clearly observed, the physics behind this clustering is not well understood. Some earthquake clusters, such as mainshock–aftershock sequences are most likely caused by triggering of events by previous events (although whether this occurs primarily as a result of static or dynamic stress changes is still debated). In other cases, such as earthquake swarms that lack an initiating mainshock, it seems more likely that the earthquakes are triggered by some underlying physical process, such as slow creep or fluid movement (e.g., Vidale and Shearer, 2006). These questions are also relevant to when foreshock sequences are observed prior to large earthquakes. Do the mainshocks occur simply because the foreshock activity itself increases the likelihood of a big earthquake, or could they be symptomatic of an underlying physical process that ultimately causes the larger event? The latter scenario provides more hope for prediction of large earthquakes, if the physical process could be understood more completely.

10.3 Searching for precursors

Short- to intermediate-term prediction (minutes to months) has proven especially problematic. Here the focus has been to search for anomalous behavior that can be observed prior to earthquakes that would provide warning that an event was imminent. Despite extensive searches for possible precursors, extremely few reliable examples have been found. This is not to say that claims for evidence of precursory phenomena have not been made. In the history of seismology such claims have occurred many times, often receiving great attention, only to be later discredited upon more careful and comprehensive study. Geller (1997) reviews much of this history.

One of the most famous examples occurred in the early 1970s, when several studies seemed to observe large (10 to 20%) changes in seismic velocity before earthquakes (e.g., Semenov, 1969; Aggarwal 1973; Aggarwal *et al.*, 1975; Whitcomb *et al.*, 1973; Robinson *et al.*, 1974). Such observations appeared to have a physical basis in laboratory studies of rock samples, which showed that when rocks are compressed until they fracture, a phenomenon termed *dilatancy* often occurs for a short time interval immediately before failure. Dilatancy is caused by microcracks forming in the sample, resulting in a slight volume increase and a change in the bulk seismic velocities. These results formed the basis of the dilatancy theory of earthquake prediction (e.g., Nur, 1972; Scholz *et al.*, 1973; Anderson and Whitcomb, 1975), which briefly was the focus of great excitement.

However, it soon became apparent that accurate measurements of changes in seismic velocity are difficult using naturally occurring events. Much greater precision can be achieved using artificial sources with nearly identical locations and waveforms. Studies of records from quarry blasts and nuclear explosions found no evidence for velocity changes before earthquakes down to levels of 1 to 2% (e.g., McEvilly and Johnson, 1974; Boore *et al.*, 1975; Kanamori and Fuis, 1976; Bolt,

1977; Chou and Crosson, 1978), providing upper limits that are an order of magnitude smaller than the changes that were claimed in the earlier studies. In addition to these observational constraints, it is also now thought that the average stress level on faults is surprisingly low (see Section 9.9), much lower than that used in laboratory experiments to fracture unbroken rocks. This implies that dilatancy, if it occurs before earthquakes, is likely to be confined to small areas of high stress concentration and not spread over significant volumes of rock where it would more readily be observed.

Other possible precursors that have received varying degrees of attention over the past few decades are changes in seismicity patterns, variations in the rate of radon gas emissions, and electromagnetic anomalies. Tantalizing suggestions of precursory behavior have often been seen for individual events, but more comprehensive studies have not been able to establish clear evidence for a link to the earthquakes. The history of these studies shows a familiar, if depressing, pattern. An apparent precursor will receive publicity as a possible method to predict earthquakes. Only rarely is there a clearly defined physical mechanism that might be causing the precursory behavior, and so the debate centers on the character of the observations and if the anomalies can indeed be correlated with earthquake occurrence. A number of papers will appear, some supporting the method and others challenging it. Typically the discussion then becomes embroiled in statistical arguments regarding the significance of the result and the exact way in which the method should be applied. These exchanges eventually become so technical as to be of little interest to anyone outside of the groups involved. The end result is that, even if the proposed method is not completely discredited, it becomes sufficiently clouded that most researchers move on to other things.

One of the best data sets to look for precursors resulted from the 2004 Parkfield earthquake. As discussed above, this region was identified in the 1980s as the likely site of a future M 6 earthquake and a large number of instruments were deployed to capture the event in detail. These included a borehole seismic array, high-resolution strain meters, electric and magnetic field sensors, and water well level meters. Although the earthquake occurred much later than anticipated, fortunately these instruments were still operating in 2004 and provided a wealth of information about the mainshock rupture and its aftershocks, as well as ongoing seismicity and creep events in the region. However, nothing unusual occurred before the mainshock – there were no notable precursory signals recorded on any of the sensors (Harris and Arrowsmith, 2006). Thus, one of the best instrumented earthquakes ever recorded did not exhibit any detectable precursors.

The only definitively established earthquake precursor is the occasional occurrence of foreshocks, events close in time and space to a subsequent mainshock. These occur far too often to be attributed to random chance; they must be related in some way to the larger event (just as the aftershocks that follow the mainshock are not randomly occurring). The existence of foreshocks made possible the most important earthquake prediction of recent times – the Chinese order to evacuate the city of Haicheng prior to the $M_S = 7.4$ earthquake of February 4, 1975. A series of small events occurred immediately prior to the mainshock, and, when the earthquake struck, most of the population had left their homes and relatively few lives were lost. At the time this was touted as a great achievement for the Chinese earthquake prediction program, but the outcome owes its success mostly to the existence of the foreshock swarm and other precursory anomalies.

Unfortunately, most large earthquakes are not signaled by easily recognized foreshock sequences. In some cases, there are no foreshocks, while in others the foreshocks are small in magnitude and not distinguishable from the many naturally occurring clusters of events that do not lead to larger earthquakes. On July 27, 1976, a $M_S = 7.8$ earthquake struck the Chinese city of Tangshan, only 200 km away from Haicheng. No foreshocks preceded this event and the population received no warning. The death toll was the greatest of any earthquake in modern times; the official count is 255 000 people killed, with unofficial estimates going much higher.

10.4 Are earthquakes unpredictable?

In California, no clearly recognizable precursor has been observed prior to any of the large earthquakes in the past few decades, despite the widespread deployment of seismometers and other instrumentation. Why should this be so? Why should such incredibly powerful events as major earthquakes apparently have no detectable precursors?

One possible explanation was described by Brune (1979), who proposed that earthquakes may be inherently unpredictable since large earthquakes start as smaller earthquakes, which in turn start as smaller earthquakes, and so on. In his model, most of the fault is in a state of stress below that required to initiate slip, but it can be triggered and caused to slip by nearby earthquakes or propagating ruptures. Any precursory phenomena will only occur when stresses are close to the yield stress. However, since even small earthquakes are initiated by still smaller earthquakes, in the limit, the region of rupture initiation where precursory phenomena might be expected is vanishingly small. Even if every small earthquake could be predicted, one is then still faced with the monumental task of deciding which of the thousands of small events will lead to a runaway cascade of rupture composing a large event.

Brune proposed his model only as a possible scenario for earthquake unpredictability, but subsequent results have tended to support his idea. The average level of shear stress on major faults is now thought to be quite low (see Section 9.9 on the heat flow paradox), far below the levels at which laboratory experiments suggest rock failure and precursory phenomena should occur. Modern ideas about *self-organized criticality* in non-linear systems suggest that it is to be expected that faults should be in a stress state such that even small events can initiate rupture to long distances. (The classic physical model for self-organized criticality is a sand pile in which individual grains of sand are continually added. The slope of the pile reaches an angle close to the maximum angle of repose. Additional sand grains may then trigger landslides of varying sizes, but it is difficult to predict in advance which grains will cause the largest slides.) Block-slider models have been devised that exhibit self-organized criticality (e.g., Bak and Tang, 1989).

Finally, studies of the beginnings of earthquakes of varying size have shown no difference between small and large events (e.g., Anderson and Chen, 1995; Mori and Kanamori, 1996). That is, there is no way to tell from the initial part of a seismogram how large the event will eventually become. This supports the notion that large earthquakes do not necessarily originate in anomalous source regions but are triggered by rupture from a smaller event (for an opposing view, see Ellsworth and Beroza, 1995; Olson and Allen, 2005). If Brune's hypothesis holds up, and the evidence for it is particularly strong in California, then short-term earthquake prediction may inherently be so difficult as to be impossible in practice. Barring dramatic new developments, an earthquake prediction program that promises timely, accurate warnings of future events, with a minimal number of false alarms, is unlikely to be achieved in the forseeable future.

In any case, seismology's most direct benefits to society are more likely to be achieved through identifying those regions most at risk from major earthquakes and encouraging suitable engineering and construction practices. People are rarely killed directly by earthquakes; rather the casualties arise from the failure of buildings and other structures. Through well-designed and rigorously enforced building codes, the death toll from earthquakes can be minimized. An example is provided by a comparison between two recent earthquakes: the $M_W = 6.8$ Armenian earthquake of December 7, 1988, and the $M_W = 6.7$ Northridge (southern California) earthquake of January 17, 1994. The Armenian event killed over 25 000 people and left 500 000 homeless, whereas only 60 were killed at Northridge. The shaking was slightly stronger in Armenia (maximum intensity of X on the modified Mercalli scale, compared to IX at Northridge), but most of the difference in the outcome can be attributed to the weaker construction practices that prevailed in Armenia. Future earthquakes are inevitable and there are few, if any, areas in the world that are completely free of earthquake risk. However, catastrophic loss of life can be minimized through sensible land use planning and the construction and maintenance of earthquake resistant structures.

10.5 Exercises

1. From Omori's law, derive an equation for the total number of aftershocks as a function of time after the mainshock. At what time will half of the total aftershocks have occurred for: (a) $c = 1$ day, (b) $c = 10$ minutes?

2. An often heard "rule of thumb" is that the largest aftershock is usually about a magnitude unit smaller than the mainshock (this was termed Bäth's law by Richter, 1958). Assuming this is true, estimate how many aftershocks a local network would record of $M \geq 1.5$, following a M 6.5 mainshock. Hint: Assume a b-value of 1.

3. (COMPUTER) Write a computer program to simulate the block-slider model of earthquake occurrence (Fig. 10.4). Consider a model with n blocks. If the displacement of the ith block is x_i and the displacement of the driving block is d, then we may use Hooke's law to express the force on the ith block as

$$F_i = k(x_{i-1} - x_i) + k(x_{i+1} - x_i) + k(d - x_i)$$

$$= k(d + x_{i-1} + x_{i+1} - 3x_i), \tag{10.5}$$

where k is the spring constant (in this case assumed constant for all springs in the model). The static and dynamic friction coefficients for the ith block may be expressed as μ_i^S and μ_i^D, respectively. At each time step, the ith block will move only if the force exceeds the resistance provided by the static friction, that is, if

$$F_i > \mu_i^S, \tag{10.6}$$

where we have assumed a unit mass for all blocks. The moving block will assume a new equilibrium position defined by the dynamic coefficient of friction

$$x_i = \tfrac{1}{3}\left[d + x_{i-1} + x_{i+1} - \mu_i^D/k\right], \tag{10.7}$$

which follows from (10.5), substituting μ^D for μ^S.

(a) Apply your program to the case where $n = 50$, $k = 1$, μ^S is randomly distributed between 0.05 and 0.55 (a different value for each block), μ^D is set to $1/3$ of the value of μ^S at each block, and the driving block moves a distance of 0.001 at each time step. Keep things simple by assuming that blocks move instantly to new equilibrium positions before they trigger any adjacent blocks. If a block moves during a time step, continue computing until no blocks are moving. Output each "quake" and plot the results for 1000 time steps. Figure 10.9 shows an example of how you may wish to display your results.

(b)   One characteristic of chaos is the extreme sensitivity of a system to small changes in initial conditions. Test your program by making a small change to one of the x values after 500 generations. How much does the output differ after 1000 generations? After 5000 generations? Is the system more sensitive to changes to high-friction blocks or to changes to low-friction blocks?

(c)   Compute a measure of the seismic moment, M_0, for each event as the sum of the slip on all of the blocks. Define event magnitudes as

$$M = \frac{2}{3}M_0 + d, \qquad (10.8)$$

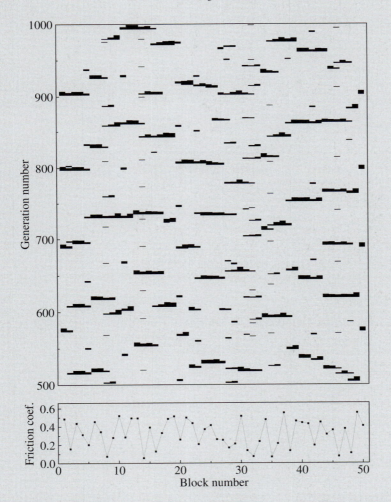

Figure 10.9 Output from a block-slider model of earthquake occurrence. The thickness of the black bars in the top plot is proportional to the amount of slip for each block during events. The lower plot shows the assumed values of the coefficient of static friction.

where the constant d is chosen so that the magnitudes have reasonable values. Make a histogram of the number of events as a function of magnitude. Can you estimate a b-value for your sequence? How could the model be changed to obtain a larger spread in event sizes?

(d)   One unrealistic aspect of this type of model is the absence of aftershock sequences following large events. Suggest some possible modifications to the model that might result in aftershock behavior.

(e)   Hint: Here is the key part of a FORTRAN program to solve this problem:

```
(set icount,d,x,xold to zero)
(initialize fric,dyfric arrays)
      n=50
      velplate=0.001
      k=1.0                   !real variable!
10    icount=icount+1
      d=d+velplate
      do i=1,n
         xold(i)=x(i)
      enddo
      idump=0
30    iquake=0
      do 40 i=1,n
         ileft=i-1
         if (ileft.lt.1) ileft=n
         iright=i+1
         if (iright.gt.n) iright=1
         force=k*(d+x(ileft)+x(iright)-3.*x(i))
         if (force.lt.fric(i)) go to 40
         iquake=1
         idump=1
         x(i)=(d+x(ileft)+x(iright)-dyfric(i)/k)/3.
40    continue
      if (iquake.eq.1) go to 30
      if (idump.eq.0) go to 10
(output details of quake, defined as difference between x and xold)
      go to 10
```

Note that "wraparound" boundary conditions are imposed at the end-points.

11

Instruments, noise, and anisotropy

Presented here are some important topics that do not easily fit into the structure of the first ten chapters.

11.1 Instruments

Throughout this book, we have often discussed Earth motion in terms of the displacement field, $\mathbf{u}(\mathbf{x}, t)$, but have not mentioned how these movements are actually measured. A device that detects seismic wave motion is termed a *seismometer*; the entire instrument package, including the recording apparatus, is called a *seismograph*. The most common type of seismometer is based on the inertia of a suspended mass, which will tend to remain stationary in response to external vibrations.

As an example, Figure 11.1 shows a simple seismometer design that will detect vertical ground motion. A mass is suspended from a spring and connected to a lever such that it can move only in the vertical direction. Motions of the lever are damped using a "dashpot" to prevent excessive oscillations near the resonant frequency of the system. The differential motion between the mass and the seismometer case (which is rigidly connected to Earth) is measured using the voltage induced in a coil by the motion of a magnet. The induced voltage is proportional to the velocity of the mass for the instrument shown in Figure 11.1. In alternative seismometer designs, the displacement or acceleration of the mass may be recorded. As we will see later, the frequency response of the seismometer is a strong function of whether the displacement, velocity, or acceleration of the mass is measured. A similar design to that shown in Figure 11.1 can be used to detect horizontal ground motion; in this case the mass is suspended as a pendulum. Both vertical and horizontal suspended mass instruments are sometimes called *inertial seismometers*.

The motion of the suspended mass is related to Earth motion through a *response function* that includes the effects of the various forces that act on the mass. This

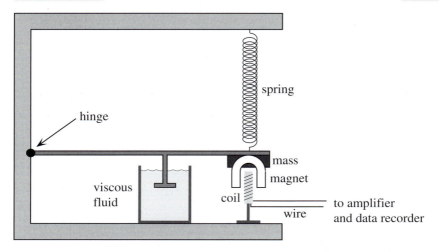

Figure 11.1 A simple inertial seismometer for measuring vertical motion. Movement of the suspended magnet induces a voltage in the coil; this signal is then amplified and recorded.

function is easily derived for simple inertial seismometers such as that shown in Figure 11.1. Let $u(t)$ be the vertical Earth displacement and $z(t)$ be the displacement of the mass with respect to Earth, each relative to their rest positions. The seismometer case is assumed to be rigidly connected to Earth, and so $z(t)$ also represents the displacement with respect to the case. The absolute displacement of the mass is given by the sum $u(t) + z(t)$. The force on the mass from the spring, F_s, will oppose the mass displacement and is given by

$$F_s = -kz, \tag{11.1}$$

where k is the spring constant. The viscous damping force, F_d, is proportional to the mass velocity and is given by

$$F_d = -D\frac{dz}{dt}, \tag{11.2}$$

where D is the damping constant. From $F = ma$, we thus have

$$-kz(t) - D\frac{dz(t)}{dt} = m\frac{d^2}{dt^2}\left[u(t) + z(t)\right], \tag{11.3}$$

where m is the mass. Rearranging, we may write

$$\ddot{z} + \frac{D}{m}\dot{z} + \frac{k}{m}z = -\ddot{u}. \tag{11.4}$$

It is convenient to define $\omega_0^2 = k/m$, where ω_0 is the resonant angular frequency of the undamped system ($D = 0$). We also define a damping parameter ϵ, such that $2\epsilon = D/m$. These substitutions give

$$\ddot{z} + 2\epsilon\dot{z} + \omega_0^2 z = -\ddot{u}. \tag{11.5}$$

This equation shows that the Earth acceleration, $\ddot{u}(t)$, can be recovered by measuring the displacement of the mass, $z(t)$, and its time derivatives.

The response function of the seismometer can also be expressed in the frequency domain. Consider harmonic Earth displacement of the form:

$$u(t) = U(\omega)e^{-i\omega t}, \tag{11.6}$$

where $\omega = 2\pi f$ is the angular frequency (see Appendix B for a description of how complex numbers are used to represent harmonic waves). The displacement response of the seismometer mass can be expressed as

$$z(t) = Z(\omega)e^{-i\omega t}. \tag{11.7}$$

We then have

$$\ddot{u} = -\omega^2 U(\omega)e^{-i\omega t}, \tag{11.8}$$

$$\dot{z} = -i\omega Z(\omega)e^{-i\omega t}, \tag{11.9}$$

$$\ddot{z} = -\omega^2 Z(\omega)e^{-i\omega t}. \tag{11.10}$$

Substituting into (11.5) and dividing by the common factor of $e^{-i\omega t}$, we obtain

$$-\omega^2 Z(\omega) - 2\epsilon i\omega Z(\omega) + \omega_0^2 Z(\omega) = \omega^2 U(\omega), \tag{11.11}$$

or

$$Z(\omega) = \frac{\omega^2}{\omega_0^2 - 2\epsilon i\omega - \omega^2}U(\omega) = \mathcal{Z}(\omega)U(\omega), \tag{11.12}$$

where $\mathcal{Z}(\omega)$ is the *frequency response function* of the sensor ($\mathcal{Z}(\omega) = Z(\omega)$ when $U(\omega) = 1$). The response function $\mathcal{Z}(\omega)$ is complex; in polar form it can be expressed as

$$\mathcal{Z}(\omega) = A(\omega)e^{i\phi(\omega)}, \tag{11.13}$$

where the amplitude, $A(\omega) = |\mathcal{Z}(\omega)|$, and the phase lag, ϕ, are real numbers.

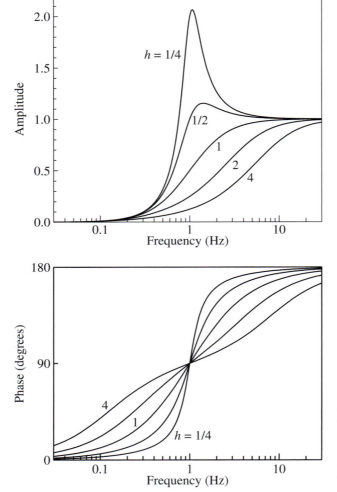

Figure 11.2 The amplitude and phase response functions for a seismometer of 1 Hz natural resonance at various levels of damping. The damping constant, h, is one for critical damping.

The strength of the damping relative to the stiffness of the spring may be described by $h = \epsilon/\omega_0$, where h is the *damping constant*. When $h = 1$ ($\epsilon = \omega_0$), the system is said to be *critically damped*. Under critical damping, a displaced mass will return to its rest position in the least possible time, without "overshooting" and oscillating about its rest position. Seismometers generally perform optimally at values of damping close to critical. Figure 11.2 plots amplitude and phase response curves for a seismometer with a natural (undamped) resonant frequency of 1 Hz (a typical short-period sensor) for values of h ranging from 1/4 to 4. At high frequencies ($\omega \gg \omega_0$), $\mathcal{Z}(\omega) \to -1$, as follows from (11.12); the amplitude response is near unity and the phase response is close to π (180°), representing a polarity reversal in the motion. In this case the motion of the mass, relative to the Earth, is simply opposite to the ground motion; the sensor stays in the same place while the Earth moves. The amplitude response falls off at frequencies below the resonant frequency and the 1-Hz sensor has little sensitivity at periods longer than 5 s (i.e., 0.2 Hz). For small values of the damping parameter (e.g., $h = 1/4$), a resonant peak occurs in the response spectrum near 1 Hz.

The frequency response function (11.12) relates the Earth displacement, u, to the sensor mass displacement, z. In the case of a seismometer that measures mass velocity, $\dot{z}$, such as that shown in Figure 11.1, the response function describes the sensor response to ground velocity, $\dot{u}$. In general, seismometers may measure the displacement, velocity, or acceleration of the sensor mass, and we may be interested in recovering the displacement, velocity, or acceleration of the ground. It is important to be aware of which combination is involved. Each time derivative introduces a factor of $-i\omega$ in the frequency domain. Thus, all other things being equal, velocity and (especially) acceleration will be enriched in high frequencies relative to displacement.

This is illustrated in Figure 11.3, which shows the $h = 1$ amplitude response for a 1-Hz sensor, as multiplied by different powers of ω. These curves provide the appropriate response for different combinations of displacement, velocity, and acceleration. Each time derivative applied to the sensor motion multiplies the response by ω^1, while each time derivative applied to the Earth motion multiplies the

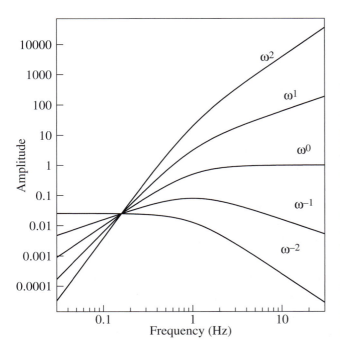

Figure 11.3 The amplitude response for a critically damped seismometer of 1 Hz natural frequency, multiplied by powers of ω from -2 to 2. The ω^0 curve is identical to the $h = 1$ curve in Figure 11.2 and shows the sensor displacement response to Earth displacement. This curve is also appropriate for the sensor velocity response to velocity and the acceleration response to acceleration. The other curves show the effect of multiplying the response by different powers of ω (see text).

response by ω^{-1}. Thus, for example, the response of sensor acceleration to ground displacement is provided by the ω^2 curve; the response of sensor velocity to ground acceleration is given by the ω^{-1} curve. Notice that all of the curves cross at $\omega = 1$ ($f = 1/2\pi$).

In two cases, the response function is flat over a wide frequency band. As discussed above, the first occurs for the ω^0 curve at high frequencies ($\omega \gg \omega_0$). The second occurs for the ω^{-2} curve at low frequencies, representing the response of a displacement sensor to ground acceleration. Defining $\ddot{U}(\omega)$ as the acceleration spectrum, the response of such a sensor to ground acceleration is given by substituting $U(\omega) = -\omega^{-2}\ddot{U}(\omega)$ into (11.12):

$$Z(\omega) = \frac{-1}{\omega_0^2 - 2\epsilon i\omega - \omega^2}\ddot{U}(\omega). \tag{11.14}$$

In the low frequency limit, we have

$$Z(\omega) = -\frac{1}{\omega_0^2}\ddot{U}(\omega) \quad \text{for } \omega \ll \omega_0. \tag{11.15}$$

This result also follows from (11.5) by considering that, at long periods, the z term will dominate over the $\dot{z}$ and $\ddot{z}$ terms, and thus

$$\omega_0^2 z = -\ddot{u}, \quad \text{for } \omega \ll \omega_0,$$

$$z = -\ddot{u}/\omega_0^2, \tag{11.16}$$

$$= -\ddot{u}T_0^2/(2\pi)^2, \tag{11.17}$$

where $T_0 = 2\pi/\omega_0$ is the undamped period of the instrument. Thus, the *sensitivity* of the sensor to long-period ground acceleration is proportional to the square of the natural period of the sensor.

In this way, measurements of the mass displacement can provide significant sensitivity to Earth acceleration at frequencies below the natural seismometer resonance frequency. Vertical-component instruments of this type can serve as *gravimeters* because they are sensitive to changes in gravity (which can be thought of as acceleration of infinitely long period). Similarly, tilt changes for horizontal component instruments can cause sensor displacements that mimic apparent horizontal accelerations.

The frequency band over which seismic waves are recorded is roughly divided into two parts by the microseism ground noise peak (see next section) that occurs at periods of approximately 6 to 8 s (0.12 to 0.17 Hz). Common usage refers to *short-period* records as those obtained at frequencies above the microseism peak, whereas *long-period* records are at frequencies below the peak. (Seismologists have become accustomed to using frequency, f, and period, $T = 1/f$, almost interchangeably. Frequency is used most often above 1 Hz, while period is typically used below 1 Hz. Remember: Short period and high frequency mean the same thing, just as long period and low frequency are equivalent.) Short-period sensors are well suited for recording high-frequency body-wave arrivals; the timing of these arrivals can be used both to locate earthquakes and to perform tomographic inversions for three-dimensional velocity structure. In addition, P first-motion data, used to compute earthquake focal mechanisms, are best obtained at short periods. However, surface wave analyses, detailed studies of source properties, and waveform modeling require data at longer periods (15 to 100 s).

Instrument sensitivity to long-period seismic waves can be achieved most simply by increasing the natural period of the sensor, either by lowering the spring constant or increasing the mass (recall that $\omega_0^2 = k/m$). Unfortunately, both approaches tend to increase the size and expense of long-period seismometers compared to their short-period counterparts. A breakthrough was achieved in 1935 when Lucien LaCoste invented the "zero-length" spring that permitted stable inertial seismometers at long natural periods. Today, typical long-period inertial seismometers have periods between 15 and 30 s.

11.1.1 Modern seismographs

The best modern instruments are more sophisticated than the simple mechanical seismograph illustrated in Figure 11.1. They are designed to achieve a linear response to Earth motions over a wide range of both amplitude and frequency.

The *dynamic range* of a seismograph is the difference between the smallest and largest amplitudes that are accurately recorded. It is desirable for the instrument to have sensitivity below typical Earth noise levels so as to record the smallest detectable events, while remaining on-scale for the largest earthquakes. The sensitivity to small motions can be improved both by the mechanical design of the instrument and through the use of low-noise amplifiers. The ultimate limit to the sensitivity is the noise due to the Brownian motion of the atoms in the sensor. For strong ground motions, the problem is to prevent non-linearity or clipping in the response. Mechanical non-linearity can arise from the finite length of the springs and levers used in the design. For example, the design shown in Figure 11.1 will be linear only for small excursions of the mass compared with the length of the lever arm. Linearity is often maintained in modern instruments through the use of *force-feedback* designs in which the mass is maintained at a fixed position. The seismograph records a measure of the force that is required to keep the mass at rest; this force is directly related to the Earth acceleration. Because the mass moves very little, this makes possible much more compact designs, including, for example, instruments sensitive to long periods that are small enough to deploy in boreholes.

Dynamic range must also be maintained in the electronics and the recording media. Analog instruments and digital systems with limited precision (e.g., 12- or 16-bit digitizers) often have insufficient range and clip at large amplitudes. The use of 24-bit digitizers has largely eliminated these problems and the need for separate channels of differing sensitivity. However, even with the best modern designs, few general-purpose seismographs can reliably record ground motion near the epicenters of shallow earthquakes with magnitudes of 7 and greater, where accelerations can exceed 1 g. Thus in many urban areas, separate networks of *strong motion* instruments have been established. These are typically low-cost accelerometers of limited sensitivity that provide on-scale records from large earthquakes. They are primarily used to provide data for the design of earthquake resistant structures. They also provide valuable near-source records for use in modeling the rupture history along the fault surface for major events.

A seismograph that provides useful sensitivity over a wide range of frequencies is termed *broadband*. The use of force-feedback techniques can greatly increase instrument bandwidth compared to purely mechanical designs and has made possible seismographs that record motion from periods of hundreds of seconds to frequencies of 10 Hz or greater. Figure 11.4 compares the frequency response of a modern broadband seismograph with some older designs. The original IDA (International Deployment of Accelerometers) network was the first digital global seismic network. IDA used gravimeters designed to record Earth's normal modes at very long periods and recorded one sample every 10 s. Data from the Global Digital Seismograph Network (GDSN) began to become available in the late 1970s. The GDSN long-period channel recorded at one sample per second; the GDSN short-period channel recorded at 20 samples per second. The GDSN response functions were designed to avoid the microseism noise peak at 5 to 8 s period (see Section 11.2). Broadband instruments began to be widely deployed in the late 1980s and early 1990s; the broadband stations in the IRIS and GEOSCOPE networks have very wide frequency responses.

Electronics have become an integral part of seismograph design; amplifiers and filters are used to modify the response and anti-aliasing filters are required for digital recording. Data may be recorded locally using a tape recorder, hard drive, or other storage medium, or they can be telemetered to a remote recording site. Modern seismic research often requires timing accuracy of 0.01 s or better to make full use of the data. Timing errors are much less common in modern instruments owing to the use of crystal oscillators with relatively constant drift rates and/or synchronization to satellite signals.

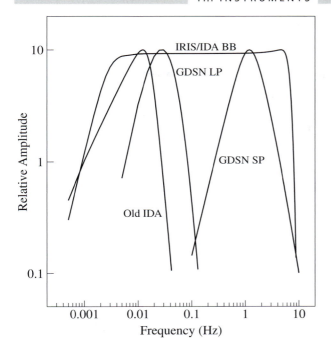

Figure 11.4 Velocity response functions for four different vertical-component instruments (old IDA station ALE, long and short-period channels for the GDSN station COL, and IRIS/IDA station ALE).

The instrument response can be defined in terms of the relationship between the digital counts in the recorded time series and the actual Earth motion. The *gain* of an instrument is the ratio between the digital counts and some measure of Earth motion; thus a high-gain instrument is more sensitive than a low-gain instrument. However, since seismograph sensitivity is frequency dependent, the concept of gain is only meaningful at a fixed frequency. A more complete description is provided by the frequency response function, $\mathcal{Z}(\omega)$, which specifies the amplitude and phase response continuously as a function of frequency. Instrument response can also be described by the *impulse response function*, which shows the seismograph output in the time domain from a delta-function input. Figure 11.5 plots the impulse response functions for four different instruments. In general, the impulse response function will more closely approximate a delta function as the instrument becomes more broadband.

There are tradeoffs in instrument design. The ideal seismograph has a flat response over a broad frequency band and sensitivity over a wide dynamic range. However, it is also sturdy, portable, low-power, and inexpensive, posing many challenges to instrument designers. For some purposes (e.g., local earthquake location),

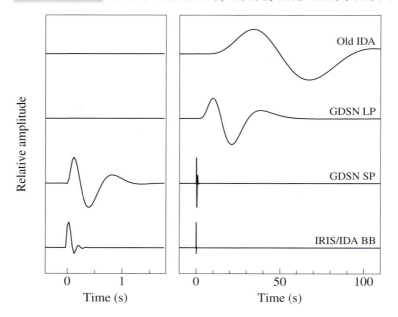

Figure 11.5 Impulse response functions for four different vertical-component instruments, showing the seismograph response to a delta-function input at zero time. The frequency response of these intruments is plotted in Figure 11.4.

large numbers of cheap instruments of limited capability may be more effective than a few state-of-the-art, but expensive, broadband seismographs. In other cases (detailed waveform modeling), there is no substitute for broadband records. Whatever the instrument design, the importance of accurate and reliable calibration information cannot be overemphasized. A raw seismogram alone is not very useful for research, without knowledge of the precise instrument location, the orientation of the horizontal sensors, the digitization rate, the time of the first sample, and the complete instrument response function.

11.2 Earth noise

Sensitive seismographs will record ground motions even in the absence of earthquakes. These motions, sometimes called *microseisms*, result from seismic waves generated primarily by wind and cultural noise at high frequencies and ocean waves and atmospheric effects at longer periods. Although microseisms are sometimes studied for their own intrinsic interest, seismologists generally consider them noise because they hamper observations of small and/or distant earthquakes. Typical noise levels will vary greatly between different sites and different frequencies. This

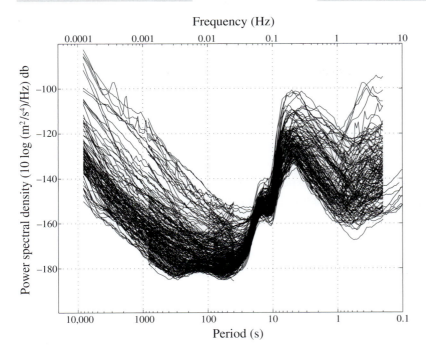

Figure 11.6 Individual acceleration spectra at over 100 global seismic stations (vertical component) computed as the average noise levels during intervals between earthquakes (adapted from Astiz, 1997). Note the microseism peak at 5 to 8 s period and the relatively low noise levels at 20 to 200 s period.

is illustrated in Figure 11.6, which plots typical noise levels for over 100 different stations from the global seismic networks (Astiz, 1997).

A large noise peak occurs at periods of about 5 to 8 s; this is termed the *microseism peak*. The source of the microseism peak involves seismic waves generated in the oceans; the peak is observed most strongly along the coast and is weaker near the middle of continents. However, the physical mechanism that generates these waves was a mystery for many years, as the typical period of surface gravity waves in the ocean is about 12 to 15 s, roughly twice the period of the main microseism peak. Thus, mechanisms such as waves breaking onto the shore cannot explain the observations (although near-shore interactions are probably responsible for the small, secondary microseism peak near 15 s period). The main microseism peak was first explained by Longuet-Higgins (1950), who showed that it results from standing waves created in the open ocean by interactions between ocean waves traveling in different directions.

These microseisms are often formed in areas of intense storm activity. During the 1940s, microseisms received considerable study and it was proposed that small arrays of seismometers could be used to determine the source locations of microseisms and potentially track the paths of hurricanes before they reached land. This scheme did not work very well, and, in any case, is now made obsolete by satellite observations. In the 1970s through the 1990s, microseism studies were relatively rare, although there was some research into their use for characterizing near-surface velocities and the site response of seismic stations. Recently, however, the field has been revitalized by the exciting discovery that cross-correlation of noise recorded at two different stations over long time periods can yield the station-to-station surface-wave Green's function (Campillo and Paul, 2003). This surprising result has now been verified by both experimental and theoretical studies and provides a new way to study Earth structure, free of the limitations imposed by the distribution of natural earthquakes. The best results are obtained when the microseisms are propagating at a variety of azimuths among the stations, i.e., not limited to a particular source region. By measuring these surface waves between multiple station pairs within an array of stations, high-resolution surface-wave tomography is possible (e.g., Shapiro *et al.*, 2005).

Current interest in microseisms also involves studying how to reduce their influence on seismic observations of earthquakes. Noise levels in seismometer installations vary greatly between sites (see Figure 11.6); a quiet site is capable of detecting and recording many more earthquakes than a noisy site. Noise levels on the continents are generally lowest at greater distances from the coast, whereas the noisiest sites are located on the seafloor and oceanic islands (for a review of seafloor noise observations, see Webb, 1998). Some improvement in noise levels can be achieved with borehole or mine shaft installations compared with near-surface vaults.

At extremely low frequencies, the largest signals are the solid Earth tides, which occur at periods of 12 and 24 hours and are seen only with sensors that are stable at these periods. As in the case of the ocean tides, they result from the gravitational effect of the Sun and the Moon on the Earth's mass. Compared with the ocean tides, they have smaller amplitudes (about 0.7 m peak-to-peak displacement) and a near-zero phase lag relative to the forcing function. Earth tides can be used to help calibrate very long period seismographs and strain meters; they can be readily predicted with standard Earth models, but care must be taken near coastlines to include the loading due to oceanic tides.

11.3 Anisotropy[†]

Most seismological modeling assumes that the Earth is isotropic, that is, seismic velocities do not vary with direction. In contrast, individual crystals and most common materials are observed to be anisotropic, with elastic properties that vary with orientation. Thus it would be surprising if the Earth was completely isotropic, but for many years seismologists were somewhat reluctant to consider the effects of anisotropy. There were several reasons for this, including the greater computational complexity required for anisotropic calculations, the difficulty in inverting data for a greater number of elastic constants, and, in many cases, the lack of compelling evidence for the existence of anisotropy. However, it has now become apparent that significant anisotropy is present in many parts of the Earth, and anisotropy studies are becoming an increasingly important part of seismology research (see Maupin and Park, 2007, for a recent review). Let us now explore some of the complications that arise when anisotropy is considered. Details on the theoretical background for this section can be found in the books by Cerveny (2001) and Chapman (2004).

Consider the wavefront generated by a point source in a homogeneous anisotropic material. The rays travel in straight lines out from the source, but the wavefront is not spherical because velocity varies as a function of ray angle (Figure 11.7). At most points on the wavefront, this means that the ray is not perpendicular to the wavefront. The implications of this are the origin of much of the complexity in anisotropic wave propagation. The ray is the direction of energy transport and represents the group velocity of the wave. In contrast, the phase velocity is the local velocity of the wavefront in the direction perpendicular to the wavefront. The group velocity is what would be measured by the wavefront arrival time at a single instrument (assuming the origin time of the pulse was known), while the phase velocity is what would be measured with travel times collected from a

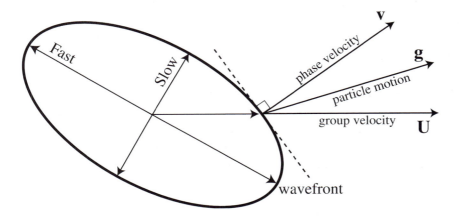

Figure 11.7 The wavefront generated by a point source in an anisotropic material, showing how the phase velocity, group velocity, and particle motion vectors point in different directions for a quasi-compressional wave (except along a symmetry plane of the anisotropy where they will coincide). The phase velocity vector **v** points in the same direction as the slowness vector **s** (see text).

small array of instruments during the local passage of the wavefront through the array. In general, the phase and group velocity directions are different for waves in anisotropic media. To complicate things still further, the particle motion direction is not simply related to either the group or phase velocity directions (see Fig. 11.7).

A more quantitative understanding of these properties can be derived from the momentum equation. Recall from Chapter 2 that the general, linear stress–strain equation is

$$\tau_{ij} = c_{ijkl}e_{kl}, \tag{11.18}$$

where τ_{ij} is the stress tensor, e_{kl} is the strain tensor, and c_{ijkl} is the elastic tensor. The tensor c_{ijkl} has the following symmetries:

$$
\begin{aligned}
c_{ijkl} &= c_{jikl} & \text{since } \tau_{ij} = \tau_{ji}, \\
&= c_{ijlk} & \text{since } e_{ij} = e_{ji}, \\
&= c_{klij} & \text{from thermodynamic considerations.}
\end{aligned}
\tag{11.19}
$$

These symmetries reduce the number of independent elastic constants in c_{ijkl} to 21 in the most general case. Often additional symmetries are considered, which further reduce the number of elastic constants. For example, crystals with orthorhombic symmetry have nine elastic constants; those with hexagonal symmetry have five elastic constants (assuming the orientation of the symmetry axis is known). Since $e_{kl} = \frac{1}{2}(\partial_k u_l + \partial_l u_k)$, we can use (11.19) to write (11.18) as

$$\tau_{ij} = c_{ijkl}\partial_l u_k. \tag{11.20}$$

Using this expression for τ_{ij}, we can write the momentum equation in the form

$$\rho \ddot{u}_i = (\nabla \cdot \tau)_i = \partial_j \tau_{ij} = \partial_j(c_{ijkl}\partial_l u_k) = c_{ijkl}\partial_j\partial_l u_k, \tag{11.21}$$

where we have assumed that c_{ijkl} is constant within a homogeneous layer.

Now assume a steady-state plane-wave solution for the displacement $\mathbf{u}$ of the form

$$\mathbf{u} = \mathbf{g}e^{-i\omega(t - \mathbf{s}\cdot\mathbf{x})}, \tag{11.22}$$

where $\mathbf{s}$ is the slowness vector, $\mathbf{x}$ is the position vector, and $\mathbf{g}$ is a polarization vector that gives the direction of particle motion. Note that $\mathbf{s}$ is orthogonal to the wavefront because $\mathbf{s}\cdot\mathbf{x}$ is unchanged for variations in $\mathbf{x}$ perpendicular to $\mathbf{s}$. Thus $\mathbf{s}$ is in the same direction as the phase velocity vector $\mathbf{v}$ and we have $u = |\mathbf{s}| = 1/c = 1/|\mathbf{v}|$, where u is the slowness and c is the phase velocity. Within a homogeneous layer, $\mathbf{s}$ and

$\mathbf{g}$ are constant. Substituting (11.22) into (11.21) and canceling the $-\omega^2 e^{-i\omega(t-\mathbf{s}\cdot\mathbf{x})}$ terms, we obtain

$$\rho g_i = g_k c_{ijkl} s_j s_l, \tag{11.23}$$

or

$$(c_{ijkl} s_j s_l - \rho \delta_{ik}) g_k = 0. \tag{11.24}$$

Now if we define the density-normalized elastic tensor

$$\Gamma_{ijkl} = \frac{c_{ijkl}}{\rho} \tag{11.25}$$

and let $\mathbf{s} = \hat{\mathbf{s}}/c$, where $\hat{\mathbf{s}}$ is the unit slowness vector and c is the phase velocity, we have

$$\left(\Gamma_{ijkl} \hat{s}_j \hat{s}_l - c^2 \delta_{ik} \right) g_k = 0. \tag{11.26}$$

Now define

$$M_{ik} = \Gamma_{ijkl} \hat{s}_j \hat{s}_l \tag{11.27}$$

so that we can write (11.26) as

$$\left(M_{ik} - c^2 \delta_{ik} \right) g_k = 0. \tag{11.28}$$

This is an eigenvalue equation for which there are three solutions for c^2 in a given direction $\hat{\mathbf{s}}$. In isotropic media, (11.28) reduces to three separate second-order equations, corresponding to P waves and two types of S waves (e.g., SV and SH). However, in the general anisotropic case, there is no simple analytic expression for $c(\hat{\mathbf{s}})$. For every direction $\hat{\mathbf{s}}$, there are three solutions that are obtained by computing the eigenvalues and eigenvectors of M_{ik}. The eigenvalues correspond to the allowed values for the phase velocity squared, c^2, while the eigenvectors, $\mathbf{g}$, specify the polarizations of the waves. Note that this eigenvalue problem must be solved for each direction for which a solution is required. There will generally be three solutions: a single *quasi-P wave* (*qP*) and two *quasi-S waves* (*qS*).

A useful aid for visualizing the allowed slownesses in anisotropic media is the *slowness surface*. This is constructed by connecting all the points in slowness space that satisfy (11.26). For isotropic media, the slowness surface consists of three concentric spheres, one with radius u_α and two, which are coincident, with

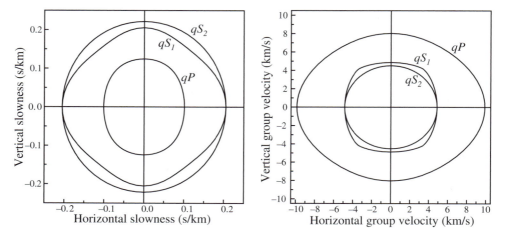

Figure 11.8 The slowness (inverse phase velocity) and group velocity surfaces computed for an example of anisotropic material with hexagonal symmetry (x_1 symmetry axis). The group velocity surfaces are the same shape as the wavefronts that would be generated by a point source. Elastic constants are $\Gamma_{1111} = 97.77$, $\Gamma_{2222} = 64.48$, $\Gamma_{1313} = 23.72$, $\Gamma_{2323} = 20.39$, and $\Gamma_{1122} = 20.83$ km^2 s^{-2}. These values are computed from the Kumazawa and Anderson (1969) measurements of olivine, assuming a fixed olivine a axis and random b and c axis orientations (see Shearer and Orcutt, 1986).

radius u_β. However, in general the three slowness surfaces are separate and non-spherical (see Fig. 11.8), and they can have many interesting and complex shapes. The slowness surface is a measure of the directional dependence of the phase velocity, and the phase velocity direction is always perpendicular to the wavefront. It can also be shown that the group velocity direction is perpendicular to the slowness surface. Thus, the wavefront and the slowness surface have an interesting reciprocal relationship (Fig. 11.9).

In principle we could compute the group velocity directions graphically from the slowness surface. However, in practice it is easier to use (e.g., Cerveny, 2001; Chapman, 2004)

$$U_i = \Gamma_{ijkl}\hat{g}_j s_k \hat{g}_l,$$

(11.29)

where $\mathbf{U}$ is the group velocity vector, $\mathbf{s}$ is the slowness vector, and $\hat{\mathbf{g}}$ is the unit normalized polarization vector. Note that $\mathbf{s}$ and $\hat{\mathbf{g}}$ can be obtained from the eigenvalue equation (11.28). The group velocity and slowness vectors are related by

$$U_i s_i = 1 .$$

(11.30)

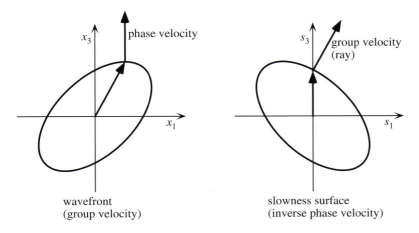

Figure 11.9 The relationship between the wavefront and slowness surface for anisotropic material.

11.3.1 Snell's law at an interface

We can determine the plane waves that couple at a horizontal interface in the same way that we did for isotropic media in Chapter 4 – by conserving horizontal slowness. If we rotate our coordinate system such that $\mathbf{s} = (p, 0, s_3)$ where p is the horizontal slowness s_1, we can then plot the slowness surfaces on both sides of the interface in the ps_3 plane (also called the *sagittal* plane). For a given horizontal slowness p there are six solutions for the vertical slowness s_3, corresponding to upgoing and downgoing qP and qS waves (Fig. 11.10a). At an interface, p must be conserved so these are the six possible solutions for the reflected and transmitted waves at an interface. As for isotropic media, values of p larger than a particular slowness surface represent waves that have turned above the layer; solutions in this case still exist but are imaginary, representing evanescent waves.

Note that the actual ray (group velocity) directions are generally not the same as the slowness directions and need not be confined to the sagittal plane. Various pathological examples are possible, such as that shown in Figure 11.10b. In this case there are four solutions for vertical slowness from a single slowness surface and the phase velocity vector can be downward pointing even while the ray direction is upward. More details concerning this example are contained in Shearer and Chapman (1989).

11.3.2 Weak anisotropy

For weakly anisotropic material, Backus (1965) used first-order perturbation theory to show that qP wave phase and group velocity variations within a plane are nearly

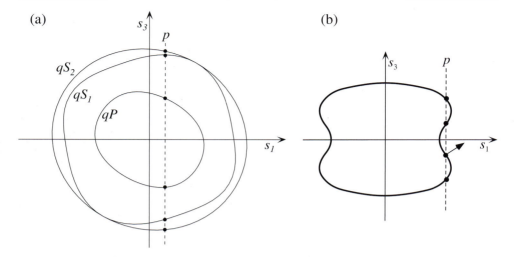

Figure 11.10 (a) For a given value of the horizontal slowness p, there are six possible values for the vertical slowness. (b) For severely anisotropic material, parts of the slowness surface can be concave outward, producing multiple solutions for the vertical slowness.

equal and can be approximated as

$$V_p^2 = A + B\cos 2\theta + C\sin 2\theta + D\cos 4\theta + E\sin 4\theta, \tag{11.31}$$

where θ is the azimuth of wave propagation. The constants A, B, C, D, and E are related to the components of the elastic tensor by

$$
\begin{aligned}
8A &= 3\Gamma_{1111} + 3\Gamma_{2222} + 2\Gamma_{1122} + 4\Gamma_{1212}, \\
2B &= \Gamma_{1111} - \Gamma_{2222}, \\
C &= \Gamma_{1112} + \Gamma_{2212}, \\
8D &= \Gamma_{1111} + \Gamma_{2222} - 2\Gamma_{1122} - 4\Gamma_{1212}, \\
2E &= \Gamma_{1112} - \Gamma_{2212}, \tag{11.32}
\end{aligned}
$$

where Γ_{ijkl} is the density-normalized elastic tensor and we have assumed that θ is measured from the x_1 axis in the plane perpendicular to the x_3 axis. There are similar equations for qS-wave velocity variations. These equations have become widely used because they provide a way to model anisotropy by fitting azimuthal travel time variations with simple 2θ and 4θ curves.

Perhaps the first definitive evidence of seismic anisotropy in the Earth was found by Hess (1964), who noticed that upper mantle velocities as measured by Pn arrival times from marine refraction experiments in the Pacific tended to be faster in directions parallel to the fossil spreading direction. Subsequent observations from

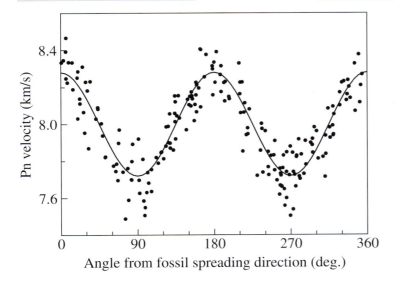

Figure 11.11 A cartoon illustrating typical observations of oceanic upper mantle *Pn* anisotropy. The azimuthal dependence of velocity is well-described with a cos 2θ curve, where θ is the angle from the original plate spreading direction.

additional experiments (e.g., Raitt *et al.*, 1969) have confirmed this relationship and *Pn* anisotropy has now been observed at a large number of sites around the world, including both oceanic and continental regions (e.g., Bamford, 1977). The anisotropic velocity variations are well modeled with the Backus theory for weak anisotropy, with the 2θ terms being dominant (Figure 11.11). Uppermost mantle anisotropy beneath the oceans most likely results from a preferred alignment of olivine crystals within the mantle that occurred during the formation of oceanic lithosphere at spreading ridges. Such alignments have also been observed in equivalent upper mantle material from ophiolites (pieces of old oceanic crust and upper mantle now exposed on land).

11.3.3 Shear-wave splitting

In isotropic media, *SV* and *SH* waves travel at the same speed. However, in anisotropic media, the two quasi-shear waves will typically travel at different speeds and arrive at slightly different times. Since the polarizations of the two *qS* waves are approximately orthogonal this can lead to *shear-wave splitting* in which a pulse will be split into two orthogonal components that will arrive at slightly different times (see Fig. 11.12).

If the time separation between the split shear waves is greater than the duration of the original pulse, then two distinct arrivals will be observed. Shear-wave splitting

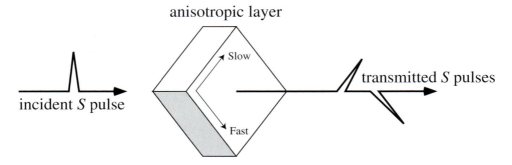

Figure 11.12 An *S* wave that travels through an anisotropic layer can split into two *S* waves with orthogonal polarizations; this is due to the difference in speed between the *qS* waves in the anisotropic material.

has been observed for both crustal and upper mantle anisotropy. Particularly good examples of shear-wave splitting are seen in teleseismic *SKS* arrivals (e.g., Kind *et al.*, 1985; Silver and Chan, 1991). In this case, the polarization of the incident *S*-wave is known to be purely *SV*, as a result of the *P* leg traveling in the fluid outer core. The delay times between the split shear waves are typically 1 to 2 seconds for *SKS* waves; the splitting is most likely caused primarily by azimuthal anisotropy within the upper mantle beneath the stations.

Shear-wave splitting is an especially valuable way to observe anisotropy, because anisotropy can often be detected on a single seismogram. In contrast, travel time analyses typically require a large number of sources and receivers at different positions in order to observe directional velocity variations, and, even with good data coverage, it can be difficult to separate the effects of anisotropy from lateral heterogeneity. Shear-wave splitting, however, is uniquely diagnostic of anisotropy because it is very difficult to produce the observed effect (two shear pulses of similar shape, observed on orthogonal components at slightly different times) without anisotropy somewhere along the ray path.

Shear-wave splitting observations typically provide two numbers – the direction of the fast *qS* polarization and the delay time between the two *qS* pulses, from which the orientation and strength of the anisotropy can be inferred. However, there is typically a tradeoff in *SKS* splitting analyses between the strength and thickness of the anisotropic layer; a thin highly anisotropic layer can produce the same delay time as a thick weakly anisotropic layer. Receiver function analysis can sometimes provide better depth resolution if splitting can be identified in *Ps* converted phases from discontinuities, because the splitting is restricted to the layer above the discontinuity. This provides a means to discriminate between anisotropy in the crust and upper mantle.

11.3.4 Hexagonal anisotropy

The most general form of anisotropy has 21 free parameters that are consistent with the required symmetries in the elastic tensor, c_{ijkl}. The number of independent elastic constants is reduced if additional symmetries are present in the elastic tensor. In seismology, the most commonly considered of these specialized cases has rotational symmetry about a vertical axis and is termed *transverse isotropy* because velocities vary only with incidence angle and not with azimuth (Figure 11.13).

In this case, c_{ijkl} contains only five independent elastic constants. Transverse isotropy is a particularly simple form of anisotropy to study because the two quasi-shear wave polarizations correspond to *SH* and *SV*, and the *qSH* and *qP/qSV* systems can be treated separately in the case of horizontally layered media. Transverse isotropy is often found in reflection seismic surveys of sedimentary layering in the crust, and this can bias estimates to reflector depths that are based on isotropic models. The upper mantle is sometimes modeled as transversely isotropic to explain discrepancies between Love and Rayleigh surface wave velocities (e.g., Forsyth, 1975). The PREM model (Dziewonski and Anderson, 1981) is transversely isotropic between 80 and 220 km depth, with *SH* waves traveling slightly faster than *SV* waves (Figure 11.14). The predicted shear-wave splitting in PREM is about 5 s at source–receiver ranges between 13° and 20° but is reduced at longer ranges where the rays pass more steeply through the anisotropic layer. No splitting is predicted for vertically traveling *S* waves propagating through transversely isotropic media. Note that this form of anisotropy does not remove the spherical symmetry in PREM, as the local symmetry axis always points in the radial direction.

The name "transverse isotropy" is somewhat unfortunate, because it only indirectly describes the nature of the anisotropy, but for historical reasons and lack of a

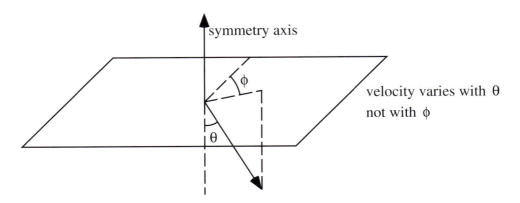

Figure 11.13 Seismic velocities in transversely isotropic materials vary only with incidence angle, not with azimuth. There is rotational symmetry about the vertical axis.

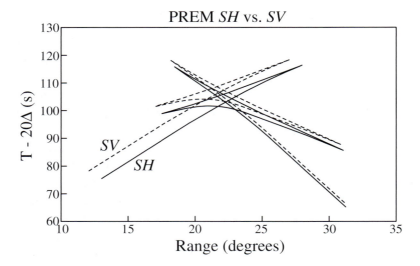

PREM *SH* vs. *SV*

Figure 11.14 *SH* and *SV* travel time curves as predicted by the transversely isotropic PREM model (plot adapted from Shearer, 1991). The triplications in the travel time curves are due to the 400 and 670-km velocity discontinuities.

better label, it is likely to continue to be used. Strictly speaking, the term transverse isotropy should be used only for the case where the symmetry axis is vertical, but this restriction is often ignored. A more general name for transverse isotropy with an arbitrary symmetry axis orientation is *hexagonally symmetric anisotropy* and this is used in the crystallography literature (e.g., Musgrave, 1970).

General hexagonal anisotropy is specified by a symmetry axis direction and five independent elastic constants (for a total of seven free parameters). Velocities vary only with angle from the symmetry axis; there is rotational symmetry about the symmetry axis. For a non-vertical symmetry axis, velocities will vary with azimuth (this is termed *azimuthal anisotropy*) and the qS polarizations in general do not correspond to *SH* and *SV*. For hexagonal anisotropy with an x_1 symmetry axis, we have the following elastic constants (e.g., Musgrave, 1970):

$$c_{1111},$$
$$c_{2222} = c_{3333},$$
$$c_{2233},$$
$$c_{1212} = c_{1313},$$
$$c_{1122} = c_{1133},$$
$$2c_{2323} = c_{3333} - c_{2233}. \tag{11.33}$$

11.3.5 Mechanisms for anisotropy

Anisotropy in rocks can arise in two fundamentally different ways, which have been termed: (1) *shape-preferred orientation* (SPO), in which the material is isotropic at very fine scales, but has anisotropic heterogeneity, which causes seismic anisotropy at long wavelengths compared to the scale of the heterogeneity, and (2) *lattice-preferred orientation* (LPO), in which the anisotropy arises from a preferred orientation of intrinsically anisotropic mineral crystals.

Examples of SPO anisotropy include thin alternating layers of fast and slow material or small aligned cracks within an isotropic rock, which can cause effective hexagonal anisotropy when the seismic wavelength is substantially greater than the layer or crack spacing (Fig. 11.15). In these cases the slow qP direction is parallel to the symmetry axis, while the fast directions are parallel to the layers or cracks. For alternating layer anisotropy (sometimes termed periodic thin-layer, or PTL, anisotropy), relationships between the layer parameters and the effective elastic constants were derived by Backus (1962). This mechanism is not very efficient at generating anisotropy because quite large velocity contrasts are required between the layers to produce significant anisotropic velocity variations at long wavelengths. However, this type of anisotropy is sometimes important in sedimentary rocks, where, for example, alternating layers of sandstone and shale might be present.

Preferred crack orientation is a more efficient way to produce anisotropy, and large porosities are not required because the cracks can be very thin. Relationships between crack parameters and the elastic constants for the bulk material are complex; theoretical results have been derived by Hudson (1980) and other authors. The strength of the anisotropy depends upon the crack density, the crack aspect

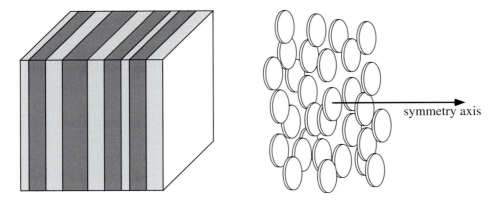

Figure 11.15 Two SPO mechanisms that can cause hexagonal anisotropy at long wavelengths in material that is isotropic at small scales: (left) alternating layers of fast and slow material, and (right) randomly distributed cracks with a single preferred orientation.

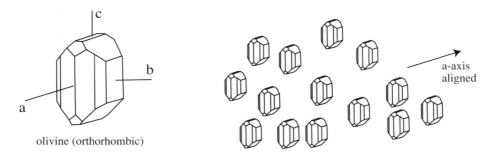

Figure 11.16 Individual olivine crystals are highly anisotropic and will cause bulk anisotropy in rocks if they crystalize with a preferred alignment of one or more of their axes. The fastest *P*-wave velocities occur in the direction of the *a* axis, which tends to align parallel to the fossil spreading direction in the oceanic upper mantle and cause *Pn* anisotropy. If the *b* and *c* axes are randomly aligned, the resulting material is hexagonally symmetric with a horizontal symmetry axis.

ratio, whether the cracks are wet or dry, and the degree of crack alignment. Cracks with no preferred orientation will not cause anisotropy but will reduce the isotropic *P* and *S* velocities of the bulk material. Aligned cracks provide the most likely explanation for many observations of azimuthal anisotropy in the shallow crust.

Individual crystals are typically highly anisotropic, with seismic velocities often differing by 20% or more among the different directions defined by the crystal symmetry axes. For example, olivine crystals are orthorhombic with an *a*-axis qP velocity of 9.89 km/s, *b*-axis velocity of 7.73 km/s, and *c*-axis velocity of 8.43 km/s (Kumazawa and Anderson, 1969). Isotropic bulk properties can nonetheless result if the crystals are randomly aligned within a rock, but if some fraction of the crystals have a preferred alignment then anisotropy is produced, as illustrated for olivine in Figure 11.16. The form of the anisotropy can be as complicated as that within the individual crystals. Preferred mineral alignments can produce hexagonal anisotropy if there is a single preferred orientation direction, with random orientations in the other two coordinates, even when the individual crystals have more general anisotropy. Azimuthal anisotropy is required to explain shear-wave splitting observed in teleseismic *S* and *SKS* arrivals and is often thought to result from a preferred alignment of olivine crystals in the upper mantle, a model that can also explain observations of *Pn* anisotropy. In this case the fast direction for horizontally traveling *P* waves corresponds to the polarization direction of the faster of the two vertically traveling qS waves.

11.3.6 Earth's anisotropy

Anisotropy is observed in Earth at a number of different depths, arising from a variety of different mechanisms (e.g., Maupin and Park, 2007). In the shallow crust, transverse anisotropy (i.e., a vertical symmetry axis) is often seen in reflection seismology data and can be caused by, for example, fine-scale horizontal layering of sandstone and shale. In this case, the velocity for vertically propagating P waves will be less than the velocity for horizontally propagating waves and the moveout equations of Chapter 7 will produce inaccurate depth estimates to the reflectors unless the anisotropy is taken into account. Azimuthal anisotropy in the shallow crust is also observed, most often in shear-wave splitting observations, and can result from preferred crack or mineral alignment. For material with thin cracks of random orientation, one should expect more "open" cracks (which affect the seismic velocities) for crack orientations perpendicular to the axis of minimum compressional stress and this has indeed been observed in laboratory experiments (Nur and Simmons, 1969). Open vertical cracks in a strike–slip regime will therefore tend to align parallel to the maximum compressive stress direction. In this case the fast polarization direction in crustal shear-wave splitting observations will align with the crack direction and this is sometime used as a measure of the direction of maximum compression in the crustal stress field. The anisotropy symmetry axis is horizontal and corresponds to the slow qP direction. Anisotropy in the mid- to lower-crust is also sometimes observed, most likely a result of preferred alignment of minerals such as biotite and hornblende.

Azimuthal upper-mantle anisotropy is observed in both Pn travel time studies and SKS splitting studies and is thought to result mainly from a preferred alignment of olivine crystals. In most cases, the observations can be largely explained with hexagonally symmetric models with a near-horizontal symmetry axis corresponding to the fast P-wave direction, although details in the observations sometimes suggest a dipping symmetry axis or a two-layer anisotropy model (Schulte-Pelkum and Blackman, 2003). It should be noted that Pn is sensitive to the very top of the mantle whereas SKS splitting delay times typically integrate the effects of azimuthal anisotropy within the top few hundreds of kilometers. Global surface-wave observations generally require transverse isotropy in the upper mantle with a vertical symmetry axis (corresponding to the slow P-wave direction) to explain both Love and Rayleigh wave data. For example, the PREM model is transversely isotropic between 24.4 km and 220 km. Transversely isotropic models predict no Pn anisotropy or SKS splitting, so at first glance there might appear to be a contradiction between the surface-wave results and other data sets. However, surface waves propagating over long paths will tend to average out the azimuthal anisotropy that may be present locally at varying orientations, resulting in globally averaged material properties that appear transversely isotropic. Many surface-wave tomography studies now solve for lateral variations in azimuthal anisotropy in the upper mantle, and these results show some agreement with SKS splitting results, although the limited lateral resolution of the surface-wave models makes detailed comparisons difficult (e.g., Montagner *et al.*, 2000). Developing upper-mantle tomography models that fully incorporate 3D velocity variations and general anisotropy is a challenging task, given the large number of additional parameters than anisotropic models require. However, resolving the orientation and strength of upper-mantle anisotropy is important because anisotropy provides valuable insights into mantle fabric and past and current deformation and geodynamic processes (see Savage, 1999; Fouch and Rondenay, 2006, for reviews of mantle anisotropy). The increasing seismic station density in many regions should help to resolve anisotropy on much finer scales than has previously been possible, at least beneath continents.

Anisotropy is observed to be relatively weak in the deeper mantle below 300 km and weak to nonexistent in the mid mantle below the 660-km discontinuity and above the D'' region near the core (e.g., Karato, 1998). The absence of mid-mantle anisotropy is consistent with some mineral physics results that predict the lack of preferred crystal alignment during deformation of silicate perovskite at mid-mantle conditions (Meade *et al.*, 1995). However, anisotropy is again observed in the bottom few hundred kilometers of the mantle in the D'' region, based on shear-wave splitting observations (e.g., Kendall and Silver, 1996; Moore *et al.*, 2004). Splitting for paths within D'' can be complicated but usually occurs with SH velocities faster than SV velocities ($V_{SH} > V_{SV}$), similar to the globally averaged anisotropy (i.e., transverse isotropy) in the upper mantle for which SH waves are also faster (see Fig. 11.14). Proposed mechanisms for D'' anisotropy include SPO models with partial melt (Kendall and Silver, 1996) or aligned perovskite plus magnesiowüstite aggregates (Karato, 1998). The fluid outer core appears to be isotropic, but anisotropy was discovered in the solid inner core from observations of PKP travel times and anomalous splitting of normal modes (Morelli *et al.*, 1986; Woodhouse *et al.*, 1986), which indicate the P velocity is fastest for directions parallel to the rotation axis. Inner-core anisotropy is believed to be due to a preferred orientation of iron crystals within the inner core, but the mechanism for causing this alignment is uncertain (see Song, 1997, for a review of inner-core anisotropy).

11.4 Exercises

1. Show that the amplitude and phase of the complex function $\mathcal{Z}(\omega)$ in equation (11.12) are given by

$$|\mathcal{Z}(\omega)| = \frac{\omega^2}{\sqrt{(\omega^2 - \omega_0^2)^2 + 4\epsilon^2\omega^2}}, \tag{11.34}$$

$$\phi(\omega) = \pi - \tan^{-1}\frac{2\epsilon\omega}{\omega^2 - \omega_0^2}. \tag{11.35}$$

2. (COMPUTER) Use equation (11.12) to compute response curves at frequencies between 0.001 and 1 Hz. Plot your results for: (a) the amplitude response curve (sensor displacement as a function of ground displacement) for a seismometer of 20 s natural period with damping constant $h = 0.7$ and (b) the mass displacement as a function of ground acceleration response curve for the same sensor.

3. What will be the displacement of a suspended mass of 20 s natural period subject to a change in gravity of 1 milligal ($= 10^{-5}$ m/s^2)?

4. Jake is driving his new sport utility vehicle when a large earthquake occurs. Assume his suspension is critically damped with a natural period of 1 Hz and a travel of ± 20 cm.

   (a) How large a vertical acceleration at 1 Hz can he experience before his suspension bottoms out? Express your answer in g ($1\,g = 9.8$ m/s^2).
   (b) How much closer to the ground would his vehicle ride if he were to travel to a planet where the gravity is 20% stronger than on Earth?

5. From Figure 11.8, what is the approximate difference in speeds between two vertically traveling qS waves in this material? Assuming that the upper mantle contains an anisotropic layer with pure olivine aligned as in this plot, how thick would the layer have to be to produce shear-wave splitting in vertical SKS waves with a 2 s delay time between the split shear waves?

6. From equation (11.31), sketch the azimuthal variation of qP waves for the following non-zero values of the harmonic constants (in km^2/s^2): (a) $A = 64$, $B = 8$, (b) $A = 64$, $D = 8$, (c) $A = 64$, $B = 4$, $D = 4$.

7. For hexagonal anisotropy, how many of the 81 components of c_{ijkl} have zero values?

8. For the values of Γ_{ijkl} listed in Figure 11.8, compute the Backus coefficients A, B, C, D, and E in equation (11.32). You will need to use the relationships for hexagonal symmetry listed in equation (11.33). Make a plot of P velocity versus azimuth.

9. (COMPUTER) Assume density-normalized elastic constants for elastic wave propagation within an iron crystal. This crystal has cubic symmetry with $\Gamma_{1111} = \Gamma_{2222} = \Gamma_{3333} = 29.64$ km^2 s^{-2}, $\Gamma_{1122} = \Gamma_{1133} = \Gamma_{2233} = 17.71$ km^2 s^{-2}, and $\Gamma_{2323} = \Gamma_{1313} = \Gamma_{1212} = 14.78$ km^2 s^{-2} (values from Musgrave, 1970).

   (a) Using the symmetry relationships (11.19), fill in all 81 values of Γ_{ijkl}. Here is a handy subroutine that does this:

```
SUBROUTINE FILLOUT(c)
real c(3,3,3,3)
do i=1,3
do j=1,3
```

```
do k=1,3
do l=1,3
   if (c(i,j,k,l).eq.0.) c(i,j,k,l)=c(j,i,k,l)
   if (c(i,j,k,l).eq.0.) c(i,j,k,l)=c(i,j,l,k)
   if (c(i,j,k,l).eq.0.) c(i,j,k,l)=c(j,i,l,k)
   if (c(i,j,k,l).eq.0.) c(i,j,k,l)=c(k,l,i,j)
   if (c(i,j,k,l).eq.0.) c(i,j,k,l)=c(l,k,i,j)
   if (c(i,j,k,l).eq.0.) c(i,j,k,l)=c(k,l,j,i)
   if (c(i,j,k,l).eq.0.) c(i,j,k,l)=c(l,k,j,i)
enddo
enddo
enddo
enddo
return
end
```

(b) Compute and plot the slowness surfaces in the s_1–s_3 plane. Label the qP and two qS waves. Note: You will need to construct the M_{ik} matrix at a range of slowness directions using equation (11.27) and then find the eigenvalues of this matrix with an appropriate subroutine or software package. Depending upon how the eigenvalues are sorted, you may find it difficult to draw lines between the points without switching between the different surfaces. If so, simply plot a symbol at each point and do not attempt to connect the points.

(c) Compute and plot the wavefronts resulting from a point source after 1 s. You will need to use equation (11.29).

(d) Test your program by repeating steps (a)–(c) for the olivine model of Figure 11.8, using the elastic constants listed in the figure caption and the hexagonal symmetry conditions given in equations (11.33).

Appendix A
The PREM model

For many years the most widely used 1-D model of Earth's seismic velocities has been the Preliminary Reference Earth Model (PREM) of Dziewonski and Anderson (1981). This model was designed to fit a variety of different data sets, including free oscillation center frequency measurements, surface wave dispersion observations, travel time data for a number of body-wave phases, and basic astronomical data (Earth's radius, mass, and moment of inertia). In addition to profiling the P and S velocities, PREM specifies density and attenuation as functions of depth. Although these parameters are known less precisely than the seismic velocities, including them is important because it makes the model complete and suitable for use as a reference to compute synthetic seismograms without requiring additional assumptions. In order to simultaneously fit Love and Rayleigh wave observations, PREM is transversely isotropic between 80 and 220 km depth in the upper mantle. This is a spherically symmetric form of anisotropy in which SH and SV waves travel at different speeds. For simplicity, the table here lists only values from an isotropic version of PREM. The true PREM model is also specified in terms of polynomials between node points; linear interpolation between the 100 km spacing of values in this table will produce only approximate results. All current Earth models have values that are reasonably close to PREM; the largest differences are in the upper mantle, where, for example, a discontinuity at 220 km is not found in most models.

Table A.1: Preliminary reference Earth model (isotropic version)							
Depth (km)	Radius (km)	V_p (km/s)	V_s (km/s)	ρ (g/cc)	Q_μ	Q_κ	P (GPa)
0.0	6371.0	1.45	0.00	1.02	0.0	57823.0	0.0
3.0	6368.0	1.45	0.00	1.02	0.0	57823.0	0.0
3.0	6368.0	5.80	3.20	2.60	600.0	57823.0	0.0
15.0	6356.0	5.80	3.20	2.60	600.0	57823.0	0.3
15.0	6356.0	6.80	3.90	2.90	600.0	57823.0	0.3
24.4	6346.6	6.80	3.90	2.90	600.0	57823.0	0.6
24.4	6346.6	8.11	4.49	3.38	600.0	57823.0	0.6
71.0	6300.0	8.08	4.47	3.38	600.0	57823.0	2.2
80.0	6291.0	8.08	4.47	3.37	600.0	57823.0	2.5
80.0	6291.0	8.08	4.47	3.37	80.0	57823.0	2.5
171.0	6200.0	8.02	4.44	3.36	80.0	57823.0	5.5
220.0	6151.0	7.99	4.42	3.36	80.0	57823.0	7.1
220.0	6151.0	8.56	4.64	3.44	143.0	57823.0	7.1
271.0	6100.0	8.66	4.68	3.47	143.0	57823.0	8.9
371.0	6000.0	8.85	4.75	3.53	143.0	57823.0	12.3
400.0	5971.0	8.91	4.77	3.54	143.0	57823.0	13.4
400.0	5971.0	9.13	4.93	3.72	143.0	57823.0	13.4
471.0	5900.0	9.50	5.14	3.81	143.0	57823.0	16.0
571.0	5800.0	10.01	5.43	3.94	143.0	57823.0	19.9
600.0	5771.0	10.16	5.52	3.98	143.0	57823.0	21.0
600.0	5771.0	10.16	5.52	3.98	143.0	57823.0	21.0
670.0	5701.0	10.27	5.57	3.99	143.0	57823.0	23.8
670.0	5701.0	10.75	5.95	4.38	312.0	57823.0	23.8
771.0	5600.0	11.07	6.24	4.44	312.0	57823.0	28.3
871.0	5500.0	11.24	6.31	4.50	312.0	57823.0	32.8
971.0	5400.0	11.42	6.38	4.56	312.0	57823.0	37.3
1071.0	5300.0	11.58	6.44	4.62	312.0	57823.0	41.9
1171.0	5200.0	11.73	6.50	4.68	312.0	57823.0	46.5
1271.0	5100.0	11.88	6.56	4.73	312.0	57823.0	51.2
1371.0	5000.0	12.02	6.62	4.79	312.0	57823.0	55.9
1471.0	4900.0	12.16	6.67	4.84	312.0	57823.0	60.7
1571.0	4800.0	12.29	6.73	4.90	312.0	57823.0	65.5
1671.0	4700.0	12.42	6.78	4.95	312.0	57823.0	70.4
1771.0	4600.0	12.54	6.83	5.00	312.0	57823.0	75.4
1871.0	4500.0	12.67	6.87	5.05	312.0	57823.0	80.4
1971.0	4400.0	12.78	6.92	5.11	312.0	57823.0	85.5
2071.0	4300.0	12.90	6.97	5.16	312.0	57823.0	90.6
2171.0	4200.0	13.02	7.01	5.21	312.0	57823.0	95.8
2271.0	4100.0	13.13	7.06	5.26	312.0	57823.0	101.1
2371.0	4000.0	13.25	7.10	5.31	312.0	57823.0	106.4
2471.0	3900.0	13.36	7.14	5.36	312.0	57823.0	111.9
2571.0	3800.0	13.48	7.19	5.41	312.0	57823.0	117.4
2671.0	3700.0	13.60	7.23	5.46	312.0	57823.0	123.0
2741.0	3630.0	13.68	7.27	5.49	312.0	57823.0	127.0
2771.0	3600.0	13.69	7.27	5.51	312.0	57823.0	128.8
2871.0	3500.0	13.71	7.26	5.56	312.0	57823.0	134.6
2891.0	3480.0	13.72	7.26	5.57	312.0	57823.0	135.8
2891.0	3480.0	8.06	0.00	9.90	0.0	57823.0	135.8
2971.0	3400.0	8.20	0.00	10.03	0.0	57823.0	144.2
3071.0	3300.0	8.36	0.00	10.18	0.0	57823.0	154.8
3171.0	3200.0	8.51	0.00	10.33	0.0	57823.0	165.2

Table A.1: Continued							
Depth (km)	Radius (km)	V_p (km/s)	V_s (km/s)	ρ (g/cc)	Q_μ	Q_κ	P (GPa)
3271.0	3100.0	8.66	0.00	10.47	0.0	57823.0	175.5
3371.0	3000.0	8.80	0.00	10.60	0.0	57823.0	185.7
3471.0	2900.0	8.93	0.00	10.73	0.0	57823.0	195.8
3571.0	2800.0	9.05	0.00	10.85	0.0	57823.0	205.7
3671.0	2700.0	9.17	0.00	10.97	0.0	57823.0	215.4
3771.0	2600.0	9.28	0.00	11.08	0.0	57823.0	224.9
3871.0	2500.0	9.38	0.00	11.19	0.0	57823.0	234.2
3971.0	2400.0	9.48	0.00	11.29	0.0	57823.0	243.3
4071.0	2300.0	9.58	0.00	11.39	0.0	57823.0	252.2
4171.0	2200.0	9.67	0.00	11.48	0.0	57823.0	260.8
4271.0	2100.0	9.75	0.00	11.57	0.0	57823.0	269.1
4371.0	2000.0	9.84	0.00	11.65	0.0	57823.0	277.1
4471.0	1900.0	9.91	0.00	11.73	0.0	57823.0	284.9
4571.0	1800.0	9.99	0.00	11.81	0.0	57823.0	292.3
4671.0	1700.0	10.06	0.00	11.88	0.0	57823.0	299.5
4771.0	1600.0	10.12	0.00	11.95	0.0	57823.0	306.2
4871.0	1500.0	10.19	0.00	12.01	0.0	57823.0	312.7
4971.0	1400.0	10.25	0.00	12.07	0.0	57823.0	318.9
5071.0	1300.0	10.31	0.00	12.12	0.0	57823.0	324.7
5149.5	1221.5	10.36	0.00	12.17	0.0	57823.0	329.0
5149.5	1221.5	11.03	3.50	12.76	84.6	1327.7	329.0
5171.0	1200.0	11.04	3.51	12.77	84.6	1327.7	330.2
5271.0	1100.0	11.07	3.54	12.82	84.6	1327.7	335.5
5371.0	1000.0	11.11	3.56	12.87	84.6	1327.7	340.4
5471.0	900.0	11.14	3.58	12.91	84.6	1327.7	344.8
5571.0	800.0	11.16	3.60	12.95	84.6	1327.7	348.8
5671.0	700.0	11.19	3.61	12.98	84.6	1327.7	352.3
5771.0	600.0	11.21	3.63	13.01	84.6	1327.7	355.4
5871.0	500.0	11.22	3.64	13.03	84.6	1327.7	358.0
5971.0	400.0	11.24	3.65	13.05	84.6	1327.7	360.2
6071.0	300.0	11.25	3.66	13.07	84.6	1327.7	361.8
6171.0	200.0	11.26	3.66	13.08	84.6	1327.7	363.0
6271.0	100.0	11.26	3.67	13.09	84.6	1327.7	363.7
6371.0	0.0	11.26	3.67	13.09	84.6	1327.7	364.0

Appendix B
Math review

This appendix is not intended to teach anyone vector calculus or complex number theory, but simply to list some of the important definitions to assist those whose skills may have become rusty. Many of the equations are expressed in both standard vector notation and the index notation used in this book.

B.1 Vector calculus

Consider a Cartesian coordinate system with x, y, and z axes. The length, or magnitude, of a vector $\mathbf{u}$ is written as $\|\mathbf{u}\|$. A vector may be expressed in terms of its components as

$$\mathbf{u} = u_x \hat{\mathbf{x}} + u_y \hat{\mathbf{y}} + u_z \hat{\mathbf{z}}, \tag{B.1}$$

where $\hat{\mathbf{x}}$, $\hat{\mathbf{y}}$, and $\hat{\mathbf{z}}$ are unit length vectors in the x, y, and z directions. The dot product of two vectors is a scalar (a single number) and is defined as

$$\lambda = \mathbf{u} \cdot \mathbf{v} = \|\mathbf{u}\| \|\mathbf{v}\| \cos\theta \tag{B.2}$$

$$= u_x v_x + u_y v_y + u_z v_z, \tag{B.3}$$

where θ is the angle between the two vectors. It follows that $\mathbf{u} \cdot \mathbf{v} = 0$ when $\mathbf{u}$ and $\mathbf{v}$ are orthogonal and that for unit vectors $\hat{\mathbf{x}} \cdot \hat{\mathbf{x}} = 1$. Note that for a unit vector $\hat{\mathbf{u}}$, the dot product $\hat{\mathbf{u}} \cdot \mathbf{v}$ gives the length of the orthogonal projection of $\mathbf{v}$ onto $\hat{\mathbf{u}}$.

The cross-product between two vectors is a third vector that points in a direction perpendicular to both (according to the right-hand rule). The cross-product can be expressed in component form as

$$\mathbf{u} \times \mathbf{v} = (u_y v_z - u_z v_y)\hat{\mathbf{x}} + (u_z v_x - u_x v_z)\hat{\mathbf{y}} + (u_x v_y - u_y v_x)\hat{\mathbf{z}}, \tag{B.4}$$

and the length of this vector may be expressed as

$$\|\mathbf{u} \times \mathbf{v}\| = \|\mathbf{u}\| \|\mathbf{v}\| \sin\theta. \tag{B.5}$$

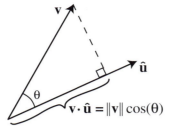

Figure B.1 The dot product of a vector with a unit vector is the length of the projection onto the unit vector.

Note that the dot product is commutative, but not the cross-product, that is

$$\mathbf{u} \cdot \mathbf{v} = \mathbf{v} \cdot \mathbf{u}, \tag{B.6}$$

$$\mathbf{u} \times \mathbf{v} = -(\mathbf{v} \times \mathbf{u}). \tag{B.7}$$

A second-order tensor, $\mathbf{U}$, is a linear operator that produces one vector from another, that is,

$$\mathbf{u} = \mathbf{U}\mathbf{v}, \tag{B.8}$$

$$u_i = U_{ij}v_j \quad \text{(sum over } j = 1, 2, 3).$$

Here we introduce the use of index notation; i and j are assumed to take on the values 1, 2, and 3 for the x, y, and z components, respectively. Notice that in a Cartesian coordinate system, the second-order tensor $\mathbf{U}$ has the form of a 3×3 matrix. We also begin using the *summation convention*; repeated indices in a product are assumed to be summed over values from 1 to 3.

The projection property of the dot product can be used to express a vector in a different (i.e., rotated) Cartesian coordinate system. If the new coordinate axes are defined by the orthogonal unit vectors $\hat{\mathbf{x}}'$, $\hat{\mathbf{y}}'$, and $\hat{\mathbf{z}}'$ (expressed in the original x,y,z coordinates), then the x' coordinate of a vector $\mathbf{v}$ is given by $\hat{\mathbf{x}}' \cdot \mathbf{v}$. In this way the vector in the new coordinate system is given by

$$\mathbf{v}' = \begin{bmatrix} \hat{x}'_1 & \hat{x}'_2 & \hat{x}'_3 \\ \hat{y}'_1 & \hat{y}'_2 & \hat{y}'_3 \\ \hat{z}'_1 & \hat{z}'_2 & \hat{z}'_3 \end{bmatrix} \mathbf{v} \equiv \mathbf{A}\mathbf{v}, \tag{B.9}$$

where $\mathbf{A}$ is the *transformation tensor* with components equal to the cosines of the angles between the primed and unprimed axes. We can express the same equation in index notation as

$$v'_i = A_{ij}v_j. \tag{B.10}$$

Because the rows of $\mathbf{A}$ are orthogonal unit vectors it follows that

$$\mathbf{A}^{\mathrm{T}}\mathbf{A} = \begin{bmatrix} 1 & 0 & 0 \\ 0 & 1 & 0 \\ 0 & 0 & 1 \end{bmatrix} = \mathbf{I}, \tag{B.11}$$

where $\mathbf{I}$ is the identity matrix.

We often will also want a way to transform a Cartesian tensor to a new coordinate system. This can be obtained by applying the transformation tensor $\mathbf{A}$ to both sides of (B.8)

$$\mathbf{u} = \mathbf{U}\mathbf{v} \tag{B.12}$$

$$\mathbf{A}\mathbf{u} = \mathbf{A}\mathbf{U}\mathbf{v} \tag{B.13}$$

$$\mathbf{A}\mathbf{u} = \mathbf{A}\mathbf{U}(\mathbf{A}^{\mathrm{T}}\mathbf{A})\mathbf{v} \tag{B.14}$$

$$\mathbf{A}\mathbf{u} = \mathbf{A}\mathbf{U}\mathbf{A}^{\mathrm{T}}(\mathbf{A}\mathbf{v}) \tag{B.15}$$

$$\mathbf{u}' = \mathbf{A}\mathbf{U}\mathbf{A}^{\mathrm{T}}\mathbf{v}' \tag{B.16}$$

and we see that the tensor operator that produces $\mathbf{u}'$ from $\mathbf{v}'$ in the primed coordinate system is given by

$$\mathbf{U}' = \mathbf{A}\mathbf{U}\mathbf{A}^{\mathrm{T}} \tag{B.17}$$

which we can use to convert $\mathbf{U}$ to $\mathbf{U}'$. In Chapter 2, we use the eigenvector matrix $\mathbf{N}$ to rotate the stress tensor into its principal axes coordinate system. The definition of $\mathbf{N}$ is similar to $\mathbf{A}$ except that the unit vectors are set to the columns rather than the rows. Thus $\mathbf{N}^{\mathrm{T}} = \mathbf{A}$ and in this case the transformation equation is

$$\mathbf{U}' = \mathbf{N}^{\mathrm{T}}\mathbf{U}\mathbf{N} \tag{B.18}$$

Useful matrix identities include

$$\mathbf{A}(\mathbf{B} + \mathbf{C}) = \mathbf{A}\mathbf{B} + \mathbf{A}\mathbf{C} \tag{B.19}$$

$$(\mathbf{A} + \mathbf{B})^{\mathrm{T}} = \mathbf{A}^{\mathrm{T}} + \mathbf{B}^{\mathrm{T}} \tag{B.20}$$

$$(\mathbf{A}\mathbf{B})^{\mathrm{T}} = \mathbf{B}^{\mathrm{T}}\mathbf{A}^{\mathrm{T}} \tag{B.21}$$

$$(\mathbf{A}\mathbf{B})^{-1} = \mathbf{B}^{-1}\mathbf{A}^{-1} \tag{B.22}$$

$$(\mathbf{A}^{-1})^{\mathrm{T}} = (\mathbf{A}^{\mathrm{T}})^{-1} \tag{B.23}$$

where for the last two we assume the existence of inverses of $\mathbf{A}$ and $\mathbf{B}$.

Functions that vary with position are termed *fields*; we can have scalar fields, vector fields, and tensor fields. In this case we may define spatial derivatives, such as the gradient, divergence, Laplacian, and curl.

The *gradient* of a scalar field, written $\nabla\lambda$, is a vector field, defined by the partial derivatives of the scalar in x, y, and z directions:

$$\mathbf{u} = \nabla\lambda = \frac{\partial\lambda}{\partial x}\hat{\mathbf{x}} + \frac{\partial\lambda}{\partial y}\hat{\mathbf{y}} + \frac{\partial\lambda}{\partial z}\hat{\mathbf{z}}, \tag{B.24}$$

$$u_i = \partial_i\lambda,$$

where ∂_i is shorthand notation for $\partial/\partial x$, $\partial/\partial y$, and $\partial/\partial z$ for $i = 1, 2$, and 3 respectively. The gradient vector, $\nabla\lambda$, is normal to surfaces of constant λ.

The gradient of a vector field is a tensor field:

$$\mathbf{U} = \nabla\mathbf{u}, \tag{B.25}$$

$$U_{ij} = \partial_i u_j.$$

The *divergence* of a vector field, written $\nabla \cdot \mathbf{u}$, is a scalar field:

$$\lambda = \nabla \cdot \mathbf{u} = \frac{\partial u_x}{\partial x} + \frac{\partial u_y}{\partial y} + \frac{\partial u_z}{\partial z} \tag{B.26}$$

$$= \partial_i u_i \quad (\text{sum over } i = 1, 2, 3).$$

The divergence of a second-order tensor field is a vector field:

$$\mathbf{u} = \nabla \cdot \mathbf{U}, \tag{B.27}$$

$$u_j = \partial_i U_{ij} \quad (\text{sum over } i = 1, 2, 3).$$

The *Laplacian* of a scalar field, written $\nabla^2\lambda$, is a scalar field:

$$\phi = \nabla^2\lambda = \nabla \cdot \nabla\lambda = \frac{\partial^2\lambda}{\partial x^2} + \frac{\partial^2\lambda}{\partial y^2} + \frac{\partial^2\lambda}{\partial z^2} \tag{B.28}$$

$$= \partial_j\partial_j\lambda \quad (\text{sum over } j = 1, 2, 3).$$

The Laplacian of a vector field is a vector field:

$$\mathbf{u} = \nabla^2\mathbf{v} = \nabla \cdot \nabla\mathbf{v}, \tag{B.29}$$

$$u_i = \partial_j\partial_j v_i \quad (\text{sum over } j = 1, 2, 3).$$

The *curl* of a vector field is a vector field:

$$\mathbf{u} = \nabla \times \mathbf{v} = \left(\frac{\partial v_z}{\partial y} - \frac{\partial v_y}{\partial z}\right)\hat{\mathbf{x}}$$

$$+ \left(\frac{\partial v_x}{\partial z} - \frac{\partial v_z}{\partial x}\right)\hat{\mathbf{y}}$$

$$+ \left(\frac{\partial v_y}{\partial x} - \frac{\partial v_x}{\partial y}\right)\hat{\mathbf{z}}. \tag{B.30}$$

The operator ∇ is distributive, that is,

$$\nabla(\lambda + \phi) \;=\; \nabla\lambda + \nabla\phi, \tag{B.31}$$

$$\nabla \cdot (\mathbf{u} + \mathbf{v}) \;=\; \nabla \cdot \mathbf{u} + \nabla \cdot \mathbf{v}, \tag{B.32}$$

$$\nabla \times (\mathbf{u} + \mathbf{v}) \;=\; \nabla \times \mathbf{u} + \nabla \times \mathbf{v}. \tag{B.33}$$

A vector field defined as the gradient of a scalar field is curl free, that is,

$$\nabla \times (\nabla\lambda) = 0. \tag{B.34}$$

A vector field defined as the curl of another vector field is divergence free, that is,

$$\nabla \cdot (\nabla \times \mathbf{u}) = 0. \tag{B.35}$$

The following identities are often useful:

$$\nabla \cdot \lambda\mathbf{u} \;=\; \lambda\nabla \cdot \mathbf{u} + \mathbf{u} \cdot \nabla\lambda, \tag{B.36}$$

$$\nabla \times \lambda\mathbf{u} \;=\; \lambda\nabla \times \mathbf{u} + \nabla\lambda \times \mathbf{u}, \tag{B.37}$$

$$\nabla \times (\nabla \times \mathbf{u}) \;=\; \nabla\nabla \cdot \mathbf{u} - \nabla^2\mathbf{u}. \tag{B.38}$$

The identity matrix $\mathbf{I}$ can be written in index notation as δ_{ij} where

$$\delta_{ij} = \begin{cases} 1 & \text{for } i = j, \\ 0 & \text{for } i \neq j. \end{cases} \tag{B.39}$$

When δ_{ij} appears as part of a product in equations, it can be used to switch the indices of other terms, that is,

$$\partial_i \delta_{ij} u_k = \partial_j u_k. \tag{B.40}$$

Of great importance in continuum mechanics is *Gauss's theorem*, which equates the volume integral of a vector field to the surface integral of the orthogonal component of the vector field:

$$\int_V \nabla \cdot \mathbf{u}\, dV = \int_S \mathbf{u} \cdot \hat{\mathbf{n}}\, dS, \tag{B.41}$$

where $\hat{\mathbf{n}}$ is the outward normal vector to the surface.

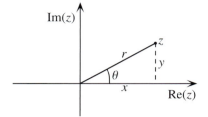

Figure B.2 The complex number z can be represented as a point in the complex plane.

B.2 Complex numbers

We use complex numbers in this book mostly as a shorthand way to keep track of the phase and amplitude of harmonic waves. The imaginary number i is defined as

$$i^2 = -1. \tag{B.42}$$

It follows that

$$\sqrt{-1} = \pm i \quad \text{and} \quad 1/i = -i. \tag{B.43}$$

A *complex number* can be written as

$$z = x + iy, \tag{B.44}$$

where $x = \text{Re}(z)$ is the real part of z and $y = \text{Im}(z)$ is the imaginary part of z (note that y itself is a real number). Complex numbers obey the commutative, associative, and distributive rules of arithmetic. The *complex conjugate* of z is defined as

$$z^* = x - iy. \tag{B.45}$$

Complex numbers may be represented as points on the complex plane (see Fig. B.2), either in Cartesian coordinates by x and y, or in polar coordinates by their *phase*, θ, and their *magnitude*, $r = |z|$. These forms are related by

$$z = re^{i\theta} = r(\cos \theta + i \sin \theta) = x + yi. \tag{B.46}$$

The magnitude $|z|$ is also sometimes referred to as the *absolute value* of z. Note that

$$y/x = \tan \theta \tag{B.47}$$

and that

$$zz^* = (x + iy)(x - iy) = x^2 + y^2 = |z|^2. \tag{B.48}$$

Now let us illustrate the convenience of complex numbers for describing wave motion. A harmonic wave of angular frequency ω is defined by its amplitude a and phase delay ϕ

(Fig. B.3), that is,

$$f(t) = a\cos(\omega t - \phi). \tag{B.49}$$

Using a trigonometric identity for $\cos(\omega t - \phi)$, this can be rewritten

$$
\begin{align}
f(t) &= a\cos\phi\cos\omega t + a\sin\phi\sin\omega t \tag{B.50}\\
&= a_1\cos\omega t + a_2\sin\omega t, \tag{B.51}
\end{align}
$$

where $a_1 \equiv a\cos\phi$ and $a_2 \equiv a\sin\phi$. This is a more convenient form because it is a linear function of the coefficients a_1 and a_2. A harmonic wave of arbitrary phase can always be expressed as a weighted sum of a sine and a cosine function. Two waves of the same frequency can be summed by adding their sine and cosine coefficients. Note that a and ϕ may be recovered from the new coefficients using

$$a^2 = a_1^2 + a_2^2 \quad \text{and} \quad \phi = \tan^{-1}(a_2/a_1). \tag{B.52}$$

We can obtain the same relationships using a single complex coefficient A by writing the function $f(t)$ as a complex exponential function

$$f(t) = \text{Re}\left[Ae^{-i\omega t}\right]. \tag{B.53}$$

Expanding this, we have

$$
\begin{align}
f(t) &= \text{Re}\left[A\left(\cos(-\omega t) + i\sin(-\omega t)\right)\right]\\
&= \text{Re}\left[A(\cos\omega t - i\sin\omega t)\right]. \tag{B.54}
\end{align}
$$

Now consider the real and imaginary parts of $A = x + iy$:

$$f(t) = \text{Re}\left[(x + iy)(\cos\omega t - i\sin\omega t)\right]. \tag{B.55}$$

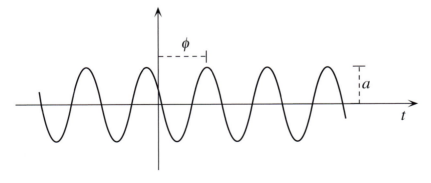

Figure B.3 The amplitude a and phase delay ϕ of a cosine function.

The real terms give

$$f(t) = x \cos \omega t + y \sin \omega t. \tag{B.56}$$

This is identical to (B.51) if we assume

$$a_1 = x = \mathrm{Re}(A), \tag{B.57}$$

$$a_2 = y = \mathrm{Im}(A). \tag{B.58}$$

In this way a single complex number can keep track of both the amplitude and phase of harmonic waves. For convenience, equations such as (B.53) usually do not explicitly include the Re function; in these cases the reader should keep in mind that the real part must always be taken before the equation has a physical meaning. This applies, for example, to equation (3.36) in Chapter 3.

Appendix C
The eikonal equation

Consider the propagation of compressional waves in heterogeneous media. From (3.31), we have

$$\nabla^2 \phi - \frac{1}{\alpha^2} \frac{\partial^2 (\phi)}{\partial t^2} = 0, \tag{C.1}$$

where the scalar potential for compressional waves, ϕ, obeys the relationship $\mathbf{u} = \nabla \phi$ where $\mathbf{u}$ is displacement. The P-wave velocity, α, is a function of position, $\alpha = \alpha(\mathbf{x})$. Now assume a harmonic solution of the form

$$\phi(t) = A(\mathbf{x}) e^{-i\omega[t - T(\mathbf{x})]}, \tag{C.2}$$

where T is a phase factor and A is the local amplitude. We can expand the spatial derivatives of ϕ as

$$\nabla \phi = \nabla A\, e^{-i\omega[t - T(\mathbf{x})]} - i\omega A\, \nabla T e^{-i\omega[t - T(\mathbf{x})]}, \tag{C.3}$$

$$\begin{aligned}
\nabla^2 \phi &= \nabla^2 A\, e^{-i\omega[t - T(\mathbf{x})]} - i\omega \nabla T \cdot \nabla A\, e^{-i\omega[t - T(\mathbf{x})]} \\
&\quad - i\omega\, \nabla A \cdot \nabla T e^{-i\omega[t - T(\mathbf{x})]} - i\omega A\, \nabla^2 T e^{-i\omega[t - T(\mathbf{x})]} \\
&\quad - \omega^2 A \nabla T \cdot \nabla T e^{-i\omega[t - T(\mathbf{x})]} \\
&= \left(\nabla^2 A - \omega^2 A |\nabla T|^2 - i[2\omega \nabla A \cdot \nabla T + \omega A\, \nabla^2 T] \right) e^{-i\omega[t - T(\mathbf{x})]}
\end{aligned} \tag{C.4}$$

and the time derivatives as

$$\frac{\partial^2 (\phi)}{\partial t^2} = -A\omega^2 e^{-i\omega[t - T(\mathbf{x})]}. \tag{C.5}$$

Substituting into (C.1) and dividing out the constant $e^{-i\omega[t - T(\mathbf{x})]}$ factor, we obtain

$$\nabla^2 A - \omega^2 A |\nabla T|^2 - i[2\omega \nabla A \cdot \nabla T + \omega A \nabla^2 T] = -\frac{A\omega^2}{\alpha^2}. \tag{C.6}$$

361

From the real part of this equation we have

$$\nabla^2 A - \omega^2 A |\nabla T|^2 = -\frac{A\omega^2}{\alpha^2} \qquad (C.7)$$

and from the imaginary part we have

$$2\omega\nabla A \cdot \nabla T + \omega A \nabla^2 T = 0$$

or

$$2\nabla A \cdot \nabla T + A\nabla^2 T = 0. \qquad (C.8)$$

Dividing (C.7) by $A\omega^2$ and rearranging, we obtain

$$|\nabla T|^2 - \frac{1}{\alpha^2} = \frac{\nabla^2 A}{A\omega^2}. \qquad (C.9)$$

We now make the high-frequency approximation that ω is sufficiently large that the $1/\omega^2$ term can be ignored. We thus have

$$|\nabla T|^2 = \frac{1}{\alpha^2}. \qquad (C.10)$$

A similar equation can be derived for S waves. Thus a more general form for this equation is

$$|\nabla T|^2 = \frac{1}{c^2}, \qquad (C.11)$$

where c is either the local P-wave speed, α, or the local S-wave speed, β. This is the standard form for the *eikonal equation* (e.g., equation 4.41 in Aki and Richards, 2002). This equation can also be expressed as

$$|\nabla T|^2 = u^2, \qquad (C.12)$$

where $u = 1/c$ is called the *slowness*. Since the velocity, c, typically appears in the denominator in ray tracing equations, we will find that it is usually more convenient to use the slowness. The phase factor, T, is also sometimes called the travel time function. We can write (C.12) in expanded form as

$$|\nabla T|^2 = (\partial_x T)^2 + (\partial_y T)^2 + (\partial_z T)^2 = u^2. \qquad (C.13)$$

Note that the phase factor T has a gradient whose amplitude is equal to the local slowness. The function $T(\mathbf{x}) = constant$ defines surfaces called wavefronts. Lines perpendicular to $T(\mathbf{x})$ or parallel to $\nabla T(\mathbf{x})$ are termed rays. The ray direction is defined by the gradient of T,

$$\nabla T = u\hat{\mathbf{k}} = \mathbf{s}, \qquad (C.14)$$

where $\hat{\mathbf{k}}$ is the unit vector in the local ray direction and $\mathbf{s}$ is the slowness vector. The function $T(\mathbf{x})$ has units of time and, because the wavefronts propagate with the local slowness in a direction parallel to the rays, it is simply the time required for a wavefront to reach $\mathbf{x}$.

The eikonal equation forms the basis for ray theoretical approaches to modeling seismic wave propagation, which are discussed in Chapter 4. It is an approximate solution, valid at high frequencies so that the terms in the wave equation that involve spatial velocity gradients in the Lamé parameters (see equation (3.18)) and the wave amplitude (C.9) can be neglected. Thus it is valid only at seismic wavelengths which are short compared to the distances in the medium over which velocity and amplitude change significantly. However, because this is often the case in the Earth, ray theoretical methods based on the eikonal equation have proven to be extremely useful.

Now recall equation (C.8), the imaginary part of (C.6):

$$2\nabla A \cdot \nabla T + A \nabla^2 T = 0, \tag{C.15}$$

where A is the wave amplitude and T is the phase factor or travel time for the wavefront. Remembering that $\nabla T = u\hat{\mathbf{k}}$ where u is the local wave slowness and $\hat{\mathbf{k}}$ is the unit vector in the ray direction, we have

$$2u\hat{\mathbf{k}} \cdot \nabla A = -\nabla \cdot (u\hat{\mathbf{k}})A, \tag{C.16}$$

or

$$A = \frac{-2u\nabla A \cdot \hat{\mathbf{k}}}{\nabla \cdot (u\hat{\mathbf{k}})}. \tag{C.17}$$

Integrating along the ray path in the direction $\hat{\mathbf{k}}$, a solution to (C.17) is provided by

$$A = \exp\left(-\frac{1}{2}\int \frac{\nabla \cdot (u\hat{\mathbf{k}})}{u} \, ds\right). \tag{C.18}$$

Substituting this expression into (C.2) for the compressional wave potential we can write

$$\phi(\omega) = Ae^{-i\omega T(\mathbf{r})} = \exp\left(-\frac{1}{2}\int_{\text{path}} \frac{\nabla \cdot (u_\alpha \hat{\mathbf{k}})}{u_\alpha} \, ds\right) \exp\left(-i\omega \int_{\text{path}} u_\alpha \, ds\right). \tag{C.19}$$

Here $u_\alpha \, ds$ is the travel time along the path and represents the usual oscillatory wave motion encountered previously. The exponent in the first exponential, however, is negative and real. Thus it represents a decay in amplitude along the ray path. This exponent can be further manipulated:

$$-\frac{1}{2}\int_{\text{path}} \frac{\nabla \cdot (u_\alpha \hat{\mathbf{k}})}{u_\alpha} \, ds = -\frac{1}{2}\int_{\text{path}} \left(\frac{\hat{\mathbf{k}} \cdot \nabla u_\alpha}{u_\alpha} + \nabla \cdot \hat{\mathbf{k}}\right) ds$$

$$= -\frac{1}{2} \int_{\text{path}} \left(\frac{1}{u_\alpha} \frac{du_\alpha}{ds} + \nabla \cdot \hat{\mathbf{k}} \right) ds$$

$$= -\frac{1}{2} \int_{\text{path}} \frac{du_\alpha}{u_\alpha} - \frac{1}{2} \int_{\text{path}} \nabla \cdot \hat{\mathbf{k}} \, ds$$

$$= -\frac{1}{2} \ln u \Big|_{u_0}^{u_\alpha} - \frac{1}{2} \int_{\text{path}} \nabla \cdot \hat{\mathbf{k}} \, ds$$

$$= -\frac{1}{2} \ln \left(\frac{u_\alpha}{u_0} \right) - \frac{1}{2} \int_{\text{path}} \nabla \cdot \hat{\mathbf{k}} \, ds. \qquad (C.20)$$

Here u_0 is the slowness at the source where the radiation first began. If we substitute (C.20) into (C.19) we find:

$$\phi(\omega) = (u_0/u_\alpha)^{1/2} e^{-\frac{1}{2} \int_{\text{path}} \nabla \cdot \hat{\mathbf{k}} \, ds} e^{-i\omega \int_{\text{path}} u_\alpha \, ds}. \qquad (C.21)$$

This equation describes the effect of geometrical spreading on wave amplitudes. $\nabla \cdot \hat{\mathbf{k}}$, the divergence of the unit vector parallel to the ray, represents the curvature of the wavefront and, when large, leads to a large amplitude reduction.

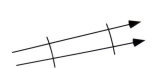   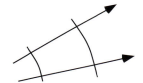

small amplitude reduction   large amplitude reduction

To illustrate the effect of the geometrical spreading term, consider a spherical wave diverging in a homogeneous whole space. In this case, wavefronts are spheres while rays are radii ($\hat{\mathbf{k}} = \hat{\mathbf{r}}$). Recalling the expression for the divergence in spherical coordinates we can write:

$$\nabla \cdot \hat{\mathbf{k}} = \frac{1}{r^2} \partial_r (r^2) = \frac{2}{r}. \qquad (C.22)$$

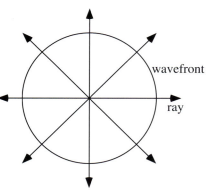

wavefront

ray

Substituting for the first exponential term in (C.21), we obtain

$$-\frac{1}{2} \int_{\text{path}} \nabla \cdot \hat{\mathbf{k}} \, ds = -\frac{1}{2} \int_{\text{path}} \frac{2ds}{r} = -\int_{r_0}^{r} \frac{dr}{r} = \ln \left(\frac{r_0}{r} \right) \qquad (C.23)$$

and thus from (C.21), and because $u_0 = u_\alpha$ from homogeneity, we obtain

$$\phi(\omega) = \left(\frac{r_0}{r}\right) e^{-i\omega \int_{\text{path}} u_\alpha \, ds} = \left(\frac{r_0}{r}\right) e^{-i\omega u_\alpha r}. \tag{C.24}$$

Thus, in a homogeneous medium (a whole space), the amplitude decays as r^{-1}. This result also follows from energy considerations (see Chapter 6) since the area of the wavefront grows as r^2.

Appendix D
Fortran subroutines

The following FORTRAN77 subroutines are required for some of the exercises.

```fortran
! LAYERXT calculates dx and dt for a ray in a layer with a linear
!   velocity gradient.  This is a highly modified version of a
!   subroutine in Chris Chapman's WKBJ program.
!
! Inputs:    p     =  horizontal slowness
!            h     =  layer thickness
!            utop  =  slowness at top of layer
!            ubot  =  slowness at bottom of layer
! Returns:   dx    =  range offset
!            dt    =  travel time
!            irtr  =  return code
!                  = -1,  zero thickness layer
!                  =  0,  ray turned above layer
!                  =  1,  ray passed through layer
!                  =  2,  ray turned in layer, 1 leg counted in dx,dt
!
       subroutine LAYERXT(p,h,utop,ubot,dx,dt,irtr)

       if (p.ge.utop) then          !ray turned above layer
          dx=0.
          dt=0.
          irtr=0
          return
       else if (h.eq.0.) then       !zero thickness layer
          dx=0.
          dt=0.
          irtr=-1
          return
       end if

       u1=utop
       u2=ubot
       v1=1./u1
       v2=1./u2
       b=(v2-v1)/h                  !slope of velocity gradient

       eta1=sqrt(u1**2-p**2)

       if (b.eq.0.) then            !constant velocity layer
          dx=h*p/eta1
          dt=h*u1**2/eta1
          irtr=1
```

```
        return
     end if

     x1=eta1/(u1*b*p)
     tau1=(alog((u1+eta1)/p)-eta1/u1)/b

     if (p.ge.ubot) then        !ray turned within layer,
        dx=x1                    !no contribution to integral
        dtau=tau1                !from bottom point
        dt=dtau+p*dx
        irtr=2
        return
     end if

     irtr=1

     eta2=sqrt(u2**2-p**2)
     x2=eta2/(u2*b*p)
     tau2=(alog((u2+eta2)/p)-eta2/u2)/b

     dx=x1-x2
     dtau=tau1-tau2

     dt=dtau+p*dx

     return
     end




! RTCOEF calculates P/SV reflection/transmission coefficients
!   for an interface between two solid layers, based on the
!   equations on p. 149-150 of Aki and Richards.  This version
!   is modified from an older routine provided by Tom Sereno.
!
!  Inputs:    vp1    =  P-wave velocity of layer 1 (top layer)
!  (real)     vs1    =  S-wave velocity of layer 1
!             den1   =  density of layer 1
!             vp2    =  P-wave velocity of layer 2 (bottom layer)
!             vs2    =  S-wave velocity of layer 2
!             den2   =  density of layer 2
!             hslow  =  horizontal slowness (ray parameter)
!  Returns:   rt(1)  =  down P to P up       (refl)
!  (complex)  rt(2)  =  down P to S up       (refl)
!             rt(3)  =  down P to P down     (tran)
!             rt(4)  =  down P to S down     (tran)
!             rt(5)  =  down S to P up       (refl)
!             rt(6)  =  down S to S up       (refl)
!             rt(7)  =  down S to P down     (tran)
!             rt(8)  =  down S to S down     (tran)
!             rt(9)  =    up P to P up       (tran)
!             rt(10) =    up P to S up       (tran)
!             rt(11) =    up P to P down     (refl)
!             rt(12) =    up P to S down     (refl)
!             rt(13) =    up S to P up       (tran)
!             rt(14) =    up S to S up       (tran)
!             rt(15) =    up S to P down     (refl)
!             rt(16) =    up S to S down     (refl)
!
! NOTE:   All input variables are real.
!         All output variables are complex!
!         Coefficients are not energy normalized.
!
```

```fortran
      SUBROUTINE RTCOEF(vp1,vs1,den1,vp2,vs2,den2,hslow,rt)
      implicit complex (a-h,o-z)
      complex rt(16)
      real vp1,vs1,den1,vp2,vs2,den2,hslow

      alpha1=cmplx(vp1,0.)
      beta1=cmplx(vs1,0.)
      rho1=cmplx(den1,0.)
      alpha2=cmplx(vp2,0.)
      beta2=cmplx(vs2,0.)
      rho2=cmplx(den2,0.)
      p=cmplx(hslow,0.)

      cone=cmplx(1.,0.)
      ctwo=cmplx(2.,0.)

      term1=(cone-ctwo*beta1**2*p**2)
      term2=(cone-ctwo*beta2**2*p**2)
      a=rho2*term2-rho1*term1
      b=rho2*term2+ctwo*rho1*beta1**2*p**2
      c=rho1*term1+ctwo*rho2*beta2**2*p**2
      d=ctwo*(rho2*beta2**2-rho1*beta1**2)
! compute signs and cosines, allowing for complex incidence angles
      si1=alpha1*p
      si2=alpha2*p
      sj1=beta1*p
      sj2=beta2*p
      ci1=csqrt(cone-si1**2)
      ci2=csqrt(cone-si2**2)
      cj1=csqrt(cone-sj1**2)
      cj2=csqrt(cone-sj2**2)

      E=b*ci1/alpha1+c*ci2/alpha2
      F=b*cj1/beta1+c*cj2/beta2
      G=a-d*ci1*cj2/(alpha1*beta2)
      H=a-d*ci2*cj1/(alpha2*beta1)
      DEN=E*F+G*H*p**2

      trm1=b*ci1/alpha1-c*ci2/alpha2
      trm2=a+d*ci1*cj2/(alpha1*beta2)
      rt(1)=(trm1*F-trm2*H*p**2)/DEN          !refl down P to P up

      trm1=a*b+c*d*ci2*cj2/(alpha2*beta2)
      rt(2)=(-ctwo*ci1*trm1*p)/(beta1*DEN)    !refl down P to S up

      rt(3)=ctwo*rho1*ci1*F/(alpha2*DEN)      !trans down P to P down

      rt(4)=ctwo*rho1*ci1*H*p/(beta2*DEN)     !trans down P to S down

      trm1=a*b+c*d*ci2*cj2/(alpha2*beta2)
      rt(5)=(-ctwo*cj1*trm1*p)/(alpha1*DEN)   !refl down S to P up

      trm1=b*cj1/beta1-c*cj2/beta2
      trm2=a+d*ci2*cj1/(alpha2*beta1)
      rt(6)=-(trm1*E-trm2*G*p**2)/DEN         !refl down S to S up

      rt(7)=-ctwo*rho1*cj1*G*p/(alpha2*DEN)   !trans down S to P down

      rt(8)=ctwo*rho1*cj1*E/(beta2*DEN)       !trans down S to S down
```

```
      rt(9)=ctwo*rho2*ci2*F/(alpha1*DEN)       !trans up P to P up

      rt(10)=-ctwo*rho2*ci2*G*p/(beta1*DEN)    !trans up P to S up

      trm1=b*ci1/alpha1-c*ci2/alpha2
      trm2=a+d*ci2*cj1/(alpha2*beta1)
      rt(11)=-(trm1*F+trm2*G*p**2)/DEN         !refl up P to P down

      trm1=a*c+b*d*ci1*cj1/(alpha1*beta1)
      rt(12)=(ctwo*ci2*trm1*p)/(beta2*DEN)     !refl up P to S down

      rt(13)=ctwo*rho2*cj2*H*p/(alpha1*DEN)    !trans up S to P up

      rt(14)=ctwo*rho2*cj2*E/(beta1*DEN)       !trans up S to S up

      trm1=a*c+b*d*ci1*cj1/(alpha1*beta1)
      rt(15)=(ctwo*cj2*trm1*p)/(alpha2*DEN)    !refl up S to P down

      trm1=b*cj1/beta1-c*cj2/beta2
      trm2=a+d*ci1*cj2/(alpha1*beta2)
      rt(16)=(trm1*E+trm2*H*p**2)/DEN          !refl up S to S down

      return
      end




! GETAUX returns auxiliary fault plane, given strike,dip,rake
! of main fault plane.
!
! Inputs:  strike1, dip1, rake1 (degrees, primary fault plane)
! Returns: strike2, dip2, rake2 (degrees, auxiliary fault plane)
!
      subroutine GETAUX(strike1,dip1,rake1,strike2,dip2,rake2)
      degrad=180./3.1415927
      s1=strike1/degrad
      d1=dip1/degrad
      r1=rake1/degrad

      d2=acos(sin(r1)*sin(d1))

      sr2=cos(d1)/sin(d2)
      cr2=-sin(d1)*cos(r1)/sin(d2)
      r2=atan2(sr2,cr2)

      s12=cos(r1)/sin(d2)
      c12=-1./(tan(d1)*tan(d2))
      s2=s1-atan2(s12,c12)

      strike2=s2*degrad
      dip2=d2*degrad
      rake2=r2*degrad

      if (dip2.gt.90.) then
         strike2=strike2+180.
         dip2=180.-dip2
         rake2=360.-rake2
      end if
      if (strike2.gt.360.) strike2=strike2-360.

      return
      end
```

Appendix E
Time series and Fourier transforms

The following is a summary of the time series concepts that are used in this book. For more details, the reader should consult Bracewell (1978) or other texts on time series analysis.

E.1 Convolution

Consider two time series $u(t)$ and $v(t)$. The *convolution* of these functions is defined as

$$u(t) * v(t) = \int_{-\infty}^{\infty} u(\tau)v(t - \tau)\, d\tau. \qquad (E.1)$$

Convolution with simple pulses generally results in a smoothing of the original time series. For example, convolution with a boxcar function will produce the same result as averaging the adjacent points (Fig. E.1).

Convolution is commutative and associative, that is,

$$u(t) * v(t) = v(t) * u(t), \qquad (E.2)$$

$$u(t) * [v(t) * w(t)] = [u(t) * v(t)] * w(t). \qquad (E.3)$$

It also follows from (E.1) that

$$\frac{\partial}{\partial t}[u(t) * v(t)] = [u(t) * v(t)]' = u'(t) * v(t) = u(t) * v'(t). \qquad (E.4)$$

Convolution provides a convenient way to describe the construction of synthetic seismograms (the predicted ground motion at a particular site as a function of the seismic source and a specified Earth model). For example, the seismogram could be written

$$u(t) = s(t) * G(t) * a(t) * r(t), \qquad (E.5)$$

371

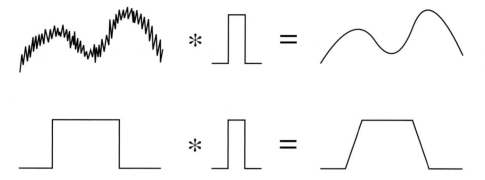

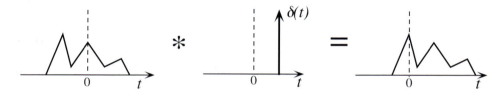

Figure E.1 Examples of convolution with a boxcar function.

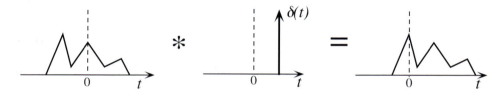

Figure E.2 Convolution with a delta function does not change the shape of a function, but shifts it in time to align with the position of the delta function.

where $s(t)$ is the effective source-time function, $G(t)$ is the elastic Green's function that connects the source and receiver (the hard part to compute!), $a(t)$ is an attenuation operator that approximates the effect of Q along the ray path, and $r(t)$ is the response of the receiver.

The *delta function*, $\delta(t)$, is often useful and is defined as

$$\delta(t) = 0 \text{ for } t \neq 0,$$

$$\int_{-\infty}^{\infty} \delta(t)\, dt = 1. \tag{E.6}$$

Convolution with a delta function leaves the original function unchanged, that is,

$$u(t) * \delta(t) = u(t). \tag{E.7}$$

The delta function may act to produce a time shift in the original time series (Fig. E.2):

$$u(t) * \delta(t - t_0) = u(t - t_0). \tag{E.8}$$

E.2 Fourier transform

There are many definitions of the Fourier transform. Here we assume the transform is defined by

$$\mathcal{F}[u(t)] = u(\omega) = \int_{-\infty}^{\infty} u(t)e^{i\omega t}\, dt \qquad (E.9)$$

and the inverse Fourier transform is

$$\mathcal{F}^{-1}[u(\omega)] = u(t) = \frac{1}{2\pi} \int_{-\infty}^{\infty} u(\omega)e^{-i\omega t}\, d\omega. \qquad (E.10)$$

The function $u(t)$ is said to be in the *time domain*; the corresponding function $u(\omega)$ is in the *frequency domain*. From these definitions, various useful relationships may be derived. The scale rule is

$$\mathcal{F}[u(t/a)] = |a|u(a\omega), \qquad (E.11)$$

$$u(t/a) = |a|\mathcal{F}^{-1}[u(a\omega)]. \qquad (E.12)$$

The differentiation rule is

$$\mathcal{F}[\dot{u}(t)] = -i\omega u(\omega). \qquad (E.13)$$

The shift theorem is

$$\mathcal{F}[u(t+a)] = u(\omega)e^{-i\omega a}. \qquad (E.14)$$

Finally, the convolution rule is

$$\mathcal{F}[u(t) * v(t)] = u(\omega)v(\omega), \qquad (E.15)$$

$$\mathcal{F}^{-1}[u(\omega)v(\omega)] = u(t) * v(t). \qquad (E.16)$$

In other words, the convolution of two functions in the time domain is equivalent to the product of the corresponding functions in the frequency domain.

E.3 Hilbert transform

A phase shift of π in the frequency domain is equivalent to a polarity reversal in the time domain (multiplying the time series by -1). In this case the pulse shapes are not changed. However, a frequency-independent phase shift that is not equal to a multiple of π will cause pulse distortion. An example that occurs frequently in seismology is the Hilbert transform, which results from a $\pi/2$ phase advance.

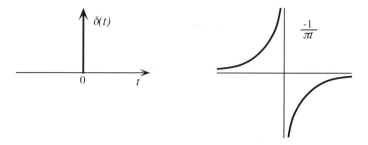

Figure E.3 The Hilbert transform of a delta function.

A forward and inverse Fourier transform will reproduce the original time series

$$u(t) = \mathcal{F}^{-1}\left(\mathcal{F}[u(t)]\right)$$

$$= \frac{1}{2\pi}\int_{-\infty}^{\infty}\left[\int_{-\infty}^{\infty}u(\xi)e^{i\omega\xi}\,d\xi\right]e^{-i\omega t}\,d\omega. \tag{E.17}$$

We may define the Hilbert transform of $u(t)$ as $\bar{u}(t)$, by inserting a $-\pi/2$ phase shift[1] in the outer integral:

$$\bar{u}(t) = \frac{1}{2\pi}\int_{-\infty}^{\infty}\left[\int_{-\infty}^{\infty}u(\xi)e^{i\omega\xi}\,d\xi\right]\frac{\operatorname{sgn}(\omega)}{i}e^{-i\omega t}\,d\omega, \tag{E.18}$$

where the $\operatorname{sgn}(\omega)$ keeps the time series real. Rearranging and evaluating the ω integral, one can show

$$\bar{u}(t) = \frac{1}{\pi}\int_{-\infty}^{\infty}\frac{u(\xi)}{\xi - t}\,d\xi, \tag{E.19}$$

in which the singularity at $\xi = t$ is handled by taking the Cauchy principle value of the integral. The Hilbert transform of a delta function is

$$\bar{\delta}(t) = \frac{1}{\pi}\int_{-\infty}^{\infty}\frac{\delta(\xi)}{\xi - t}\,d\xi = -\frac{1}{\pi t} \tag{E.20}$$

and is illustrated in Figure E.3.

Thus the Hilbert transform may also be expressed as a convolution

$$\bar{u}(t) = u(t) * \left(-\frac{1}{\pi t}\right). \tag{E.21}$$

---

[1] The sign of the phase shift depends upon the sign of $i\omega$ in the Fourier transform; here we assume the sign convention of (E.9) and (E.10).

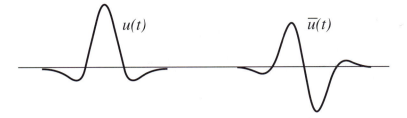

Figure E.4 The Hilbert transform of a typical seismic pulse.

Note also that

$$\overline{f * g} = f * \bar{g} = \bar{f} * g. \tag{E.22}$$

Equations (E.18), (E.19), and (E.21) are all equivalent definitions of the Hilbert transform of $u(t)$. In practice, Hilbert transforms are most easily calculated with a computer by using (E.18) and a Digital Fourier Transform.

Any frequency-independent phase shift may be expressed as a linear combination of a function and its Hilbert transform (an extension to all frequency components of the fact that a phase shift for a harmonic wave can be expressed as a linear combination of a sine and cosine wave; the Hilbert transform of a sine wave is a cosine wave). Applying the Hilbert transform twice produces a polarity reversal in the time series (a π phase shift); applying the Hilbert transform four times reproduces the original time series. The Hilbert transform does not change the amplitude spectrum of a time series; only the phase of the different frequency components is affected.

The Hilbert transform of a delta function is acausal in the sense that a precursory tail extends to $-\infty$ in time. In seismology, of course, energy cannot arrive prior to the time of source initiation (and in most cases cannot arrive before the time of the direct P-wave arrival). This is explained by the fact that the Hilbert transformed pulses occur on secondary arriving branches (such as PmP, PP, etc.) and that the Hilbert transform is predicted only as a high-frequency approximation to the true pulse shape. In practice, with bandlimited data, the Hilbert transform does not produce notably acausal pulse shapes (Fig. E.4).

Another useful function is the *analytic time series*, defined as:

$$U(t) = u(t) + i\bar{u}(t). \tag{E.23}$$

In a sense, $U(t)$ bears the same resemblance to $u(t)$ as does $e^{i\omega t}$ to $\cos(\omega t)$. We also have the *envelope time function*, defined as:

$$E(t) = \left[u^2(t) + \bar{u}^2(t) \right]^{1/2}. \tag{E.24}$$

The envelope time function is useful as a local estimate of the amplitude that is not sensitive to individual zero crossings in the seismogram.

BIBLIOGRAPHY

Abercrombie, R. E. (1995). Earthquake source scaling relationships from -1 to 5 M_L using seismograms recorded at 2.5-km depth, *J. Geophys. Res.*, **100**, 24 015–36.

Aggarwal, Y. P. (1973). Premonitory changes in seismic velocities and prediction of earthquakes, *Nature*, **241**, 101–4.

Aggarwal, Y. P., Sykes, L. R., Simpson, D. W., and Richard, P. G. (1975). Spatial and temporal variations in t_s/t_p and in P wave residuals at Blue Mountain Lake, New York: Application to earthquake prediction, *J. Geophys. Res*, **80**, 718–32.

Agnew, D. C. (2002). History of seismology, in *International Handbook of Earthquake and Engineering Seismology, Part A*, ed. W. H. Lee, San Diego: Academic Press, 3–11.

Agnew, D. C. and Jones, L. M. (1991). Prediction probabilities from foreshocks, *J. Geophys. Res.*, **96**, 11 959–71.

Akcelik, V., Bielak, J., Biros, G., Epanomeritakis, I., Fernandez, A., Ghattas, O., Kim, E., Lopez, J., O'Halloron, D., Tu, T., and Urbanic, J. (2003). High resolution forward and inverse modeling on terrascale computers, SC2003, Phoenix, AZ.

Aki, K. (1966). Estimation of earthquake moment, released energy, and stress-strain drop from G wave spectrum, *Bull. Earthquake Res. Inst. Tokyo U.*, **44**, 23–88.

Aki, K. (1967). Scaling law of seismic spectrum, *J. Geophys. Res.*, **72**, 1217–31.

Aki, K. (1981). A probabilistic synthesis of precursory phenomena, in *Earthquake Prediction*, ed. D. W. Simpson and P. G. Richards, Washington: American Geophysical Union, 566–74.

Aki, K. and Richards, P. G. (1980). *Quantitative Seismology: Theory and Methods*, San Francisco: W. H. Freeman.

Aki, K. and Richards, P. G. (2002). *Quantitative Seismology*, 2nd edn, Sausalito, CA: University Science Books.

Allmann, B. P. and Shearer, P. M. (2009). Global variations of stress drop for moderate to large earthquakes, *J. Geophys. Res.*, doi:10.1029/2008JB005821, **114**.

Anderson, D. and Whitcomb, J. (1975). Time-dependent seismology, *J. Geophys. Res.*, **80**, 1497–1503.

Anderson, J. and Chen, Q. (1995). Beginnings of earthquakes in the Mexican subduction zone on strong-motion accelerograms, *Bull. Seismol. Soc. Am.*, **85**, 1107–1116.

Astiz, L. (1997). Characteristic low and high noise models from robust PSE of seismic noise of broadband stations, presented at IASPEI 1997 meeting.

Astiz, L., Earle, P., and Shearer, P. (1996). Global stacking of broadband seismograms, *Seismol. Res. Lett.*, **67**, 8–18.

Backus, G. E. (1962). Long-wave elastic anisotropy produced by horizontal layering, *J. Geophys. Res.*, **67**, 4427–40.

Backus, G. (1965). Possible forms of seismic anisotropy of the uppermost mantle under oceans, *J. Geophys. Res.*, **70**, 3429–39.

Backus, G. and Gilbert, F. (1967). Numerical applications of a formalism for geophysical inverse problems, *Geophys. J. Roy. Astron. Soc.*, **13**, 247–76.

Backus, G. and Gilbert, F. (1968). The resolving power of gross Earth data, *Geophys. J. Roy. Astron. Soc.*, **16**, 169–205.

Backus, G. and Gilbert, F. (1970). Uniqueness in the inversion of inaccurate gross Earth data, *Phil. Trans. Roy. Soc. A*, **266**, 123–92.

Baig, A. M. and Dahlen, F. A. (2004). Travel time biases in random media and the S-wave discrepancy, *Geophys. J. Int.*, **158**, 922–38.

Bak, P. and Tang, C. (1989). Earthquakes as a self-organized critical phenomenon, *J. Geophys. Res.*, **94**, 15 635–7.

Bakun, W. H. (1999). Seismic activity of the San Francisco Bay region, *Bull. Seismol. Soc. Am.*, **89**, 764–84.

Bamford, D. (1977). Pn velocity anisotropy in a continental upper mantle, *Geophys. J. Roy. Astron. Soc.*, **49**, 29–48.

Berckhemer, H., Kampfmann, W., Aulbach, E., and Schmeling, H. (1982). Shear modulus and q of forsterite and dunite near partial melting from forced-oscillation experiments, *Phys. Earth. Planet. Int.,* **29**, 30–41.

Bessonova, E. N., Fishman, V. M., Schnirman, M. G., and Sitnikova, G. A. (1974). The tau method for inversion of travel times – I. Deep seismic sounding data, *Geophys. J. Roy. Astron. Soc.*, **36**, 377–98.

Bessonova, E. N., Fishman, V. M., Schnirman, M. G., Sitnikova, G. A., and Johnson, L. R. (1976). The tau method for inversion of travel times – II. Earthquake data, *Geophys. J. Roy. Astron. Soc.*, **46**, 87–108.

Boatwright, J. (1980). A spectral theory for circular seismic sources: simple estimates of source dimension, dynamic stress drop and radiated energy, *Bull. Seismol. Soc. Am.*, **70**, 1–27.

Boatwright, J. and J.B. Fletcher (1984). The partition of radiated energy between P and S waves, *Bull. Seismol. Soc. Am.*, **74**, 361–76.

Bolt, B. (1977). Constancy of P travel times from Nevada explosions to Oroville Dam station 1970–1976, *Bull. Seismol. Soc. Am.*, **67**, 27–32.

Bolt, B. A. (1993). *Earthquakes and Geological Discovery*, New York: Scientific American Library.

Boore, D. M., Lindh, A. G., McEvilly, T. V., and Tolmachoff, W. W. (1975). A search for travel time changes associated with the Parkfield earthquake of 1966, *Bull. Seismol. Soc. Am.*, **65**, 1407–18.

Bostock, M. G. (1998). Mantle stratigraphy and evolution of the Slave province, *J. Geophys. Res.*, **103**, 21 183–200.

Bostock, M. G., Rondenay, S., and Shragge, J. (2001). Multiparameter two-dimensional inversion of scattered teleseismic body waves 1. Theory for oblique incidence, *J. Geophys. Res.*, **106**, 30 771–82.

Bracewell, R. N. (1978). *The Fourier Transform and Its Applications*, New York: McGraw-Hill.

Brodsky, E. D. and Kanamori, H. (2001). Elastohydrodynamic lubrication of faults, *J. Geophys. Res.*, **106**, 16 357–74.

Brune, J. (1969). Heat flow, stress, and the rate of slip along the San Andreas Fault, California, *J. Geophys. Res.*, **74**, 3821–7.

Brune, J. (1970). Tectonic stress and the spectra of seismic shear waves from earthquakes, *J. Geophys. Res.*, **75**, 4997–5009.

Brune, J. N. (1979). Implications of earthquake triggering and rupture propagation for earthquake prediction based on premonitory phenomena, *J. Geophys. Res.*, **84**, 2195–8.

Brune, J. N., Brown, S., and Johnson, P. A. (1993). Rupture mechanism and interface separation in foam rubber models of earthquakes: a possible solution to the heat flow paradox and the paradox of large overthrusts, *Tectonophysics*, **218**, 59–67.

Brune, J. N. and W. Thatcher (2002). Strength and energetics of active fault zones, in *International Handbook of Earthquake and Engineering Seismology, Part A*, ed. W. H. Lee, San Diego: Academic Press, 569–88.

Bullen, K. E. and Bolt, B. A. (1985). *An Introduction to the Theory of Seismology*, Cambridge: Cambridge University Press.

Burridge, R. and Knopoff, L. (1967). Model and theoretical seismicity, *Bull. Seismol. Soc. Am.* **57**, 341–71.

Cao, A., Romanowicz, B., and Takeuchi, N. (2005). An observation of PKJKP: Inferences on inner core shear properties, *Science*, **308**, 1453–5, doi: 10.1126/science.1109134.

Campillo, M. and Paul, A. (2003). Long-range correlations in the diffuse seismic coda, *Science*, **299**, 547–9, doi: 10.1126/science.1078551.

Cerveny, V. (2001). *Seismic Ray Theory*, Cambridge: Cambridge University Press.

Chapman, C. (2004). *Fundamentals of Seismic Wave Propagation*, Cambridge: Cambridge University Press.

Chapman, C. H., Jen-Yi, C., and Lyness, D. G. (1988). The WKBJ seismogram algorithm, in *Seismological Algorithms*, ed. D. Doornbos, London: Academic Press, 341–71.

Chou, C. W. and Crosson, R. S. (1978). Search for time-dependent seismic P travel times from mining explosions near Centralia, Washington, *Geophys. Res. Lett.*, **5**, 97–100.

Choy, G. L. and Boatwright, J. (2007). The energy radiated by the 26 December 2004 Sumatra-Andaman earthquake estimated from 10-minute P-wave windows, *Bull. Seismol. Soc. Am.*, **97**, doi: 10.1785/0120050623, S18–S24.

Choy, G. L. and Richards, P. G., (1975). Pulse distortion and Hilbert transformation in multiply reflected and refracted body waves, *Bull. Seismol. Soc. Am.*, **65**, 55–70.

Claerbout, J. F. (1976). *Fundamentals of Geophysical Processing*, New York: McGraw-Hill.

Claerbout, J. F. (1985). *Imaging the Earth's Interior*, Oxford: Blackwell Scientific Publications.

Custodio, S., Liu, P. C., and Archuleta, R. J. (2005). The 2004 Mw 6.0 Parkfield, California, earthquake: inversion of near-source ground motion using multiple data sets, *Geophys. Res. Lett.*, **32**, doi:10.1029/2005GL024417.

Dahlen, F. A., Hung, S.-H., and Nolet, G. (2000). Frechet kernels for finite-frequency travel times – I. Theory, *Geophys. J. Int.*, **141**, 157–74.

Dahlen, F. and Nolet, G. (2005). Comment on the paper on sensitivity kernels for wave equation transmission tomography by de Hoop and van der Hilst, *Geophys. J. Int.*, **163**, 949–51.

Dahlen, F. A. and Tromp, J. (1998). *Theoretical Global Seismology*, Princeton: Princeton University Press.

de Hoop, M. and van der Hilst, R. (2005a). On sensitivity kernels for wave equation transmission tomography, *Geophys. J. Int.*, **160**, 621–33.

de Hoop, M. and van der Hilst, R. (2005b). Reply to a comment by F.A. Dahlen and G. Nolet on: On sensitivity kernels for wave equation tomography, *Geophys. J. Int.*, **163**, 952–5.

Deuss, A., Woodhouse, J., Paulssen, H., and Trampert, J. (2000). The observation of inner core shear waves, *Geophys. J. Int.*, **142**, 67–73.

Dewey, J. and Byerly, P. (1969). The early history of seismometry (to 1900), *Bull. Seismol. Soc. Am.*, **59**, 183–227.

Di Toro, G., Goldsby, D. L., and T. E. Tullis (2004). Friction falls towards zero in quartz rock as slip velocity approaches seismic rates, *Nature*, **427**, 436–9.

Dieterich, J. (1994). A constitutive law for rate of earthquake production and its application to earthquake clustering, *J. Geophys. Res.*, **99**, 2601–18.

Doornbos, D. (1974). The anelasticity of the inner core, *Geophys. J. Roy. Astron. Soc.*, **38**, 397–415.

Doornbos, D.J. (1983). Observable effects of the seismic absorption band in the Earth, *Geophys. J. Roy. Astron. Soc.*, **75**, 693–711.

Dorman, L. M. and Jacobson, R. S. (1981). Linear inversion of body wave data, part 1: velocity structure from travel-times and ranges, *Geophysics*, **46**, 138–50.

Dziewonski, A. M. and Anderson, D. L. (1981). Preliminary reference Earth model, *Phys. Earth Planet. Inter.*, **25**, 297–356.

Dziewonski, A. M. and Woodhouse, J. H. (1983). An experiment in systematic study of global seismicity: Centroid-moment tensor solutions for 201 moderate and large earthquakes of 1981, *J. Geophys. Res.*, **88**, 3247–71.

Ekström, G. and Dziewonski, A. M. (1988). Evidence of bias in estimations of earthquake size, *Nature*, **332**, 319–23.

Ekström, G., Nettles, M., and Abers, G.A. (2003). Glacial earthquakes, *Science*, **302**, 622–4.

Ekström, G., Tromp, J., and Larsen, E.W. (1997). Measurements and global models of surface wave propagation, *J. Geophys. Res.*, **102**, 8137–57.

Ellsworth, W. L. and Beroza, G. C. (1995). Seismic evidence for an earthquake nucleation phase, *Science*, **268**, 851–5.

Eshelby, J. D. (1957). The determination of the elastic field of an ellipsoidal inclusion, and related problems, *Prec. Roy. Soc. London*, **A241**, 376–96.

Felzer, K. R. and Brodsky, E. E. (2005). Testing the stress shadow hypothesis, *J. Geophys. Res.*, **110**, doi:10.1029/2004JB003277.

Felzer, K. R. and Brodsky, E. E. (2006). Decay of aftershock density with distance indicates triggering by dynamic stress, *Nature*, **441**, 735–8.

Forsyth, D. W. (1975). The early structural evolution and anisotropy of the oceanic upper mantle, *Geophys. J. Roy. Astron. Soc.*, **43**, 103–62.

Fouch, M. J. and Rondenay, S. (2006). Seismic anisotropy beneath stable continental interiors, *Phys. Earth Planet. Inter.*, **158**, 292–320, doi: 10.1016/j.pepi.2006.03.024.

Freed, A. M. (2005). Earthquake triggering by static, dynamic, and postseismic stress transfer, *Annu. Rev. Earth Planet. Sci.*, **33**, 335–67, doi: 10.1146/annurev.earth.33.092203.122505.

Frohlich, C. (1979). An efficient method for joint hypocenter determination for large groups of earthquakes, *Comput. Geosci.*, **5**, 387–9.

Fuchs, K. and Müller, G. (1971). Computation of synthetic seismograms with the reflectivity method and comparison with observations, *Geophys. J. Roy. Astron. Soc.*, **23**, 417–33.

Gardner, J. K. and Knopoff, L. (1974). Is the sequence of earthquakes in southern California, with aftershocks removed, Poissonian?, *Bull. Seismol. Soc. Am.*, **64**, 1363–7.

Garmany, J., Orcutt, J. A., and Parker, R. L. (1979). Travel time inversion: a geometrical approach, *J. Geophys. Res.*, 3615–22.

Geller, R. J. (1997). Earthquake prediction: a critical review, *Geophys. J. Int.*, **131**, 425–50.

Gephart, J. W. and Forsyth, D. W. (1984). Improved method for determining the regional stress tensor using earthquake focal mechanism data: Application to the San Fernando earthquake sequence, *J. Geophys. Res.*, 89, 9305–20.

Gilbert, F. and Dziewonski, A. M. (1975). An application of normal mode theory to the retrieval of structural parameters and source mechanisms from seismic spectra, *Phil. Trans. R. Soc. Lond. A*, **278**, 187–269.

Goins, N. R., Dainty, A. M., and Toksöz, M. N. (1981). Lunar seismology: the internal structure of the moon, *J. Geophys. Res.*, **86**, 5061–74.

Grant, L. (2002). Paleoseismology, in *International Handbook of Earthquake and Engineering Seismology, Part A*, ed. W. H. Lee, San Diego: Academic Press, 475–89.

Green, H. W. (1994). Solving the paradox of deep earthquakes, *Sci. Am.*, September, 64–71.

Hanks, T. C. and Kanamori, H. (1979). A moment magnitude scale, *J. Geophys. Res.*, **84**, 2348–50.

Harris, R. (2002). Stress triggers, stress shadows, and seismic hazard, in *International Handbook of Earthquake and Engineering Seismology, Part A*, ed. W. H. Lee, San Diego: Academic Press, 1217–32.

Harris, R. A. and Arrowsmith, J. R. (2006). Introduction to the special issue on the 2004 Parkfield earthquake and the Parkfield Earthquake Prediction Experiment, *Bull. Seismol. Soc. Am.*, **96**, S1–S10, doi: 10.1785/0120050831.

Harris, R. A. and Simpson, R. W. (1992). Changes in static stress on southern California faults after the 1992 Landers earthquake, *Nature*, **360**, 251–4.

Harris, R. A., Simpson, R. W., and Reasenberg, P. A. (1995). Influence of static stress changes on earthquake locations in southern California, *Nature*, **375**, 221–224.

Harvey, J. (1995). Helioseismology, *Physics Today*, October, 32–8.

Heaton, T. H. (1990). Evidence for and implications of self-healing pulses of slip in earthquake rupture, *Phys. Earth Planet. Inter.*, **64**, 1–20.

Helmstetter, A. and Sornette, D. (2002). Diffusion of epicenters of earthquake aftershocks, Omori's law, and generalized continuous-time random models, *Phys. Rev. E*, **66**, 061104.

Helmstetter, A., Sornette, D., and Grasso, J.-R. (2003). Mainshocks are aftershocks of conditional foreshocks: How do foreshock statistical properties emerge from aftershock laws, *J. Geophys. Res.*, **108**, doi: 10.1029/2002JB001991.

Hess, H. H. (1964). Seismic anisotropy of the uppermost mantle under oceans, *Nature*, **203**, 629–31.

Hough, S. E. (1997). Empirical Green's function analysis: Taking the next step, *J. Geophys. Res.*, **102**, 5369–84.

Hough, S. E., Armbruster, J. G., Seeber, L., and Hough, J. F. (2000). On the modified Mercalli intensities and magnitudes of the 1811–1812 New Madrid earthquakes, *J. Geophys. Res.*, **105**, 23 839–64.

Houston, H. and Kanamori, H. (1986). Source spectra of great earthquakes: teleseismic constraints on rupture process and strong motion, *Bull. Seismol. Soc. Am.*, **76**, 19–42.

Hudson, J. A. (1980). Overall properties of a cracked solid, *Math. Proc. Cambridge Phil. Soc.*, **88**, 371–84.

Hung, S.-H., Dahlen, F., and Nolet, G. (2000). Frechet kernets for finite-frequency travel times – I. Examples, *Geophys. J. Int.*, **141**, 175–203.

Ide, S. and Beroza, G. C. (2001). Does apparent stress vary with earthquake size?, *Geophys. Res. Lett.*, **28**, 3349–52.

Igel, H. and Weber, M. (1995). SH-wave propagation in the whole mantle using high-order finite differences, *Geophys. Res. Lett.*, **22**, 71–4.

Ishii, M., Shearer, P. M., Houston, H., and Vidale, J. E. (2005). Extent, duration and speed of the 2004 Sumatra-Andaman earthquake imaged by the Hi-Net array, *Nature*, 435, doi:10.1038/nature03675.

Iyer, H. M. and Hirahara, K. (eds.) (1993). *Seismic Tomography: Theory and Practice*, London: Chapman and Hall.

Jackson, I., Paterson, M. S., and Fitz Gerald, J. D. (1992), Seismic wave dispersion and attenuation in Aheim dunite: an experimental study, *Geophys. J. Int.*, **108**, 517–34.

Jackson, I., Fitz Gerald, J. D. Faul, U. H., and Tan, B. H. (2002), Grain-size-sensitive seismic wave attenuation in polycrystalline olivine, *J. Geophys. Res.*, **107**, doi:10.1029/2001JB001225.

Jahnke, G., Thorne, M. S., Cochard, A., and Igel, H. (2008). Global SH-wave propagation using a parallel axisymmetric spherical finite-difference scheme: application to whole mantle scattering, *Geophys. J. Int.*, **173**, 815–26, doi: 10.1111/j.1365-246X.2008.03744.x.

Jordan, T. H. and Sverdrup, K. A. (1981). Teleseismic location techniques and their application to earthquake clusters in the south-central Pacific, *Bull. Seismol. Soc. Am.*, **71**, 1105–30.

Julian, B. R., Miller, A. D., and Foulger, G. R. (1998). Non-double-couple earthquakes – 1. Theory, *Rev. Geophys.*, **36**, 525–49.

Kagan, Y. Y. and Houston, H. (2005). Relation between mainshock rupture process and Omori's law for aftershock moment release rate, *Geophys. J. Int.*, **163**, 1039–1048, doi: 10.1111/j.1365-246X.2005.02772.x.

Kagan, Y. Y. and Jackson, D. D. (1991). Seismic gap hypothesis: ten years after, *J. Geophys. Res.*, **96**, 21 419–31.

Kampmann, W. and Müller G. (1989). PcP amplitude calculations for a core–mantle boundary with topography, *Geophys. Res. Lett.*, **16**, 653–56.

Kanamori, H. (1977). The energy release in great earthquakes, *J. Geophys. Res.*, **82**, 2981–7.

Kanamori, H. (2006). The radiated energy of the 2004 Sumatra-Andaman earthquake, in *Earthquakes: Radiated Energy and the Physics of Faulting, Geophysical Monograph Series 170*, Washington: American Geophysical Union, doi: 10.1029/170GM07, 59–68.

Kanamori, H. and Allen, C. R. (1986). Earthquake repeat time and average stress drop, in *Earthquake Source Mechanics*, ed. S. Das and C. H. Scholz, Washington: American Geophysical Union, C.C., 227–35.

Kanamori, H. and Anderson, D. L. (1975). Theoretical basis of some empirical relations in seismology, *Bull. Seismol. Soc. Am.*, **65**, 1073–95.

Kanamori, H., Anderson, D. L., and Heaton, T. H. (1998). Frictional melting during the rupture of the 1994 Bolivian earthquake, *Science*, **279**, 839–42.

Kanamori, H. and Brodsky, E. (2004). The physics of earthquakes, *Rep. Prog. Phys.*, **67**, 1429–96.

Kanamori, H., Ekstrom, E., Dziewonski, A., Barker, J. S., and Sipkin, S. A. (1993). Seismic radiation by magma injection – an anomalous seismic event near Tori Shima, Japan, *J. Geophys. Res.*, **98**, 6511–22.

Kanamori, H. and Fuis, G. (1976). Variation of P wave velocity before and after the Galway Lake Earthquake ($M_L = 5.2$) and the Goat Mountain earthquakes ($M_L = 4.7, 4.7$), 1975, in the Mojave Desert, California, *Bull. Seismol. Soc. Am.*, **66**, 2027–37.

Kanamori, H. and Given, J. W. (1982). Analysis of long-period seismic waves excited by the May 18, 1980, eruption of Mt. St. Helens: a terrestrial monopole, *J. Geophys. Res.*, **87**, 5422–32.

Kanamori, H., Given, J. W., and Lay, T. (1984). Analysis of seismic body waves excited by the Mount St. Helens eruption of May 18, 1980, *J. Geophys. Res.*, **89**, 1856–66.

Kanamori, H., Mori, J., Hauksson, E., Heaton, T. H., Hutton, K. L., and Jones, L. M. (1993). Determination of earthquake energy release and M_L using Terrascope, *Bull. Seismol. Soc. Am.*, **83**, 330–46.

Kanamori, H. and Rivera, L. (2006). Energy partitioning during an earthquake, in *Earthquakes: Radiated Energy and the Physics of Faulting*, Geophysical Monograph Series 170, ed. R. Abercro A. McGair, H. Kanamori, and G. di Toro, Washington: American Geophysical Union, 3–13, doi: 10.1029/170GM03.

Karato, S.-I. (1998). Seismic anisotropy in the deep mantle, boundary layers and the geometry of mantle convection, *Pure Appl. Geophys.*, **151**, 565–87.

Kawakatsu, H. (1991). Insignificant isotropic component in the moment tensor of deep earthquakes, *Nature*, **351**, 50–3.

Kendall, J. M. and Silver, P. G. (1996). Constraints from seismic anisotropy on the nature of the lowermost mantle, *Nature*, **381**, 409–12.

Kennett, B. N. L. (1974). Reflections, rays and reverberations, *Bull. Seismol. Soc. Am.*, **64**, 1685–96.

Kennett, B. N. L. (1983). *Seismic Wave Propagation in Stratified Media*, New York: Cambridge University Press.

Kennett, B. N. L. (1991). *IASPEI 1991 Seismological Tables*, Res. School of Earth Science, Australian National University, Canberra, Australia.

Kennett, B. N. L. (1991). The removal of free surface interactions from three-component seismograms, *Geophys. J. Int.*, **104**, 153–63.

Kennett, B. L. N. (2001). *The Seismic Wavefield, Vol.1: Introduction and Theoretical Development*, Cambridge: Cambridge University Press.

Kennett, B. N. L. and Engdahl, E. R. (1991). Travel times for global earthquake location and phase identification, *Geophys. J. Int.*, **106**, 429–65.

Kent, G. M., Harding, A. J., Orcutt, J. A., Detrick, R. S., Mutter, J. C., and Buhl, P. (1994). Uniform accretion of oceanic crust south of the Garrett transform at 14°15′S on the East Pacific Rise, *J. Geophys. Res.*, **99**, 9097–116.

Kilb, D., Gomberg, J., and Bodin, P. (2000). Triggering of earthquake aftershocks by dynamic stresses, *Nature*, **408**, 570–4, doi: 10.1038/35046046.

Kind, R., Kosarev, G. L., Makeyeva, L. I., and Vinnik, L. P. (1985). Observations of laterally inhomogeneous anisotropy in the continental lithosphere, *Nature*, **318**, 358–61.

Kirby, S. H., Durham, W. B., and Stern, L. A. (1991). Mantle phase changes and deep-earthquake faulting in subducting lithosphere, *Science*, **252**, 216–25.

Knopoff, L. (1958). Energy release in earthquakes, *Geophys. J.*, **1**, 44–52.

Knopoff, L. and Randall, M. (1970). The compensated vector linear dipole: a possible mechanism for deep earthquakes, *J. Geophys. Res.*, **75**, 4957–63.

Komatitsch, D., Ritsema, J., and Tromp, J. (2002). The spectral-element method, Beowulf computing, and global seismology, *Science*, **298**, 1737–42.

Komatitsch, D., Tsuboi, S., and Tromp, J. (2005). The spectral-element method in seismology, in *Seismic Earth: Array Analysis of Broadband Seismograms*, Geophysical Monograph Series, **157**, ed. A. Levander and G. Nolet, Washington: American Geophysical Union, 207–27.

Kostrov, B. (1974). Seismic moment and energy of earthquakes and seismic flow of rock, *Izv. Acad. Sci., USSR, Phys. Solid Earth (Engl. Transl.)*, **1**, 23–40.

Kumazawa, M. and Anderson, D. L. (1969). Elastic moduli, pressure derivatives, and temperature derivatives of single-crystal olivine and single-crystal fosterite, *J. Geophys. Res.*, **74**, 5961–72.

Lachenbruch, A. H. (1980). Frictional heating, fluid pressure, and the resistance to fault motion, *J. Geophys. Res.*, **85**, 6097–112.

Lachenbruch, A. H. and Sass, J. H. (1980). Heat flow and energetics of the San Andreas Fault zone, *J. Geophys. Res.*, **85**, 6185–222.

Langston, C. A. (1977). Corvallis, Oregon, crustal and upper mantle receiver structure from teleseismic body waves, *Bull. Seismol. Soc. Am.*, **67**, 713–24.

Latham, G., Ewing, M., Press, F., and Sutton, G. (1969). The Apollo passive seismic experiment, *Science*, **165**, 241–50.

Lawrence, J. F. and Shearer, P. M. (2006). A global study of transition zone thickness using receiver functions, *J. Geophys. Res.*, **111**, doi: 10.1029/ 2005JB003973.

Lay, T. and Wallace, T. C. (1995). *Modern Global Seismology*, San Diego: Academic Press.

Levander, A. R. (1988). Fourth-order finite-difference P-SV seismograms, *Geophysics*, **53**, 1425–36.

Lin, G. and Shearer, P. (2005). Tests of relative earthquake location techniques using synthetic data, *J. Geophys. Res.*, **110**, B04304, doi: 10.1029/2004JB003380.

Lin, G. and Shearer, P. (2006). The COMPLOC earthquake location package, *Seismol. Res. Lett.*, **77**, 440–44.

Lin, G., Shearer, P., and Hauksson, E. (2007). Applying a three-dimensional velocity model, waveform cross-correlation, and cluster analysis to locate southern California seismicity from 1981 to 2005, *J. Geophys. Res.*, **112**, B12309, doi: 10.1029/2007JB004986.

Liu, H., Anderson, D. L., and Kanamori, H. (1976). Velocity dispersion due to anelasticity: Implications for seismology and mantle composition, *Geophys. J. R. Astron. Soc.*, **47**, 41–58.

Liu, Q. and Tromp, J. (2006). Finite-frequency kernels based on adjoint methods, *Bull. Seismol. Soc. Am.*, **96**, 2383–97.

Longhurst, R. S. (1967). *Geometrical and Physical Optics*, New York: John Wiley.

Longuet-Higgins, M. S. (1950). A theory of the origin of microseisms, *Phil. Trans. R. Soc. London A*, **243**, 1–35.

Lundquist, G. M. and Cormier, V. C. (1980). Constraints on the absorption band model of *Q. J. Geophys. Res.*, **85**, 5244–56.

Ma, S. and Archuleta, R. J. (2006). Radiated seismic energy based on dynamic rupture models of faulting, *J. Geophys. Res.*, **111**, doi: 10.1029/2005JB004055.

Madariaga, R. (1976). Dynamics of an expanding circular fault, *Bull. Seismol. Soc. Am.*, **66**, 639–66.

Magistrale, H., Day, S., Clayton, R. W., and Graves, R. (2000). The SCEC southern California reference three-dimensional seismic velocity model Version 2, *Bull. Seismol. Soc. Am.*, **90**, S65–S76.

Mallick, S. and Frazer, L. N. (1987). Practical aspects of reflectivity modeling, *Geophysics*, **52**, 1355–64.

Malvern, L. E. (1969). *Introduction to the Mechanics of a Continuous Medium*, Englewood Cliffs, NJ: Prentice-Hall.

Manners, U. J., and Masters, G. (2008). Analysis of core-mantle boundary structure using S and P diffracted waves, *Geophys. J. Int.*, submitted.

Masters, T. G., Johnson, S., Laske, G., and Bolton, H. (1996). A shear-velocity model of the mantle, *Phil. Trans. R. Soc. Lond. A*, **354**, 1385–411.

Masters, G., Laske, G., Bolton, H., and Dziewonski, A. (2000). The relative behavior of shear velocity, bulk sound speed, and compressional velocity in the mantle: implications for chemical and thermal structure, in *Earth's Deep Interior*, Geophysical Monograph Series 117, ed. S. Karato, Washington: American Geophysical Union, 63–87.

Maupin, V. and Park, J. (2007). Theory and observations – Wave propagation in anisotropic media, in *Treatise on Geophysics, Volume 1: Deep Earth Structure*, ed. G. Schubert, Oxford: Elsevier, pp. 289–321.

Mayeda, K. and Walter, W. R. (1996). Moment, energy, stress drop, and source spectra of western United States earthquakes from regional coda envelopes, *J. Geophys. Res.*, **101**, 11 195–208.

McCann, W. R., Nishenko, S. P., Sykes, L. R., and Krause, J. (1979). Seismic gaps and plate tectonics: seismic potential for major boundaries, *Pure Appl. Geophys.*, **117**, 1082–47.

McEvilly, T. V. and Johnson, L. R. (1974). Stability of P and S velocities from central california quarry blasts, *Bull. Seismol. Soc. Am.*, **64**, 343–53.

Meade, C., Silver, P. G., and Kaneshima, S. (1995). Laboratory and seismological observations of lower mantle isotropy, *Geophys. Res. Lett.*, **22**, 1293–6.

Michael, A. J. (1987). Use of focal mechanisms to determine stress: A control study, *J. Geophys. Res.*, **92**, 357–69.

Miller, A., Foulger, G. R. and Julian, B. R. (1998). Non-double-couple earthquakes 2. Observations, *Rev. Geophys.*, **36**, 551–68.

Montagner, J.-P., Griot-Pommera, D.-A., and Lave, J. (2000). How to relate body wave and surface wave anisotropy?, *J. Geophys. Res.*, **105**, 19 015–27.

Montelli, R., Nolet, G., Dahlen, F. A., Masters, G., Engdahl, E. R., and Hung, S.-H. (2004). Finite-frequency tomography reveals a variety of plumes in the mantle, *Science*, **303**, 338–43.

Moore, M. M., Garnero, E. J., Lay, T., and Williams, Q. (2004). Shear wave splitting and waveform complexity for lowermost mantle structures with low-velocity lamellae and transverse isotropy, *J. Geophys. Res.*, **109**, doi: 10.1029/2003JB002546.

Morelli, A., Dziewonski, A. M., and Woodhouse, J. H. (1986). Anisotropy of the inner core inferred from *PKIKP* travel times, *Geophys. J. Roy. Astron. Soc.*, **13**, 1545–8.

Mori, J. and Kanamori, H. (1996). Initial rupture of earthquakes in the 1995 Ridgecrest, California sequence, *Geophys. Res. Lett.*, **23**, 2437–40.

Moser, T. J. (1991). Shortest path calculation of seismic rays, *Geophysics*, **56**, 59–67.

Müller, G. (1971). Approximate treatment of elastic body waves in media with spherical symmetry, *Geophys. J. Roy. Astron. Soc.*, **23**, 435–49.

Müller, G. (1985). Source pulse enhancement by deconvolution of an empirical Green's function, *Geophys. Res. Lett.*, **12**, 33–6.

Müller, G. (2007). *Theory of Elastic Waves*, ed. M. Weber, G. Rümpker, and D. Gajewski, Potsdam (http://bib.gfz-putsdam.de/pub/str0703/0703.htm).

Musgrave, M. J. P. (1970), *Crystal Acoustics*, San Francisco: Holden-Day.

Narkounskaia, G. and Turcotte, D. L. (1992). A cellular-automata, slider-block model for earthquakes I. Demonstration of chaotic behavior for a low-order system, *Geophys. J. Int.*, **111**, 250–8.

Nettles, M. and Ekström, G. (1998). Faulting mechanism of anomalous earthquakes near Bardarbunga Volcano, Iceland, *J. Geophys. Res.*, **103**, 17 973–84.

Nolet, G. (ed.) (1987). *Seismic Tomography*, Holland: D. Reidel.

Nolet, G. and Moser, T.-J. (1993). Teleseismic delay times in a 3-D earth and a new look at the S discrepancy, *Geophys. J. Int.*, **114**, 185–95.

Nur, A. (1972). Dilatancy, pore fluids and premonitory variations of t_s/t_p travel times, *Bull. Seismol. Soc. Am.*, **62**, 1217–22.

Nur, A. and Simmons, G. (1969). Stress-induced velocity anisotropy in rock: An experimental study, *J. Geophys. Res.*, **74**, 6667–74.

Ogata, Y. (1999). Seismicity analysis through point-process modeling: a review, *Pure Appl. Geophys.*, **155**, 471–507.

Okada, Y. (1992). Internal deformation due to shear and tensile faults in a half-space, *Bull. Seismol. Soc. Am.*, **82**, 1018–40.

Okal, E. A. and Cansi, Y. (1998). Detection of PKJKP at intermediate periods by progressive multi-channel correlation, *Earth Planet. Sci. Lett.*, **164**, 23–30.

Olsen, K. B., Day, S. M., Minster, J. B., Chi, Y., Chourasia, A., Faerman, M., Moore, R., Maechling, P., and Jordan, T. (2006). Strong shaking in Los Angeles expected from southern San Andreas Fault earthquake, *Geophys. Res. Lett.*, **33**, L07305, doi:10.1029/2005GL025472.

Olson, E. L. and Allen, R. M. (2005). The deterministic nature of earthquake rupture, *Nature*, **438**, 212–215, doi:10.1029/2004JB003277.

Omori, F. (1894). On the aftershocks of earthquakes, *J. Coll. Sci. Imp. Univ. Tokyo*, **7**, 111–216.

Orcutt, J. A. (1980). Joint linear, extremal inversion of seismic kinematic data, *J. Geophys. Res.*, **85**, 2649–60.

Orowan, E. (1960). Mechanism of seismic faulting, *Geol. Soc. Am. Bull.*, **79**, 323–345.

Peng, Z., Vidale, J. E., Ishii, M., and Helmstetter, A. (2007). Seismicity rate immediately before and after main shock rupture from high-frequency waveforms in Japan, *J. Geophys. Res.*, **112**, doi: 10.1029/2006JB004386.

Pérez-Campos, X. and Beroza, G. C. (2001). An apparent mechanism dependence of radiated seismic energy, *J. Geophys. Res.*, **106**, 11,127–11,136.

Prieto, G. A., Shearer, P. M., Vernon, F. L. and Kilb, D. (2004). Earthquake source scaling and self-similarity estimation from stacking P and S spectra, *J. Geophys. Res.*, **109**, doi: 1029/2004JB003084.

Raitt, R. W., Shor, G. G., Francis, T. J. G., and Morris, G. B. (1969). Anisotropy of the Pacific upper mantle, *J. Geophys. Res.*, **74**, 3095–109.

Reasenberg, P. A. and Simpson, R. W. (1992). Response of regional seismicity to the static stress change produced by the Loma Prieta earthquake, *Science*, **255**, 1687–90.

Rice, J. R., and Ben-Zion, Y. (1996). Slip complexity in earthquake fault models, *Proc. Natl. Acad. Sci.*, **93**, 3811–18.

Richards-Dinger, K. and Shearer, P. (2000). Earthquake locations in southern California obtained using source-specific station terms, *J. Geophys. Res.*, **105**, 10 939–60.

Richter, C. F. (1958). *Elementary Seismology*, San Francisco: W. H. Freeman.

Robinson, R., Wesson, R. L., and Ellsworth, W. L. (1974). Variations of P wave velocity before the Bear Valley, California, earthquake of 24 February 1972, *Science*, **184**, 1281–3.

Rong, Y., Jackson, D. D., and Kagan, Y. Y. (2003). Seismic gaps and earthquakes, *J. Geophys. Res.*, **108**, doi:10.1029/2002JB002334.

Savage, M. K. (1999). Seismic anisotropy and mantle deformation: What have we learned from shear-wave splitting?, *Rev. Geophys.*, **37**, 65–106.

Schmalzle, G., Dixon, T. H., Malservisi, R., and Govers, R. (2006). Strain accumulation across the Carrizo segment of the San Andreas Fault, California: Impact of laterally varying crustal properties, *J. Geophys. Res.*, **111**, B05403, doi 10.1029/2005JB003843.

Scholz, C. H. (1982). Scaling laws for large earthquakes: Consequences for physical models, *Bull. Seismol. Soc. Am.*, **72**, 1–14.

Scholz, C. H. (1987). Size distributions for large and small earthquakes, *Bull. Seismol. Soc. Am.*, **87**, 1074–7.

Scholz, C. (2000). Evidence for a strong San Andreas fault, *Geology*, **28**, 163–6.

Scholz, C., Sykes, L., and Aggarwal, Y. (1973). Earthquake prediction: a physical basis, *Science*, **181**, 803–10.

Schulte-Pelkum, V. and Blackman, D. K. (2003). A synthesis of seismic P and S anisotropy, *Geophys. J. Int.*, **154**, 166–78, doi: 10.1046/j.1365-246X.2003.01951.x.

Schulte-Pelkum, V., Monslave, G., Sheehan, A., Pandey, M. R., Sapkota, S., Bilham, R., and Wu, F. (2005). Imaging the Indian subcontinent beneath the Himalaya, *Nature*, **435**, 1222–5, doi: 10.1038/nature03678,

Scott, P. and Helmberger, D. (1983). Applications of the Kirchhoff–Helmholtz integral to problems in seismology, *Geophys. J. Roy. Astron. Soc.*, **72**, 237–54.

Semenov, A. M. (1969). Variation in the travel time of transverse and longitudinal waves before violent earthquakes, *Izv. Acad. Sci. USSR Phys. Solid Earth*, **4**, 245–8.

Shapiro, N. M., Campillo, M., Stehly, L., and Ritzwoller, M. H. (2005). High-resolution surface-wave tomography from ambient seismic noise, *Science*, **307**, 1615–18, doi: 10.1126/science.1108339.

Shearer, P. M. (1991). Imaging global body wave phases by stacking long-period seismograms, *J. Geophys. Res.*, **96**, 20 353–64.

Shearer, P. M. (1994). Imaging Earth's seismic response at long periods, *Eos Trans. Am. Geophys. Union*, **75**, 451–2.

Shearer, P. M. and Chapman, C. H. (1988). Ray tracing in anisotropic media with a linear gradient, *Geophys. J. Int.*, **94**, 575–80.

Shearer, P. M. and Chapman, C. H. (1989). Ray tracing in azimuthally anisotropic media – I. Results for models of aligned cracks in the upper crust, *Geophys. J. Int.*, **96**, 51–64.

Shearer, P. M. and Orcutt J. A. (1986). Compressional and shear wave anisotropy in the oceanic lithosphere – the Ngendei seismic refraction experiment, *Geophys. J. Roy. Astron. Soc.*, **87**, 967–1003.

Shearer P. M., Prieto, G, and E. Hauksson (2006). Comprehensive analysis of earthquake source spectra in southern California, *J. Geophys. Res.*, **111**, doi: 10.1029/2005JB003979.

Sheriff, R. E. and Geldart L. P. (1995). *Exploration Seismology*, 2nd edn, Cambridge: Cambridge University Press.

Shimazaki, K. and Nakata, T. (1980). Time-predictable recurrence model for large earthquakes, *Geophys. Res. Lett.*, **7**, 279–82.

Sieh, K. (1996). The repetition of large-earthquake ruptures, *Proc. Natl. Acad. Sci.*, **93**, 3764–71.

Silver, P. G. and Chan, W. W. (1991). Shear wave splitting and sub-continental mantle deformation, *J. Geophys. Res.*, **96**, 16 429–54.

Singh, S. K. (1994). Seismic energy release in Mexican subduction zone earthquakes, *Bull. Seismol. Soc. Am.*, **84**, 1533–50.

Sipkin, S. A. and Jordan, T. H. (1979), Frequency dependence of Q_{ScS}, *Bull. Seismol. Soc. Am.*, **69**, 1055–79.

Song, X. (1997). Anisotropy of the Earth's inner core, *Rev. Geophys.*, **35**, 297–313.

Stark, P. B. and Parker, R. L. (1987). Smooth profiles from $\tau(p)$ and $X(p)$ data, *Geophys. J. Roy. Astron. Soc.*, **89**, 997–1010.

Stark, P. B., Parker, R. L., Masters, G., and Orcutt, J. A. (1986). Strict bounds on seismic velocity in the spherical earth, *J. Geophys. Res.*, **91**, 13 892–902.

Stein, R. S. (1999). The role of stress transfer in earthquake occurrence, *Nature*, **402**, 605–9, doi: 10.1038/45144.

Stein, R. S., King, G. C. P., and Lin, J. (1992). Changes in failure stress on the southern San Andreas fault system caused by the 1992 magnitude $= 7.4$ Landers earthquake, *Science*, **258**, 1328–32.

Stein, R. S., King, G. C. P., and Lin, J. (1994). Stress triggering of the 1994 $M = 6.7$ Northridge, California, earthquake by its predecessors, *Science*, **265**, 1432–1435.

Stein, S. and Wysession (2003). *Introduction to Seismology, Earthquakes and Earth Structure*, Oxford: Blackwell.

Suteua, A. M. and Whitcomb, J. H. (1979). A local earthquake coda magnitude and its relation to duration, moment M_0, and local Richter magnitude M_L, *Bull. Seismol. Soc. Am.*, **69**, 353–68.

Tanimoto, T. (1987). The three-dimensional shear wave structure in the mantle by overtone waveform inversion, I. Radial seismogram inversion, *Geophys. J. Roy. Astron. Soc.*, **89**, 713–40.

Thatcher, W. (1975). Strain accumulation and release mechanism of the 1906 San Francisco earthquake, *J. Geophys. Res.*, **80**, 4862–872.

Thorne, M. S., Lay, T., Garnero, E. J., Jahnke, G., and Igel, H. (2007). Seismic imaging of the laterally varying D″ region beneath the Cocos Plate, *Geophys. J. Int.*, **170**, 635–48, doi: 10.1111/j.1365-246X.2006.03279.x.

Turcotte, D. L. (1997). *Fractals and Chaos in Geology and Geophysics*, 2nd edn, Cambridge: Cambridge University Press.

Utsu, T. (2002a). Statistical features of seismicity, in *International Handbook of Earthquake and Engineering Seismology, Part A*, ed. W.H. Lee, San Diego: Academic Press, 719–32.

Utsu, T. (2002b). Relationships between magnitude scales, in *International Handbook of Earthquake and Engineering Seismology, Part A*, ed. W. H. Lee, San Diego: Academic Press, 733–46.

Veith, K. F. and Clawson, G. E. (1972). Magnitude from short-period P-wave data, *Bull. Seismol. Soc. Am.*, **62**, 435–52.

Venkataraman, A., Beroza, G.C., and Boatwright, J. (2006). A brief review of techniques used to estimate radiated seismic energy, in *Earthquakes: Radiated Energy and the Physics of Faulting*, Geophysical Monograph Series 170, ed. R. Abercrombie, McGair, H. Kanamori, and G. di Toro, Washington: American Geophysical Union, 15–24, doi: 10.1029/170GM04.

Venkataraman, A. and Kanamori, H. (2004). Observational constraints on the fracture energy of subduction zone earthquakes, *J. Geophys. Res.*, **109**, doi: 10.1029/2003JB002549.

Vidale, J. (1988). Finite-difference calculation of travel times, *Bull. Seismol. Soc. Am.*, **78**, 2062–76.

Vidale, J. E., Agnew, D. C., Johnston, M. J. S., and Oppenheimer, D. H. (1998). Absence of earthquake correlation with Earth tides: an indication of high preseismic fault stress rate, *J. Geophys. Res.*, **103**, 24 567–72.

Vidale, J. E. and Shearer, P. M. (2006). A survey of 71 earthquake bursts across southern California: Exploring the role of pore fluid pressure fluctuations and aseismic slip as drivers, *J. Geophys. Res.*, **111**, doi: 10.1029/2005JB004034.

Virieux, J. (1986). P-SV wave propagation in heterogeneous media: velocity-stress finite difference method, *Geophysics*, **51**, 889–901.

Waldhauser, F. (2001). hypoDD: A program to compute double-difference hypocenter locations, US Geological Survey Open File Report, 01-113.

Waldhauser, F. and Ellsworth, W. L. (2000). A double-difference earthquake location
  algorithm: method and application to the northern Hayward fault, *Bull. Seismol. Soc.
  Am.*, **90**, 1353–68.

Walter, W. R., Mayeda, K., Gok, R., and Hofstetter, A. (2006). The scaling of seismic
  energy with moment: simple models compared with observations, in *Earthquakes:
  Radiated Energy and the Physics of Faulting*, Geophysical Monograph Series 170, ed.
  R. Abercrombie, McGair, H. Kanamori, and G. di Toro, Washington: American
  Geophysical Union, 25–41, doi: 10.1029/170GM05.

Warren, L. M., and P. Shearer (2000), Investigating the frequency dependence of mantle Q
  by stacking P and PP spectra, *Geophys. Res. Lett.*, 105, **25** 391–402.

Webb, S. C. (1998). Broadband seismology and noise under the ocean, *Rev. Geophys.*, **36**,
  105–42.

Whitcomb, J. H., Garmany, J. E., and Anderson, D. L. (1973). Earthquake prediction:
  variation of seismic velocities before the San Francisco (sic) Earthquake, *Science*, **180**,
  632–5.

Woodhouse, J. H., Giardini, D., and Li, X.-D. (1986). Evidence for inner core anisotropy
  from free oscillations, *Geophys. J. Roy. Astron. Soc.*, **13**, 1549–52.

Wookey, J. and Helffrich, G. (2008). Inner-core shear-wave anisotropy and texture from
  an observation of PKJKP waves, *Nature*, **454**, 873–7, doi: 10.1038/nature07131.

Yilmaz, O. (1987). *Seismic Data Processing*, Tulsa: Society of Exploration Geophysicists.

Zhao, L., Jordan, T. H., and Chapman, C. H. (2000). Three-dimensional Frechet
  differential kernels for seismic delay times, *Geophys. J. Int.*, **141**, 558–76.

Zhao, L., Jordan, T. H., Olsen, K. B., and Chen, P. (2005). Fréchet kernels for imaging
  regional earth structure based on three-dimensional reference models, *Geophys. J. Int.*,
  95, 2066–80.

Zoback, M. D., Zoback, M. L., Mount, V. S., Suppe, J., Eaton, J. P., Healy, J. H.,
  Oppenheimer, D., Reasenberg, P., Jones, L., Raleigh, C. B., Wong, I. G., Scotti, O., and
  Wentworth, C. (1987). New evidence on the state of stress of the San Andreas fault
  system, *Science*, **238**, 1105–11.

INDEX

Abel transform, 104
absorption band model, 168–71
Airy phase, 225
Aki, Keiiti, 13
analytic time series, 375
Andersonian faulting theory, 294
anisotropy, 30, 332–46
  aligned crack, 343
  azimuthal, 342
  crustal, 344–5
  hexagonal, 341–4
  inner core, 346
  lower mantle, 346
  mechanisms for, 343–4
  mineral aligned, 344
  observations in Earth, 344–6
  periodic-thin layer, 343
  phase vs. group velocity, 334
  Pn, 338
  shear-wave splitting, 339–40
  slowness surfaces, 335
  upper mantle, 338, 345–6
  wavefronts, 333
  weak, 337–9
antipodal phase advance, 240
antipode, 226
apparent stress, 276
Armenian earthquake, 317
arrival time, 90
attenuation, 163–77
  absorption band model, 168–71
  global politics, 177
  in Earth, 173–5
  non-linear, 176–7
  nuclear test sites, 177
  observing, 175–6
  velocity dispersion, 167
autocorrelation, 191, 211
auxiliary fault plane, 248, 370

b-value, 288–90
banana-doughnut kernels, 126
baseline correction, 175

beach balls, 251–60
Bessonova method, 113
block-slider model, 304, 317–20
body forces, 19
broadband seismographs, 328
Brune fault model, 270
Brune, James, 266, 270, 296, 316–17
Brune-type stress drop, 271
bubble pulse, 189
bulk modulus, 31
Bullen, K. E., 3, 5
Bullen, K.E., 90
Byerlee's law, 294, 295

catalog completeness, 289
Cauchy tetrahedron, 19
caustics, 72, 144
Cecchi, Filippo, 2
centroid moment tensor (CMT), 245, 297
chaotic behavior, 305, 306
characteristic earthquake hypothesis, 303, 306
checkerboard test, tomography, 123
chi-squared statistics, 130–1
Chilean earthquake, 8
co-seismic strain field, 301
Coalinga earthquake, 304
coda Q, 176
common-midpoint stacking, 5, 184–8
compensated linear vector dipole (CLVD), 249, 251
complex numbers, review, 358–60
complex plane, 358
compressional quadrant, 253
computers, importance in seismology, 10
constitutive relationships, 30–3
continental drift, 5
convolution, 188, 191, 210, 371–3
core, 3
  seismic phases, 87
core–mantle boundary, 3
corner frequencies, 266
Coulomb failure function, 37, 310
critical angle, 154
critical damping, seismometer, 324

critical slip, 278
cross-correlation, 190–1
　of noise, 332
cross-product, vector, 353
crossover point, 86
crust, 3
　seismic phases, 86
curl, vector field, 27, 356

damped least squares, 121
damping, seismometer, 321–2
deaths from earthquakes, 13
Debye peak, 172
deconvolution, 184, 188–91, 200
deep earthquakes, 6, 296
delay time, 73–6, 216, 218, 252
delta functions, 51, 154, 188, 240, 329, 372, 374, 375
density structure, Earth's, 5
depth phases, 88
Did You Feel It website, 291
diffraction, 193
diffraction hyperbolas, 193–5
digital seismographs, 10, 328
dilatancy, as earthquake predictor, 314–15
dilatation, 27
dilatational quadrant, 253
dinner party, Los Angeles, 299
dip, fault plane, 245
dip-slip faults, 245
directivity, source, 262
discrete modeling methods, 53–6
dispersion, 167, 216, 219, 224–5, 238–9
displacement field, 25
divergence, 356
dot product, vector, 353
double-couple source, 6, 243–60
　major and minor decomposition, 250
double-difference location method, 135
dynamic friction, 302
dynamic range, seismograph, 327

Earth
　internal structure, 3, 173–5, 349
　pressure of interior, 24
　seismic velocities, 3, 349
Earth-flattening transformation, 80, 82–3
earthquake energy partitioning, 277–80
earthquake triggering, 309–14
earthquakes
　deep, 296
　distribution of, 5–7
　how to locate, 127–35, 138
　magnitude, 280–8
　precursors, 314–6
　prediction of, 301–17
　table of large, 287
　unpredictability of, 316–17
East Pacific Rise, 12
eikonal equation, 65, 361–5
elastic moduli, 31–3
　units, 32

elastic rebound theory, 13, 301
elastic tensor, 30, 334
　density normalized, 335
　symmetries, 334, 342
elastohydrodynamic lubrication, 295
empirical Green's functions (EGF), 267–8
energy
　in seismic waves, 139–42
　partitioning in earthquakes, 277–80
　radiated from source, 273–7
envelope time function, 375
epicenter, 127
equation of motion, 42
equivalent body forces, 243
error ellipse, earthquake location, 130–1
Eshelby circular fault stress drop, 268
ETAS model, 311, 312
Eulerian description of motion, 25
evanescent waves, 154, 222, 337
Ewing, James, 2
exploding reflector model, 193

far-field term, 252
faults
　as seismic sources, 245–60
　fractal distribution of, 305
　stress level on, 315
Fermat's principle, 67, 119
finite-difference methods, 43, 52, 54–62
finite-difference ray tracing, 86, 119
finite-element methods, 52, 55
finite-frequency tomography, 125–7
finite-slip modeling, 291–3
focal mechanisms, 246–60, 297, 298
　plotting, 259–60
focal sphere, 255–7
fold, reflection seismology, 188
foot wall, fault, 245
force couple, 243
force-feedback seismometers, 327, 328
foreshocks, 311, 314–6
Fortran examples, 63, 240, 320, 347, 367–70
Fourier transforms, 155, 165, 189, 237, 265, 373
Fréchet derivatives, 125
fractal scaling, 305
fracture energy, 277
free oscillations, 8, 231–7
　sun, 9
frequency response function, 329, 347
frequency response, seismograph, 323–30
frictional energy, 277

gain, instrument, 329
Galitzen, B. B., 3
Garmany method, 110
gather, reflection seismology, 187
Gauss's theorem, 357
geometrical spreading, 141–4, 364
GETAUX subroutine, 370
Global CMT catalog, 245, 257–9, 287, 297
Global Digital Seismograph Network (GDSN), 328

Global Positioning System (GPS), 308
gradient operator, 356
graph theory, 86, 119
gravimeters, 326
Green's functions, 52, 242, 244–5, 332, 372
    empirical, 267–8
Green's theorem, 203
ground roll, 188
group velocity, 216, 217, 333, 336
Gutenberg, B., 3
Gutenberg, Beno, 13
Gutenberg–Richter
    energy relation, 286
    magnitude–frequency relation, 288

Haicheng earthquake, 316
hanging wall, fault, 245
harmonic waves, 47, 358–60
    table of parameters, 47
Haskell fault model, 264
Haskell source, 262
Haskell stress-displacement vector, 160
heat-flow paradox, 293–6
Hector Mine earthquake, 308
helioseismology, 9
Helmholtz decomposition theorem, 46
Herglotz–Wiechert formula, 105–6
hexagonal anisotropy, 341–4
Hilbert transform, 154, 156, 158–9, 373–5
Himalayan cross-section, 202
history of seismology, 2–15
Holy Grail, 88
homogeneous equation of motion, 42
horizontal slowness, 66
Huygens' principle, 192–3, 213
hydrostatic strain, 28
hydrostatic stress, 22–4, 30
hyperbolic equation, 40
hypocenter, 127
hypocentroidal decomposition, 134

identity matrix, 355, 357
impedance, 142, 150
Imperial Valley earthquake locations, 135
impulse response function, 329
impulse response test, tomography, 123
incremental stress, 30
index notation, 29, 354
inertial seismometers, 321
infinitesimal strain theory, 26
inhomogeneous waves, 154
inner-core anisotropy, 346
inner-core boundary, 4
intensity scale, 290–1
International Deployment of Accelerometers (IDA), 328
International Seismological Centre (ISC) picks, 90
inverse theory, 8, 110, 116
isotropy, 30
iterative inversions, 122
iterative location methods, 133

JB tables, 3, 90
Jeffreys, Harold, 3, 90
joint hypocenter velocity (JHV) inversions, 124

Kirchhoff applications, 208
Kirchhoff migration, 210
Kirchhoff theory, 202–11
Kronecker delta, 31

L1 norm, 129
L2 norm, 129
LaCoste, Lucien, 327
Lagrangian description of motion, 25
Lamé parameters, 31–3
Landers earthquake, 36, 308
Laplacian operator, 121, 356
lattice-preferred orientation (LPO), 343
layer matrix, 160
layer-cake model, 107, 111
LAYERXT subroutine, 100, 367–8
left-lateral motion, 245
Lehmann, Inge, 3
Leighton, Robert, 9
linear programming, 115
linear superposition, 263
linear velocity gradients, 71, 98
linearly elastic material, 30
lithostatic stress, 294
locating earthquakes, 127–35, 138
Loma Prieta earthquake, 33, 308
Love waves, 215–9
low-velocity zones, 76–7, 105–6
lunar seismology, 8

Madariaga fault model, 270, 275
magnitude, 13, 280–8
    body wave (m_b), 282
    local (M_L), 281
    moment (M_W), 284
    saturation of, 283, 285
    surface wave (M_S), 283
Mallet, Robert, 2
mantle, 3
    seismic phases, 87
MARMOD model, 99
Martian seismology, 9
master event location methods, 134
math review, 353–60
matrix identities, 355
matrix methods, layered models, 159–63
Mercalli scale, 290
microseism peak, 331
microseisms, 326, 330
mid-ocean ridges, 5
migration, 191–7
    poststack, 197
    prestack, 197
Mode III cracks, 279
Moho, 3, 86
    reflections off, 150, 152
Mohorovičić, Andrija, 3

moment magnitude, 284
moment tensor, 243–60
   decomposition of, 248–50, 296–7
   isotropic and deviatoric, 249
moment, scalar (M_0), 246
   from spectra, 262
momentum equation, 40–2
Moon
   pressure of interior, 24
   quakes, 8
   seismometers, 8
moveout, 186

Nakano, H., 6
Naples earthquake, 2
near-field term, 252
New Madrid earthquakes, 291
Newton's law, 40
nodal lines, 253
nodal points, 254
noise
   cross-correlation of, 332
   Earth, 330–2
non-double-couple sources, 248–51
non-linear material behavior, 30
normal faults, 245
normal modes, 8, 52, 231–7, 240
   Sun, 9
normal moveout, 186
Northridge earthquake, 90, 308, 309, 317
nuclear explosion seismology, 7

oceanic crust, 107
Oldham, Richard, 3
olivine crystal alignment, 344
olivine elastic constants, 336
Omori's law, 309
origin time, earthquake, 127
Orowan fault model, 277
Orowan stress drop, 278
outliers, in data, 129
overthrust faults, 245

P waves
   polarization, 48
   radiation pattern, 253
   velocity, 32, 45
paleoseismology, 306
Parkfield earthquake, 292, 315
Parkfield earthquake sequence, 303
Pascals, 24
PcS, 88
periodic-thin layer (PTL) anistropy, 343
Pg, 86
phase advance, 155
phase velocity, 216, 333
Piñon Flat Observatory, 35
PKIKP, 88
PKJKP, 88
PKP, 88

plane waves, 46–50, 62, 65, 140, 144, 151, 156–63,
   193, 215, 220, 337
   anisotropic media, 334
plate tectonics, 5
PmP, 86
Pn, 86
point sources, 243
Poisson's ratio, 32
polarizations, seismic waves, 48–50
potentials, 46, 220–1
PP, 88
pP, 88
PPP, 88
precritical reflections, 154
precursors, earthquake, 314–16
Preliminary Determination of Epicenters (PDE), 127
PREM model, 4, 341, 345, 349
pressure, 22–5
pressure axis, 248, 257
primary fault plane, 248
principal axes
   computing, 22–3
   strain, 28
   stress, 21
prograde branch, 72
propagator matrix, 161
Ps, 199
pseudotachylytes, 295
pulse shapes, far-field, 260–5

Q, 163–77
quality factor, 164
quasi-*P* waves, 335
quasi-*S* waves, 335

radial component, 89
radiated seismic energy, 273–7
   ratio to moment, 276
radiation efficiency, 279
radiation patterns, 251–60
Radon transform, 113
rake, fault plane, 245
rate-and-state friction, 312
ray names, 86–9
ray parameter
   definition, 66
   spherical, 81
ray tracing
   finite difference, 86
   flat Earth, 67–72
   spherical Earth, 80–3
   summary of equations, 77
   three-dimensional, 83–6
   two-point, 119
ray-theoretical methods, 52
Rayleigh function, 223, 238
Rayleigh waves, 219–24
   image of, 230
Receiver functions, 199–202
recurrence intervals, earthquake, 302, 306
reduction velocity, 73

reflection and transmission coefficients, 144–56, 161
  energy normalized, 151–2
  inhomogeneous waves, 154–5
  P/SV, 149, 368–70
  SH, 145–9
  vertical incidence, 149–50
reflection seismology, 5, 181–214
reflectivity method, 162
regularization, 115
Reid, H.F., 13, 301
relative event location methods, 134–5
relaxed modulus, 172
representation theorem, 242
residuals, travel time, 117
resonant frequency, 324
retrograde branch, 72
reverse faults, 245
Richter, Charles, 13, 281
Ricker wavelet, 213–14
right-lateral motion, 245
rise time, 263
rotation tensor, 26
row action methods, 123
RTCOEF subroutine, 178, 368–70

S discrepancy, 175
S-P time, 132
S waves
  polarization, 50
  radiation pattern, 253
  velocity, 32, 45
sagittal plane, 337
San Andreas Fault, 5, 35, 246, 259, 295, 305
San Francisco earthquake, 13, 301
San Simeon earthquake, 291, 308
scalar seismic moment, 246
  from spectra, 262
scaled energy, 276
secular strain rate, 301
seismic efficiency, 280, 295
seismic gap hypothesis, 307–8
seismic moment
  from spectra, 262, 299
  scalar, 246
seismographs, 321–30
  broadband, 328
  electromagnetic, 3
  history, 2–3
  modern, 327–30
  response functions, 321–30
  strong-motion, 328
seismometers, 321
self-organized criticality, 317
self-similar faults, 305
self-similarity of earthquakes, 271–3, 276
SH and *SV* polarizations, 145
SH reflection and transmission coefficients, 148
shadow zones, 77
shape-preferred orientation (SPO), 343
shear modulus, 31
shear-wave splitting, 339–40

SKS, 340, 347
sinc function, 265
singular value decomposition, 120
SKKS, 88
SKS, 88
  splitting from anisotropy, 340, 345
slant stack, 113
Slichter, L.B., 107
slip predictable model, 302
slip vector, 245
slip-weakening model, 278
slowness
  definition, 66
  surface, 335
  vector, 68
SmS, 86
Snell's law, 65–7, 98, 337
solar seismology, 9
solution matrix, 160
sound channel, 77
source spectra, 265–73, 298, 299
source-specific station terms, 135
SP, 88
sP, 88
spectral-element method, 55
spherical harmonics, 233–5
spherical waves, 50–1
spheroidal modes, 233
splitting, normal modes, 236
SS, 88
sS, 88
stacks, global seismograms, 90
staggered-grid finite differences, 59
standard linear solid, 171–3
standing waves, 232
static equilibrium equation, 42
static friction, 302
statics corrections, 198–9
station corrections, 124
station terms, 134
stick-slip behavior, 302
straight-line fitting, 106–9
strain, 25–9
  computing, 29
  energy, 139
  extensional, 25
  hydrostatic, 28
  tensor, 25–9
  units, 29
stress, 17–25
  deviatoric, 23
  hydrostatic, 22–4, 30
  incremental, 30
  orientation from focal mechanisms, 259
  tensor, 17–22
  units, 24
stress drop, 268–71, 299
  total, 280
stress shadow hypothesis, 308
stress-displacement vector, 160
strike, fault plane, 245

strike–slip faults, 245
strong-motion instruments, 328
subduction zones, 5
Sumatra earthquake, 13, 236
summary rays, 122
summation convention, 30, 354
super shear rupture, 264
superposition principle, 242, 263
surface waves, 215–31
   global, 226–8
   observing, 228–31
swarms, earthquake, 314
Synthetic Aperture Radar (SAR), 309
synthetic seismograms, 42, 51–3, 162

t^*, 165
takeoff angle, 67
Tangshan earthquake, 316
tau-p transform, 113
teleseism, 2
Temblovia, Republic of, 299
tension axis, 248, 253, 257
tensor, second-order, 354
Thomson-Haskell method, 162
thrust faults, 245
tides, solid Earth, 332
time predictable model, 302
time series analysis, 371–5
tomography, 10, 118–27
Tonga subduction zone, 7
toroidal modes, 233
traction, 17
   continuity across interface, 146
traction vector, computing, 19–20
transformation tensor, 354
transition zone, 3
transverse component, 89
transverse isotropy, 341–2, 345, 349
travel time curves, 68, 72–3
   crustal, 86

global, 90
travel time tables, 3
triplications, 72
Turner, H. H., 6
turning point, 67

uniqueness theorem, 242
unrelaxed modulus, 172

variance reduction, 128
vector calculus, review, 353–7
vector dipole, 243
vector identities, 357
vector spherical harmonics, 235
velocity analysis, 197–8
velocity inversion, 103–27
   one-dimensional, 103–17, 135
   three-dimensional, 117–27
vertical slowness, 69
vibroseis, 5, 190–2, 211

Wadati, Kiyoo, 6
Wadati–Benioff zones, 6
Wallace Creek, 306
wave equation
   one-dimensional, 39
   seismic, 39–61
wave guides, 77
waveform cross-correlation, 135
wavenumber, 47, 145
Wegener, Alfred, 5
Wiechert, E., 3
Wood–Anderson seismograph, 281
Worldwide Standardized Seismograph Network
   (WWSSN), 8, 259

Young's modulus, 31

zero-length spring, 327
zero-offset sections, 182–3